Biodiversity Conservation and Land Degradation Neutrality

Sanjay Singh • Mukesh K. Gautam
Editors

Biodiversity Conservation and Land Degradation Neutrality

From Loss to Restoration

Springer

Editors
Sanjay Singh
Centre of Excellence on Sustainable Land Management
Indian Council of Forestry Research and Education
Dehradun, Uttarakhand, India

Mukesh K. Gautam
Biology Department
Medgar Evers College
City University of New York
NY, USA

ISBN 978-981-96-7209-7 ISBN 978-981-96-7210-3 (eBook)
https://doi.org/10.1007/978-981-96-7210-3

This Springer imprint is published by the registered company Springer Nature Singapore Pte Ltd.
The registered company address is: 152 Beach Road, #21-01/04 Gateway East, Singapore 189721, Singapore

Preface

Centuries of deforestation in tropical region including Amazonia and south-east Asia, Acidification of European ecosystems in the 1900s, destruction of prairies in the United States in 1930s, Mekong Delta and Indo-Gangetic Plains soil erosion and salinization, desert expansion or desertification in many parts of the world, especially in China and India, and forest fires in Australia and Amazon, in the past and present, are some of the cautionary tales how land degradation and its drivers have profoundly shaped our land and their productivity.

In the present time, increased domination of Earth's ecosystems by humans has tipped the balance off natural ecological processes, leading to widespread land degradation. As a multifaceted global crisis, it disrupts natural systems, making it one of the most pressing ecological issues affecting the health, structure, and function of ecosystems, cascading into significant socioeconomic and cultural challenges. The intricate link between land degradation and biodiversity (loss) is complex, each exacerbating the other in a feedback loop that undermines ecosystem health and resilience. Land degradation, caused by deforestation, overgrazing, soil erosion, desertification, and unsustainable agricultural practices, destroys habitats and reduces the variety of species that can thrive in a given area. As ecosystems degrade, they lose key species that play vital roles in nutrient cycling, soil fertility, pollination, and water regulation, leading to further land deterioration. Conversely, the loss of biodiversity weakens ecosystems' ability to recover from disturbances, making them more vulnerable to degradation. For instance, the loss of vegetation diversity exposes soil to erosion, while the decline of decomposers reduces soil quality and fertility. Although land degradation is a global phenomenon, these processes are particularly evident in tropical ecosystems areas, for example, rainforests of Amazon and South-East Asia, where deforestation and habitat destruction not only depletes biodiversity but also alters biogeochemical cycles, further exacerbating land degradation.

Biodiversity conservation and Land Degradation Neutrality are closely linked. Achieving Land Degradation Neutrality is crucial for restoring and reinforcing biodiversity—its structure and function, which ultimately are vital for resilience. Halting land degradation is crucial in halting biodiversity loss because it directly addresses one of the primary drivers of biodiversity decline—habitat destruction. To prevent further land degradation to ensure that ecosystems remain resilient against environmental stressors, Land Degradation Neutrality (LDN) was introduced by the

United Nations Convention to Combat Desertification (UNCCD) and endorsed in 2015 as part of the UN's Sustainable Development Goals (SDGs), specifically Goal 15.3 that states "To combat desertification, restore degraded land and soil, including land affected by desertification, drought, and floods, and strive to achieve a land degradation-neutral world." The aim of LDN is to maintain or improve the amount and quality of land resources necessary to support ecosystem functions and services while meeting human needs. Achieving LDN is a fundamental step toward stopping the ongoing destruction of habitats and drain of biodiversity and ensuring long-term ecological and socioeconomic stability.

This book presents original research and synthesis studies addressing key issues in biodiversity conservation, land degradation, and global efforts to achieve land degradation neutrality. At its core, it emphasizes ecosystem restoration as a crucial strategy for mitigating land degradation and promoting biodiversity conservation. The chapters provide practical examples and best practices in conservation and restoration, highlighting the complex drivers of land-use change and degradation. As a whole it examines the direct impact of land degradation on biodiversity and explores global frameworks, strategies, and policies essential for effective conservation through land degradation neutrality. It presents a holistic and multidisciplinary approach underscoring the importance of collaboration among academia, policymakers, and subject matter experts in safeguarding terrestrial ecosystems. It further accentuate the significance of terrestrial ecosystems to achieve biodiversity goals outlined in the SDG-15—Life on Land through a multidisciplinary approach and contributions from academia, policymakers, and subject matter specialists. Designed as a comprehensive resource, this book will help professionals, trainees, and students with valuable insights into land degradation, its consequences, and the pathways toward biodiversity conservation.

Chapters 1, 2, 3, and 4 establish the foundation by highlighting the urgency of biodiversity conservation and land degradation neutrality. Chapters 5 and 6 highlight the importance of RS-GIS tools for monitoring and assessment of land degradation. Chapters 7 and 8 discuss the drivers behind land use change and degradation while highlighting the complex interplay between human activities and shifting biodiversity. Chapters 9, 10, and 11 emphasize the importance of sustainable practices while discussing the impact of land degradation on biodiversity and biodiversity loss. These chapters provide an overview of global efforts, frameworks, and concepts related to biodiversity conservation and land degradation neutrality, providing readers with a broader context about significance of global LDN initiatives. Chapters 12, 13, 14, 15, 16, and 17 highlight the restoration strategies and case studies, highlighting the importance of Indigenous traditional knowledge and community involvement. The case studies, best practices, and strategies for land restoration and conservation presented in these chapters provide a tangible model for implementing ecosystem restoration and conservation, presenting practical and actionable practices for achieving LDN.

We thank the contributors, for accepting our invitation, sharing their knowledge, and research and their consistent support during the editing of this book, for finalizing the chapters that made this book possible. We express our gratitude to the Springer International team for their kind assistance throughout the entire book production process.

Dehradun, Uttarakhand, India — Sanjay Singh
NY, USA — Mukesh K. Gautam

Contents

Editors and Contributors

About the Editors

Sanjay Singh obtained a PhD in Botany from Hemvati Nandan Bahuguna Garhwal University, India. At present, he works at the Center of Excellence on Sustainable Land Management (CoE-SLM), Indian Council of Forestry Research and Education, India. The establishment of the CoE-SLM was announced during the 14th Conference of Parties (COP-14) of the United Nations Convention to Combat Desertification, India, and Dr. Singh played a pivotal role as one of the key scientists who worked toward the establishment of the Center at the Indian Council of Forestry Research and Education, India. As a member of Team India for the Environmental Sustainability Working Group of the G20, he was also involved in the publishing of G20 Presidency documents. Before this, he was a Scientist at the Tropical Forest Research Institute-Jabalpur (India), where he worked extensively in the forests of Central India, with a primary focus on issues related to biodiversity characterization and its conservation. He has over 50 research publications in the field of biodiversity, ecosystem ecology, plant ecology, climate change, and land degradation. His current work is aligned with the Global Biodiversity Framework and Sustainable Development Goal 15- UNEP (SDG-15).

Mukesh K. Gautam is an ecosystem ecologist and biogeochemist, with a PhD in ecology and environmental science from Forest Research Institute-India, followed by postdoctoral experiences from Indian Institute-Ahmedabad (India) and Korea Basic Science Institute-South Korea. In 20+ years of academic career, he has conducted collaborative research on ecosystem ecology and vegetation ecology, biogeochemistry of forested ecosystem, stable isotope ecology, ecohydrology, soil science, and element dynamics with the aim to understand and reinforce the knowledge and prediction about the ability of ecosystems to provide important services and maintain biodiversity. In his research and academic journey, he has authored and published 38 research articles in various influential and prestigious peer-reviewed journals. As an integral part of the science dissemination and peer-review system, he regularly serves as a guest reviewer for various scientific journals across a wide range of scientific disciplines. Currently, he is an assistant professor at the Biology Department, Medgar Evers College (CUNY), NY, New York, where he teaches different courses related to general biology including ecology and conduct research in broader framework of ecosystem analysis.

Contributors

Bikas Adhikari Dhankuta Municipality, Dhankuta, Nepal

Rabindra Adhikari University of Tubingen, Department of Geosciences, Chair of Soil Science, Tubingen, Germany

Juhi Agrawal University of Petroleum and Energy Studies, Dehradun, Uttarakhand, India

Ram Avtar Hokkaido University, Sapporo, Hokkaido, Japan

Tserennadmid Bataa School of Agroecology and Business, Mongolian Agricultural University of Life Science, Ulaanbaatar, Mongolia

Nabin Bhattarai Armale Innovation Pvt. Ltd, Balkot, Bhaktapur, Nepal

Seema Bhatt Food and Agriculture Organization of the United Nations (FAO), New Delhi, India

Jigyasa Bisaria Centre for Climate Change Studies, Indian Institute of Forest Management, Bhopal, Madhya Pradesh, India

Arghya Chakrabarty ICFRE-Rain Forest Research Institute, Jorhat, Assam, India

Wetlands International South Asia, New Delhi, India

Harish Bahadur Chand Forest Ecology and Climate Change Division, Forest Research Institute, Dehradun, Uttarakhand, India

Mohan Chand Cryospheric Society of Nepal, Kathmandu, Nepal

Konda Chavva Food and Agriculture Organization of the United Nations (FAO), New Delhi, India

Amarmend Chuluunbaatar School of Agroecology and Business, Mongolian Agricultural University of Life Science, Ulaanbaatar, Mongolia

Suruchi Devi Centre for Climate Change Studies, Indian Institute of Forest Management, Bhopal, Madhya Pradesh, India

Devesh Dubey Arahas Technologies, Gurugram, Haryana, India

Moses Fayiah Department of Forestry, School of Natural Resources Management, Njala University, Njala, Sierra Leone

Mukesh K. Gautam Biology Department, Medgar Evers College, City University of New York, NY, USA

Krishna Giri Centre of Excellence on Sustainable Land Management, Indian Council of Forestry Research and Education, Dehradun, Uttarakhand, India

Suraj Goswami Centre for Climate Change Studies, Indian Institute of Forest Management, Bhopal, Madhya Pradesh, India

Thilanka Gunaratne Department of Botany, Faculty of Science, University of Peradeniya, Peradeniya, Sri Lanka

R. S. C. Jayaraj UT Environmental Impact Assessment Authority, Andaman and Nicobar Islands, India

Fariha Jesmin Department of Environmental Science and Disaster Management, Daffodil International University, Dhaka, Bangladesh

Padam Raj Joshi Chang'an University, Xi'an, Shaanxi, China

A. B. M. Kamal Pasha Department of Environmental Science and Disaster Management, Daffodil International University, Dhaka, Bangladesh

Bikalpa Karki National Trust for Nature Conservation, Biodiversity Conservation Center, Chitwan, Nepal

Bhaskar Singh Karky Asian Development Bank, Kathmandu, Nepal

Jasmeet Kaur Food and Agriculture Organization of the United Nations (FAO), New Delhi, India

Chanda Khadka Hetauda Sub-Metropolitan City, Hetauda, Nepal

Aakash Kumar Himalayan Institute for Learning and Leadership (HILL), University of Petroleum and Energy Studies (UPES), Dehradun, Uttarakhand, India

Forest Research Institute (Deemed to be) University, Dehradun, Uttarakhand, India

Abhishek Kumar Department of Biotechnology, Graphic Era (Deemed to be University), Dehradun, Uttarakhand, India

Forest Ecology and Climate Change Division, Forest Research Institute, Dehradun, Uttarakhand, India

Anil Kumar University of Petroleum and Energy Studies, Dehradun, Uttarakhand, India

Manoj Kumar Centre of Excellence on Sustainable Land Management (CoE-SLM), Indian Council of Forestry Research & Education, Dehradun, Uttarakhand, India

Sanjeev Kumar Centre of Excellence on Sustainable Land Management, Indian Council of Forestry Research and Education, Dehradun, Uttarakhand, India

S. M. Mahmudur Rahman Department of Environmental Science and Disaster Management, Daffodil International University, Dhaka, Bangladesh

Kambam Boxen Meetei ICFRE-Rain Forest Research Institute, Jorhat, Assam, India

Gaurav Mishra Centre of Excellence on Sustainable Land Management, Indian Council of Forestry Research and Education, Dehradun, Uttarakhand, India

Sagar Mozumder Department of Environmental Science and Disaster Management, Daffodil International University, Dhaka, Bangladesh

Aishwarya Negi Forest Ecology and Climate Change Division, Forest Research Institute, Dehradun, Uttarakhand, India

Odbayar Otgonbaatar School of Agroecology and Business, Mongolian Agricultural University of Life Science, Ulaanbaatar, Mongolia

Hitendra Padalia Forestry and Ecology Department, Indian Institute of Remote Sensing (IIRS), Indian Space Research Organisation (ISRO), Dehradun, Uttarakhand, India

Divesh Pangtey Centre of Excellence on Sustainable Land Management (CoE-SLM), Indian Council of Forestry Research & Education, Dehradun, Uttarakhand, India

Mahfuza Parveen Department of Environmental Science and Disaster Management, Daffodil International University, Dhaka, Bangladesh

Center for Environment and Sustainability, Daffodil International University, Dhaka, Bangladesh

Rajan Prasad Paudel National Trust for Nature Conservation, Biodiversity Conservation Center, Chitwan, Nepal

Prawesh Poudel National Trust for Nature Conservation, Biodiversity Conservation Center, Chitwan, Nepal

Kangujam Premkumar Forest Ecology and Climate Change Division, Forest Research Institute, Dehradun, Uttarakhand, India

Department of Forestry, North-Eastern Hill University, Tura, Meghalaya, India

Sodgerel Purevee School of Agroecology and Business, Mongolian Agricultural University of Life Science, Ulaanbaatar, Mongolia

Ranjit Singh Rawat Indian Council for Forestry Research and Education (ICFRE), Dehradun, Uttarakhand, India

Trisha Roy ICAR-Indian Institute of Soil and Water Commission, Dehradun, Uttarakhand, India

Tulasi Sangraula Federation of Community Forestry Users Nepal, Changunarayan, Nepal

Rachana Shah National Trust for Nature Conservation, Biodiversity Conservation Center, Chitwan, Nepal

Kripa Shrestha Montclair State University, Montclair, NJ, USA

Sanjay Singh Centre of Excellence on Sustainable Land Management, Indian Council of Forestry Research and Education (ICFRE), Dehradun, Uttarakhand, India

Shreyansh Singh Forest Research Institute, Dehradun, Uttarakhand, India

Bhaskar Sinha Centre for Climate Change Studies, Indian Institute of Forest Management, Bhopal, Madhya Pradesh, India

Shivang Sinha Hokkaido University, Sapporo, Hokkaido, Japan

Kamani Sylva Department of Engineering Management, Faculty of Engineering, University of Peradeniya, Peradeniya, Sri Lanka

Rajesh Bahadur Thapa International Centre for Integrated Mountain Development, Lalitpur, Kathmandu, Nepal

Meribeni Tsopoe ICFRE-Rain Forest Research Institute, Jorhat, Assam, India

Vatsala Forest Research Institute (Deemed to be) University, Dehradun, Uttarakhand, India

Teiji Watanabe Hokkaido University, Sapporo, Hokkaido, Japan

Land Degradation Neutrality: Concept and Strategies

1

Suruchi Devi, Suraj Goswami, Jigyasa Bisaria, and Bhaskar Sinha

Abstract

Land Degradation Neutrality (LDN) is the concept that can be defined as the amount of land resources in a balanced state in an ecosystem. It is an important component of Agenda 2030 for Sustainable Development, notably addressed in Target 15.3 under the sustainable development goal (SDG 15), "Life on Land." Land degradation continues to worsen globally due to climate change and anthropogenic activities, affecting over 3.2 billion people, and humans have already altered 70% of ice-free land. LDN seeks to reverse this trend by maintaining or improving the status of land resources to support ecosystems and promote food security. The major causes of land degradation are salinization, alkalization, acidification, pollution, nutrient depletion, reduction in total and biomass carbon, wind erosion, waterlogging, lowering of the water table, mining and quarrying, and decline in land biodiversity, which are further accelerated by climate change. Sustainable land management is the primary strategy for stopping, reducing, and reversing land degradation. Achieving LDN can lead to a healthier ecosystem, improved food security, and a more sustainable future. Furthermore, long-term development policies can address this by promoting efficiency, resilience, and social inclusion in the food, water, and energy sectors.

Keywords

Climate change · Land degradation neutrality · UNCCD

S. Devi · S. Goswami · J. Bisaria · B. Sinha (✉)
Centre for Climate Change Studies, Indian Institute of Forest Management, Bhopal, Madhya Pradesh, India
e-mail: bsinha@iifm.ac.in

S. Singh, M. K. Gautam (eds.), *Biodiversity Conservation and Land Degradation Neutrality*, https://doi.org/10.1007/978-981-96-7210-3_1

1.1 Introduction

Land Degradation Neutrality, as defined by the United Nations Convention to Combat Desertification (UNCCD), is a state whereby the amount and quality of land resources necessary to support ecosystem functions and services, and to enhance food security, remain stable or increase across specified temporal and spatial scales and ecosystems. The concept of LDN is focused on "no net loss" and maintaining healthy and productive land through two main approaches: avoiding or reducing land degradation and reversing existing degradation so that gains counterbalance losses (Cowie et al. 2018). This concept was introduced by the UN in 2012 and emphasized by the UNEP Global Environmental Outlook to continue striving toward the target of land degradation; however, achieving neutrality remains challenging.

About 75% of the world's land is degraded (IPBES 2018) and land degradation continues to grow worldwide, affecting over 3.2 billion people, and humans have already modified almost two-thirds of ice-free land (Olsson et al. 2019). The major drivers of land degradation are water erosion, wind erosion, waterlogging, lowering of the water table, mining and quarrying, and urban and industrial waste salinization, alkalization, acidification, pollution, nutrient depletion, reduction in total and biomass carbon, and decline in land biodiversity (Sivakumar and Stefanski 2007). All these unsustainable land use practices are contributing to increased pressure on land resources and complicating efforts to reverse degradation (Hannam 2022). Additionally, climate change is further accelerating land degradation and increasing vulnerabilities. If current trends continue, almost 90% of all land could be changed by 2050. This degradation alters rainfall patterns, exacerbates severe weather, and leads to climate change, resulting in societal instability, poverty, war, and migration.

Achieving land degradation neutrality is a key component of Agenda 2030 for Sustainable Development, specifically mentioned in Target 15.3, under the sustainable development goal (SDG 15), "Life on Land." The UNCCD's goal of achieving LDN by 2030 aims to reverse this trend by stabilizing or enhancing the condition of land resources to support ecosystems and increase food security (Kumar et al. 2021). This could be achieved by combating desertification and restoring degraded land and soil, including land affected by desertification, drought, and floods (IUCN 2015). To achieve this target, the Conference of Parties (COP 12) to UNCCD has integrated land degradation neutrality (LDN) in their process and directed the UNCCD Science-Policy Interface (SPI) to create a scientific conceptual framework for LDN (Gilbey et al. 2019), which can be applied to all types of land use and ecosystem services.

Rapid growth in the human population is another challenge to achieving land degradation neutrality (Conacher 2009). It is estimated that the demand for food will grow by 50% and water and energy by 40% by 2030 (Gnacadja 2012; Sharma et al. 2023). The rapid increase of the global population demands long-term development strategies that can extensively address emerging challenges in food, water, and energy sectors. To effectively address these intertwined needs, strategies should prioritize efficiency, resilience, and social inclusion (UNCCD 2012). The issues of desertification, land degradation, and drought can be resolved through LDN by promoting an ecosystem-based approach with two primary pathways of action: first, by addressing current and future land degradation (avoiding, preventing, and minimizing land degradation processes), such as transitioning to Sustainable Land Management (SLM) and second, by redressing past degradation, such as rehabilitating working landscapes and restoring the natural ecosystem. SLM is the main approach for preventing, mitigating, and reversing land degradation, and it has been promoted through various land use projects across different countries. However, the problem of land degradation is still a major global threat (Kust et al. 2017; Kumar et al. 2024). This chapter introduces concepts and strategies designed to combat land degradation in the context of these challenges.

1.2 Understanding Land Degradation

Land degradation is a global challenge and a major threat to soil, with ~24 billion tonnes of fertile soil lost annually to water erosion and 75 billion tonnes eroded annually from arable lands worldwide, resulting in an estimated financial loss of US $400 billion per year (Global Soil Partnership 2017). Land degradation is described as a temporary or permanent reduction of the biological and economic productivity of land due to natural pressure as well as anthropogenic activities (Jinger et al. 2023). This leads to a reduction in food production, water storage capacity, biodiversity, and carbon sequestration in soils and vegetation (IUCN 2015). It is a complex process encompassing physical, chemical, biological, and anthropogenic forms. Physical degradation includes vegetation degradation, primarily caused by deforestation and overgrazing, leading to soil erosion and a decline in organic matter (SAC 2021). Soil erosion, both water and wind erosion, is a major contributor to global land degradation (Reddy et al. 2018). Furthermore, waterlogging is a physical process that occurs when soil becomes saturated with water for extended periods, reducing agricultural productivity due to insufficient soil aeration. Frost heaving and frost shattering are additional physical degradation processes, with frost heaving involving the formation of ice lenses beneath the soil during freezing conditions and frost shattering involving the mechanical weathering of rocks due to temperature fluctuations around 0 °C, leading to the formation of talus or scree (SAC 2021). Acidification of ecosystems, leaching, salinization, and soil salinity pose significant threats to cation retention capacity and fertility depletion of soils (Mohamed et al. 2019). Anthropogenic degradation processes like mining, quarrying, brick kilns, industrial effluents, city waste, and urban agglomeration contribute significantly to land degradation. Unsustainable practices are accelerating land degradation, posing serious environmental, ecological, and socio-economic challenges (SAC 2021).

In India, approximately 120.7 million hectares (Mha) of land suffer from various forms of degradation (NAAS 2010). The primary cause is soil erosion, responsible for ~71% (85.7 Mha) of degraded land, with water erosion alone contributing ~60.7% (73.3 Mha) and wind erosion accounting for 10.3% (12.4 Mha).

Deforestation (2.07 Mha from 2001 to 2021), intensive rainfall, uncontrolled grazing (5.65 Mha), excessive fertilizer use (32 million tonnes per year), and shifting cultivation (7.6 Mha) further exacerbate land degradation (Down To Earth 2019; Fertilizer Association of India 2021; Global Forest Watch 2022; Bhat et al. 2022; Jinger et al. 2023). Degradation results in loss of adaptive capacity, increased environmental risks, and insecurity in food, water, and energy supplies. It can cause human displacement due to natural hazards like droughts, flash floods, heat waves, and dust storms. Additionally, it significantly contributes to biodiversity loss and changes in species abundance, leading to a decline in ecosystem services. Reduced infiltration and groundwater recharge, coupled with increased runoff and siltation, exacerbate downstream drought and flood impacts (IUCN 2015).

1.3 The Concept of Land Degradation Neutrality

The concept of land degradation neutrality was first introduced to the global community in 2012 by the document "Zero Net Land Degradation," a new SDG for Rio+20. Further, utilizing this document, the UNCCD advocated the inclusion of a reference to the LDN in the Rio+20 outcomes document, "The Future We Want" (UN 2012; Chasek et al. 2019). In COP 11, the Intergovernmental Working Group (IWG) developed the definition of LDN. This definition was subsequently endorsed by the UNCCD COP 12 in 2015. COP-12 then requested the UNCCD Science-Policy Interface (SPI) to develop the "Scientific Conceptual Framework for Land Degradation Neutrality" (LDN-SCF) based on definition and guidance. The LDN-SCF was designed to provide a scientifically sound basis for understanding and implementing LDN and to inform the development of practical guidance for pursuing LDN and monitoring progress toward the LDN target (IUCN 2015). The LDN concept implies achieving a neutral state of land where land restoration efforts balance ongoing land degradation. To attain a state of zero net loss of productive land, LDN aimed to promote the implementation of the ideal combination of activities that are designed to avoid, reduce, and reverse land degradation (Hannam 2022). In this way, LDN can be seen as a tool for policy to balance the land degradation process with restoration, rehabilitation, and reclamation at local, national, regional, and global levels (Kust et al. 2017). To maintain food security, LDN aims to:

- Maintain the sustainable delivery of ecosystem services
- Improve land productivity
- Increase the resilience of the land and populations that depend on it
- Look for opportunities with other social, economic, and environmental objectives
- Reinforce responsible and inclusive land governance (Orr et al. 2017).

SGD 15.3 has documented 19 LDN principles to guide the LDN implementation process. These principles are considered pillars of the scientific conceptual framework for LDN, and the three important principles are to maintain national sovereignty, protect human rights and enhance human well-being, and respect or enhance land-based natural capital (Boer and Hannam 2021). The framework is designed to achieve neutrality, which projects and seeks to balance expected positive and negative changes. It promotes adaptive management during the planning, implementation, monitoring, and interpretation of LDN (Cowie et al. 2018). This framework provides an approach for decision-makers to vision, frame of reference, mechanism for neutrality, achieving neutrality, and monitoring neutrality.

The LDN conceptual framework can be applied to different land types, land uses, and ecosystem services, and every country can use this according to its situation (Orr et al. 2017). Keesstra et al. (2018) introduced four concepts: (i) systems thinking, (ii) connectivity, (iii) nature-based solutions, and (iv) regenerative economics. Systems thinking is vital for understanding the impact of land management and land use changes, optimizing sustainable use of the system. Connectivity illustrates the impact of land management and land use changes within one natural system, as well as the impact of spatially or geographically connected natural systems such as river basins and groundwater bodies. Nature-based solutions utilize the characteristics and dynamics of the natural system to develop resilient solutions and regenerative economics that integrates environmental, social, and economic practices to create a thriving world where both people and nature can flourish. These concepts can support comprehensive decision-making for LDN across different scales. Monitoring and assessment are also crucial for achieving land degradation neutrality.

1.4 Monitoring and Assessment of Land Degradation Neutrality

The assessment of LDN can use indicators and questionnaire methods. Land degradation is a multidimensional socioeconomic and environmental problem, and no single indicator can fully assess the health of the land. Because of the different definitions, diverse approaches to land degradation assessment have been used (Feng et al. 2022). Several studies have suggested a need for an indicator-based assessment and monitoring framework for effective natural resource management (Singh et al. 2022, 2024a; Chaki et al. 2022). The UNCCD secretariat and other international initiatives proposed a set of three global biophysical indicators for reporting on land degradation as part of the monitoring and evaluating framework (UNCCD 2013). The SDG 15.3.1 indicator is derived from a binary classification of land condition (degraded or not degraded) with the corresponding set of three measurable sub-indicators of land degradation: (1) Trends in land cover; (2) Trends in land productivity; and (3) Trends in carbon stocks (above and below ground) are currently represented by soil organic carbon (SOC) stocks. These sub-indicators are calculated using a one-out-all-out (1OAO) method, in which a significant reduction or negative change in any of the three sub-indicators is considered to comprise land degradation (Bernoux and Chevallier 2014; Sims et al. 2021). Other new indicators, such as wind erosion, water erosion, timber stocks, forest cover, and biodiversity levels, have been included in the LDN assessment globally (Chappell et al. 2019; Ptichnikov and Martynyuk 2020; Wuepper et al. 2021). Whereas, at the regional and local level, indicators like land contamination, gully erosion, soil loss rate, total soil loss, and atmospheric humidity are being considered for assessment (Speranza et al.

2019; Tsymbarovich et al. 2020; Kuderina et al. 2020). The questionnaire methods, such as participatory mapping, farmer interviews, field surveys, mapping questionnaires, and expert empirical data triangulation, are suggested by various studies, and it is an important supplement to the indicator method (Crossland et al. 2018; Von Maltitz et al. 2019; Garcia et al. 2019; Nzuza et al. 2021; Feng et al. 2022).

1.5 Strategies for Achieving Land Degradation Neutrality

1.5.1 Sustainable Land Management Practices

Sustainable land management (SLM) practices are crucial for achieving LDN, as they can help to prevent, reduce, or reverse degradation depending on the specific conditions of the intervention and site. A global network of SLM specialists established in 1992, the World Overview of Conservation Approaches and Technologies (WOCAT) aims to combat land degradation and support knowledge sharing and evidence-based decision-making to promote SLM and contribute to LDN. According to WOCAT, LDN is the sustainable use of land resources, including soil, water, animals, and plants, for producing goods to meet evolving human needs while ensuring the long-term productive potential and environmental functions of these resources (Liniger et al. 2011; Gonzalez-Roglich et al. 2019). SLM practices provide important local, regional, and global benefits. SLM practices, such as agroforestry, mulching, and water harvesting, can help capture and store carbon dioxide (CO_2) in vegetation and soil, reducing CO_2 emissions, mitigating climate change impacts and helping to preserve agrobiodiversity. Effective management of land resources based on indicators of agricultural zoning can optimize capital investments and maximize returns while ensuring the rational use of land and preserving natural resources. It aims to promote environmental conservation, enhance agricultural productivity, and support the overall well-being of communities while maintaining the health and resilience of land ecosystems (Critchley et al. 2023; Lukhno 2023; Abab et al. 2023). FAO launched the GEF-supported project, "Decision Support for Scaling up and Mainstreaming Sustainable Land Management (DS-SLM)" in September 2015. The project has 15 participating countries in Africa (Lesotho, Morocco, Nigeria and Tunisia), East and South Asia (Bangladesh, China, the Philippines and Thailand), Europe and Central Asia (Bosnia and Herzegovina, Turkey, and Uzbekistan), and South and Central America (Argentina, Colombia, Ecuador, and Panama). Several Indian government schemes are contributing to the achievement of land degradation neutrality. In the forestry sector, initiatives include the Green India Mission, Compensatory Afforestation, Nagar Van Scheme, Eco-development Forces, Intensification of Forest Management Scheme, Green Skill Development Programme, and National Coastal Mission. In the agriculture sector, efforts include the National Mission on Sustainable Agriculture and Soil Health Cards. Watershed management schemes such as the National Bamboo Mission (NBM), Integrated Watershed Management Programme (under PMKSY), and

CADWM Initiatives by State Governments under IWMP are also instrumental (Singh et al. 2023).

According to FAO, SLM works based on four principles, which include: (i) targeted policy and institutional support, including the development of incentive mechanisms for SLM adoption and income generation at the local level; (ii) land user-driven and participatory approaches; (iii) the integrated use of natural resources on farms and at the ecosystem scale; and (iv) multilevel, multistakeholder involvement and partnerships at all levels—land users, technical experts, and policymakers.

1.5.2 Socio-ecological System Approach

A socio-ecological system (SES) approach suggests that promoting and maintaining healthy land ecosystems relies on politically driven initiatives to avoid, reduce, or reverse land degradation. Land managers and institutions must ensure people care for and value ecosystems while efficiently managing rights and privileges over time and locations (Orr et al. 2017). This concept, also known as coupled human and natural systems or coupled human-environment systems, provides a robust framework for understanding the interconnectedness of human and ecological systems. It highlights people's dependence on nature and their ethical responsibilities toward it (Fischer et al. 2015). It offers a novel interdisciplinary platform to integrate different views and dimensions of global land system changes (Leslie et al. 2015). It is now necessary to draw insights from the humanities, social sciences, and natural sciences to promote a new land stewardship culture, such as emphasizing humans as nature (Braito et al. 2017). The integration of SES in LDN can help to foster LDN within a rapidly re-aligning global environmental change and sustainable development policy context (Okpara et al. 2018).

1.5.3 Agroforestry Approaches

Agroforestry is an inventive way to restore soil fertility and achieve LND. It is a simple biological method involving the cultivation of trees, shrubs, crops, or livestock, is cost-effective, and can reduce pressure on conventional agricultural lands and forests. Among various biological approaches, agroforestry stands out for its ability to increase productivity while maintaining ecological and economic stability. Furthermore, agroforestry systems provide a wide range of ecosystem benefits, including provisioning, regulation, support, and cultural services, directly contributing to land restoration efforts (Ghosh and Sinha 2016; Paramesh et al. 2019; Jinger et al. 2022). Many studies highlighted agroforestry as a soil conservational method that holds carbon (C) in the soil profile and enhances carbon storage. It could retain around 83.6 t C ha^{-1} up to 30 cm soil depth (Jha et al. 2001). However, Maikhuri et al. (2000) assessed the carbon capture potential of trees at around 3.9 t ha^{-1} y^{-1} for abandoned agricultural land and around 1.79 t ha^{-1} y^{-1} for degraded forest land

(Edrisi et al. 2018). Agroforestry systems have the potential to address a wide range of land degradation issues, including flooded areas, wastewater-affected sites, mining areas, wind-eroded lands, degraded slopes, ravines, and drought-prone regions. This adaptability shows agroforestry's ability to provide suited solutions for all kinds of degraded land, bringing out its promising role in sustainable land management practices (Dagar 2013; Uthappa et al. 2015; Jinger et al. 2023; Goswami et al. 2024). Combating land degradation is critical for food and water security and climate change adaptation and mitigation (IPBES 2018; Mehta et al. 2023; Hazarika et al. 2024). Considering the current challenges associated with land degradation, agroforestry appears as a better strategy for achieving land degradation neutrality (Jinger et al. 2023; Hazarika et al. 2024). Most developing and least developed countries depend solely on agriculture for their livelihood, so agroforestry has become a more reliable approach for land degradation neutrality in these areas. In this regard, developing robust institutional support and market linkages is key to the successful scaling of agroforestry, especially in developing countries (Ghosh and Sinha 2018, 2019).

1.5.4 Sustainable Agronomic Practices

The different strategies, such as sustainable agronomic practices, manage cropping systems, enhance soil quality, use more water, and protect the environment. Crop rotations and intercropping with nitrogen-fixing crops (groundnuts, beans, and cowpeas) can enhance soil fertility and enrich the nutrient supply for subsequent crops, resulting in increased crop yield (Woodfine 2009). Studies showed that crop rotation in the North Rift and western regions of Kenya increased maize yields by 71% and bean yields by 158% in 2005 (Hine and Pretty 2008), and in Brazil, intercropping maize with legumes (*Vignaunguiculata* and *Canavalisensiformis*), and plowing these back in as green manures showed significant yield (Parrott and Marsden 2002; Branca et al. 2013). Strip cropping combines contouring and crop rotation by alternating strips of row crops and soil-conserving crops on the same slope to help combat wind and water erosion in drylands and hilly regions. Intercropping cowpea with maize (2 rows of cowpea, 1 row of maize) reduced runoff by 10% and soil loss by 28% compared to pure maize. Minimum runoff was recorded under barnyard millet (36% of rainfall), followed by black soybean (37%) and maize (42%). Black soybean and maize alone had the highest soil loss (7.1 and 6.7 tons/ha, respectively), while barnyard millet had 4.8 tons/ha. Line sowing wheat and mustard (8:1 row ratio) optimized space and soil moisture use, increasing wheat equivalent yield by 14% and net returns by 30% compared to mixed sowing (Bhattacharyya et al. 2015). Thus, this implied that appropriate agronomic practices significantly enhance the restoration of degraded land. On degraded sites, Jatropha plants are reported to perform better when organic manure is applied as opposed to mineral fertilizers (Kumar and Sharma 2008). The Jatropha seed cake has high nitrogen content, making it a great source of plant nutrients that boost biomass content and plant growth (Behera et al. 2010; Edrisi et al. 2018).

1.5.5 Application of Organic Fertilizers

Extensive studies on different combinations of amendments, including fly ash, vermicompost, farmyard manure, chemical fertilizers, and bio-fertilizers, show significant improvements in the physicochemical and biological qualities of soil. Using organic manures and compost enhances the soil's organic carbon pool more effectively than applying the same amount of nutrients as inorganic fertilizers. Additionally, residue mulching improves soil structure, reduces bulk density, and increases both infiltration capacity and soil organic carbon pool (Gregorich et al. 2001; Shaver et al. 2002; Branca et al. 2013). Studies indicate that maize yields doubled in Kenya in 2005 (Hine and Pretty 2008), millet yields increased by 75–195%, and groundnut yields rose by 100–200% in Senegal in 2001 (Parrott and Marsden 2002). Similarly, in Bolivia, potato yields increased by 250–375% between the early 1980s and 2000 by using organic fertilizers. In regions where essential nutrients like N, P, and K are often limited, the use of bacterial strains known as plant growth-promoting bacteria (PGPBs), isolated from a variety of habitats, can favorably affect a wide range of plant growth and yield (de-Bashan et al. 2012). Additionally, Philippot et al. (2013) highlighted that plant association with arbuscular mycorrhiza contributes to the microbial-mediated enhancement of soil quality. This symbiosis is crucial for land restoration, facilitated by glycoproteins like glomalin that contribute to stable soil aggregate formation (Edrisi et al. 2018).

1.5.6 Water Management

Terracing and contour farming significantly enhance yields by reducing erosion and soil loss, as well as mitigating flooding and runoff. Studies show that restored Incan terraces increased upland crop yields by 150% (Altieri 2001), while contour hedgerows improved maize yields by up to 15% in the Philippines (Shively 1999). In Burkina Faso and Niger, millet yields increased significantly with the use of bunds, though high rainfall areas saw decreased yields (Dutilly-Diane et al. 2003). In Tanzania, terraces improved soil composition and boosted maize and bean yields (Branca et al. 2011). Water harvesting and conservation techniques, crucial for boosting agricultural production, have notably increased crop yields in various regions, including a 100% increase in cereal yields in Zimbabwe (Branca et al. 2013). Mechanical soil and water conservation measures in India's hilly regions are vital for controlling erosion and managing runoff. These include bunding to collect runoff and prevent erosion, graded bunds for high rainfall areas, contour bunds to reduce runoff, and bench and half-moon terraces for deeper soils supporting agroforestry. Additionally, grassed waterways manage natural drainage, and water harvesting ponds provide irrigation and fish farming resources (Bhattacharyya et al. 2015).

1.6 Challenges and Barriers to Achieve LDN

Land degradation neutrality (LDN) is a complex and multidimensional aim that entails protecting and enhancing land resources while balancing economic, social, and environmental needs. Despite its critical importance, achieving LDN presents numerous challenges and barriers. LDN faces major financial and economic challenges. Funding for sustainable land management is frequently insufficient, relying heavily on erratic donor support. Economic systems favor short-term profits from land exploitation, discouraging investment in environmentally friendly practices. Poverty forces communities to engage in unsustainable land use, which is exacerbated by a scarcity of viable alternatives. Therefore, Land Degradation Neutrality and meeting global goals require effective land governance, stable institutions, adequate financial resources, and stakeholder collaboration (Allen et al. 2020; Singh et al. 2024b). Significant investments are required to address the trillion-dollar challenge of land degradation, as public resources alone are insufficient (Dallimer and Stringer 2018). Hybrid governance models such as public-private partnerships (PPPs) can help to attract large-scale investments, but private sector participation remains low due to uncertainty about LDN viability (Gilbey et al. 2019). Promoting LDN's ability to provide multiple benefits from investments is critical to overcoming these barriers.

LDN is hampered by inconsistent land use policies across sectors such as agriculture, forestry, and urban development. Poor institutional capacities, a lack of coordination among government agencies, corruption, and transparency all impede effective land management. Unclear land tenure rights cause conflicts and discourage sustainable practices, while indigenous peoples' marginalized participation exacerbates non-compliant land use. To effectively promote LDN, these challenges must be overcome by improving policy coherence, strengthening institutions, ensuring transparent governance, and establishing clear land tenure rights. Implementing SLM policies and practices to minimize current and prevent future land degradation and restoring degraded and abandoned lands (UNDP 2017). Technical and knowledge barriers to achieving LDN include a lack of understanding of local SLM practices, limited access to modern monitoring technology in developing regions due to cost and expertise constraints, and insufficient data on land degradation trends and restoration outcomes, exacerbated by a lack of standardized assessment methodologies. There are broader knowledge and technological gaps among the developing and developed countries. These challenges impede the effective implementation of LDN strategies, necessitating increased research, technology access, and data collection efforts. Limited community participation in decision-making is one of the social and cultural barriers to achieving LDN, which leads to opposition to sustainable land practices. Cultural traditions that conflict with these practices, as well as a lack of awareness about LDN, further impede progress. Inadequate educational programs exacerbate these issues, emphasizing the importance of inclusive decision-making, cultural sensitivity, and improved education in promoting sustainable land management that aligns with LDN goals. Community-based initiatives improve land management to meet the needs and well-being of communities in a sustainable manner, thereby enhancing livelihoods. Thus, they may contribute to LDN (Akhtar-Schuster et al. 2017).

Land degradation is influenced by both climatic and non-climatic factors (e.g., unsustainable grazing and land use), which have an inverse effect on soil physical, chemical, and biological properties, particularly soil organic matter such as nitrogen content. Climatic factors accelerate the droughts and extreme weather events exacerbate land degradation, while non-climatic factors such as unsustainable agricultural practices and urbanization contribute directly to land degradation, making achieving LDN difficult. Restoration efforts have shown great promise in grassland ecosystems. According to studies, 70–90% of soil nitrogen can be restored after grazing is halted in China and Mongolia. This restoration not only improves grassland ecosystem services but also helps to increase soil carbon sequestration (Chen et al. 2022; Yu et al. 2023). Deserts in Central Asia have expanded rapidly in recent decades, exacerbated by desertification caused primarily by unsustainable water use practices, particularly the over-extraction of freshwater from rivers and lakes for activities such as cotton farming (e.g., the Aral Sea crisis). This has resulted in severe water scarcity and dwindling arable land availability. Land degradation on the Arabian Peninsula is primarily driven by water scarcity, which is exacerbated by extensive oil extraction and a scarcity of cultivable land (Chan et al. 2023; Westerberg and Myint 2016). Land degradation in Namibia and the broader southwest African region is caused by natural factors such as persistent drought and severe wind erosion, as well as human-induced factors such as over-cultivation, mining activities, and deforestation. Rapid population growth in the North African Sahel region has exacerbated land degradation through overgrazing, uncontrolled land reclamation, and widespread exploitation of mineral resources like oil and gas (Zhao et al. 2007). These interconnected factors highlight the complex and multifaceted nature of land degradation in these regions, emphasizing the critical need for SLM practices and effective environmental policies to mitigate its consequences (Chan et al. 2023). Climate change worsens droughts and floods, reducing land productivity and complicating restoration efforts. Overexploitation of natural resources such as water and forests results in depletion of biodiversity and critical ecosystem services required for land's health and productivity, posing significant environmental and ecological barriers to achieving LDN. These challenges and barriers must be considered when creating LDN plans (Fig. 1.1). By implementing best practices in governance, financing, and stakeholder engagement, we can effectively overcome these barriers and promote SLM around the world.

1.7 Conclusion

The increasing pressure on land due to the unsustainable use of land resources is leading to the degradation of productive land. To address this issue, achieving land degradation neutrality (LDN) has become part of the global agenda of the United Nations Convention to Combat Desertification (UNCCD). The scientific conceptual framework provided by the UNCCD supports this goal, offering a foundation for

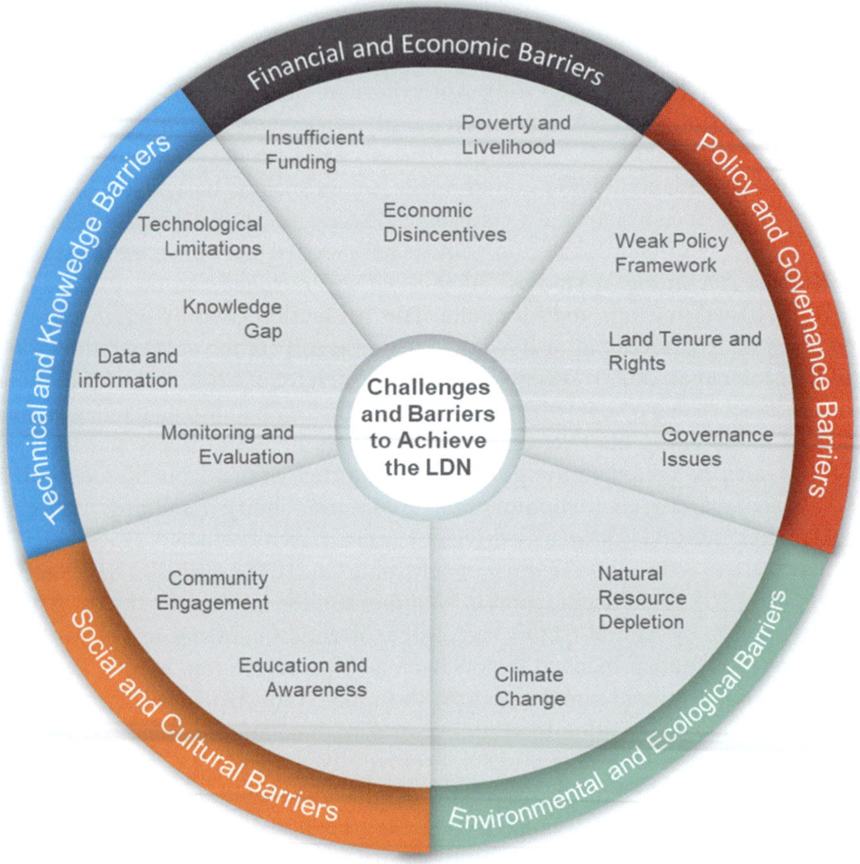

Fig. 1.1 Challenges and barriers to achieving Land Degradation Neutrality

restoring all types of land. The LDN framework suggests monitoring progress using three key indicators: trend in land cover, trend in land productivity, and trend in carbon stocks (Cowie et al. 2018). To achieve land degradation neutrality, various strategies have been proposed in numerous studies. Four key approaches, such as systems thinking, connectivity, nature-based solutions, and a regenerative economy, are identified as essential and strongly interrelated to each other for achieving LDN (Gonzalez-Roglich et al. 2019). Sustainable Land Management (SLM) is a crucial approach for achieving LDN. SLM practices are designed to avoid, reduce, and reverse land degradation through the restoration and rehabilitation of degraded land. These practices include agroforestry, agronomic practices, socio-ecological system approaches, water management, and the use of organic fertilizers.

References

Abab SA, Senbeta F, Negash TT (2023) The effect of land tenure institutional factors on small landholders' sustainable land management investment: evidence from the highlands of Ethiopia. Sustainability 15(12):9150

Akhtar-Schuster M, Stringer LC, Erlewein A, Metternicht G, Minelli S, Safriel U, Sommer S (2017) Unpacking the concept of land degradation neutrality and addressing its operation through the Rio Conventions. J Environ Manag 195:4–15

Allen C, Metternicht G, Verburg P, Akhtar-Schuster M, Inacio da Cunha M, Sanchez Santivañez M (2020) Delivering an enabling environment and multiple benefits for land degradation neutrality: stakeholder perceptions and progress. Environ Sci Policy 114:109–118

Altieri M (2001) Applying agroecology to enhance the productivity of peasant farming systems in Latin America. Paper presented at the reducing poverty through sustainable agriculture. University of Essex, London

Bhattacharyya R, Ghosh BN, Mishra PK, Mandal B, Rao CS, Sarkar D, Franzluebbers AJ (2015) Soil degradation in India: challenges and potential solutions. Sustainability 7(4):3528–3570

Behera SK, Srivastava P, Tripathi R, Singh JP, Singh N (2010) Evaluation of plant performance of Jatropha curcas L. under different agro-practices for optimizing biomass–a case study. Biomass Bioenergy 34:30–41

Bernoux M, Chevallier T (2014) Carbon in dryland soils: multiple essential functions. Les dossiers thematiques du CSFD, France

Bhat Y, Nandy S, Padalia H, Nath AJ, Pebam R (2022) Role of geospatial technology in shifting cultivation studies in India: a systematic review. J Indian Soc Remote Sens 50:2359–2379

Boer B, Hannam I (2021) Restoration of ecosystems and land degradation neutrality. In: Fulfilling the sustainable development goals. Routledge, pp 392–404

Braito MT, Böck K, Flint C, Muhar A, Muhar S, Penker M (2017) Human-nature relationships and linkages to environmental behaviour. Environ Values 26(3):365–389

Branca G, Lipper L, McCarthy N, Jolejole MC (2013) Food security, climate change, and sustainable land management. A review. Agron Sustain Dev 33:635–650

Branca G, Lipper L, Neves B, Lopa D, Mwanyoka I (2011) Payments for watershed services supporting sustainable agricultural development in Tanzania. J Environ Dev 20(3):278–302 https://doi.org/10.1177/1070496511415645 (Original work published 2011)

Chaki A, Sinha B, Bisaria J, Srivastava P (2022) Need for a comprehensive monitoring framework of Indian forests. Econ Polit Wkly 57(2):25–28

Chan FKS, Chen J, Li P, Wang J, Wang J, Zhu Y (2023) The cross-boundary of land degradation in Mongolia and China and achieving its neutrality-challenges and opportunities. Ecol Indic 151:110311

Chappell A, Webb NP, Leys JF, Waters CM, Orgill S, Eyres MJ (2019) Minimising soil organic carbon erosion by wind is critical for land degradation neutrality. Environ Sci Policy 93:43–52

Chasek P, Akhtar-Schuster M, Orr BJ, Luise A, Ratsimba HR, Safriel U (2019) Land degradation neutrality: the science-policy interface from the UNCCD to national implementation. Environ Sci Policy 92:182–190

Chen Q, Shang Y, Zhu R, Bao Q, Lin S (2022) Long-term enclosure at heavy grazing grassland affects soil nitrification via ammonia-oxidizing bacteria in Inner Mongolia. Sci Rep 12(1):21464

Conacher A (2009) Land degradation: a global perspective. N Z Geogr 65:91–94

Cowie AL, Orr BJ, Sanchez VMC, Chasek P, Crossman ND, Erlewein A, Welton S (2018) Land in balance: the scientific conceptual framework for Land Degradation Neutrality. Environ Sci Policy 79:25–35

Critchley W, Harari N, Mollee E, Mekdaschi-Studer R, Eichenberger J (2023) Sustainable land management and climate change adaptation for small-scale land users in sub-Saharan Africa. Land 12(6):1206

Crossland M, Winowiecki LA, Pagella T, Hadgu K, Sinclair F (2018) Implications of variation in local perception of degradation and restoration processes for implementing land degradation neutrality. Environ Dev 28:42–54

Dagar JC (2013) Greening salty and waterlogged lands through agroforestry systems for livelihood security and better environment. In: Agroforestry systems in India: livelihood security and ecosystem services. Springer India, New Delhi, pp 273–332

Dallimer M, Stringer LC (2018) Informing investments in land degradation neutrality efforts: a triage approach to decision making. Environ Sci Policy 89:198–205

de-Bashan LE, Hernandez JP, Bashan Y (2012) The potential contribution of plant growth promoting bacteria to reduce environmental degradation – a comprehensive evaluation. Appl Soil Ecol 61:171–189

Down To Earth (2019) Available at: https://www.downtoearth.org.in/news/agriculture/india-lost-31-of-grasslands-in-a-decade-66643

Dutilly-Diane C, Sadoulet E, de Janvry A (2003) Household behaviour under market failures: how natural resource management in agriculture promotes livestock production in the Sahel. J Afr Econ 12:343–370

Edrisi SA, Tripathi V, Abhilash PC (2018) Towards the sustainable restoration of marginal and degraded lands in India. Trop Ecol 59(3):397–416

Feng S, Zhao W, Zhan T, Yan Y, Pereira P (2022) Land degradation neutrality: a review of progress and perspectives. Ecol Indic 144:109530

Fertilizer Association of India (2021) Available at: https://www.faidelhi.org/general/connpk.pdf. Accessed 11 Jan 2023

Fischer J, Gardner T, Bennett E, Balvanera P, Biggs R, Carpenter S, Daw T, Folke C et al (2015) Advancing sustainability through mainstreaming a social-ecological systems perspective. Curr Opin Environ Sustain 14:144–149

Garcia CL, Teich I, Gonzalez-Roglich M, Kindgard AF, Ravelo AC, Liniger H (2019) Land degradation assessment in the Argentinean Puna: comparing expert knowledge with satellite-derived information. Environ Sci Pol 91:70–80

Ghosh M, Sinha B (2016) Impact of forest policies on timber production in India: a review. Nat Res Forum 40(1–2):62–76

Ghosh M, Sinha B (2018) Policy analysis for realizing the potential of timber production from trees outside forests (TOF) in India. Int For Rev 20(1):89–103

Ghosh M, Sinha B (2019) Institutional imperatives for promoting trees outside forests (TOFs) to enhance timber production in India. Small-scale For 18:57–79

Gilbey B, Davies J, Metternicht G, Magero C (2019) Taking land degradation neutrality from concept to practice: early reflections on LDN target setting and planning. Environ Sci Policy 100:230–237

Global Forest Watch (2022) Available at: https://www.globalforestwatch.org/

Global Soil Partnership (2017) Endorses guidelines on sustainable soil management. Available at: http://www.fao.org/global-soil-partnership/resources/highlights/detail/en/c/416516/

Gnacadja L (2012) Moving to zero-net rate of land degradation. Statement by Executive Secretary. UN Convention to Combat Desertification, Rio de Janeiro

Gonzalez-Roglich M, Zvoleff A, Noon M, Liniger H, Fleiner R, Harari N, Garcia C (2019) Synergizing global tools to monitor progress towards land degradation neutrality: trends. Earth and the World Overview of Conservation Approaches and Technologies sustainable land management database. Environ Sci Policy 93:34–42

Goswami S, Reddy BV, Yadav S, Adhruj A, Dash U, Rathore A (2024) Rice–fish-based agroforestry system: a climate smart way to reconcile sustainable livelihood options. In: Agroforestry to combat global challenges: current prospects and future challenges. Springer Nature Singapore, Singapore, pp 551–568

Gregorich EG, Drury CF, Baldock JA (2001) Changes in soil carbon under long-term maize in monoculture and legume-based rotation. Can J Soil Sci 81(1):21–31

Hannam I (2022) Soil governance and land degradation neutrality. Soil Secur 6:100030

Hazarika A, Kurmi B, Francaviglia R, Sileshi GW, Paramesh V, Das AK, Nath AJ (2024) The transition from shifting cultivation to indigenous agroforestry as nature-based solution for land restoration in the Indian Eastern Himalayas. Ecol Indic 162:112031
Hine R, Pretty J (2008) Organic agriculture and food security in Africa. United Nations Conference on Trade and Development and United Nations Environment Programme, Geneva
IUCN (2015) Land degradation neutrality: implications and opportunities for conservation, techni cal brief, 2nd edn. IUCN, Nairobi, p 19
Lukhno A (2023) The land resources management according to agrarian land zoning. Mod Manag Rev 28(2):39–49
Jha MN, Gupta MK, Raina AK (2001) Carbon sequestration: forest soil and land use management. Ann For 9:249–256
Jinger D, Khatri P, Kumari K, Kumar D, Dinesh D (2022) Agroforestry-based ecosystem services for livelihood resilience. Food Sci Rep 3:50–55
Jinger D, Kaushal R, Kumar R, Paramesh V, Verma A, Shukla M, Kumawat S (2023) Degraded land rehabilitation through agroforestry in India: achievements, current understanding, and future prospectives. Front Ecol Evol 11:1088796
Keesstra S, Mol G, Leeuw J, Okx J, Molenaar C, Cleen M, Visser S (2018) Soil-related sustainable development goals: four concepts to make land degradation neutrality and restoration work. Land 7(4):133
Kuderina TM, Suslova SB, Lunin VN, Kudikov AV (2020) Atmospheric moisture as a factor of land degradation neutrality in forest–steppe landscapes. Arid Ecosyst 10:156–160
Kumar A, Sharma S (2008) An evaluation of multipurpose oil seed crop for industrial uses (Jatropha curcas L.): a review. Ind Crop Prod 28:1–10
Kumar A, Singh S, Chand HB, Kewat AK, Singh KP (2021) Road towards restoring land degradation in India: a UNCCD perspective. RJCES 9(4):27–38
Kumar A, Singh S, Kumar R, Kumar M, Rani K, James J (2024) Balancing conservation and development for achieving land degradation neutrality. Horizon 2(4):17–24
Kust G, Andreeva O, Cowie A (2017) Land degradation neutrality: concept development, practical applications and assessment. J Environ Manag 195:16–24
Leslie HM, Basurto X, Nenadovic M, Sievanen L, Cavanaugh KC, Cota-Nieto JJ, Aburto-Oropeza O (2015) Operationalizing the social-ecological systems framework to assess sustainability. Proc Natl Acad Sci 112(19):5979–5984
Liniger HP, Mekdaschi R, Hauert C, Gurtner M (2011) Sustainable land management in practice: guidelines and best practices for sub-Saharan Africa. FAO, Rome
Maikhuri RK, Semwal RL, Rao KS, Singh K, Saxena KG (2000) Growth and ecological impacts of traditional agroforestry tree species in Central Himalaya, India. Agrofor Syst 48:257–271
Mehta D, Pandey R, Gupta AK, Juhola S (2023) Nature-based solutions in Hindu Kush Himalayas: IUCN global standard-based synthesis. Ecol Indic 154:110875
Mohamed E, Belal AA, Ali RR, Saleh A, Hendawy EA (2019) Land degradation. In: The soils of Egypt. Springer, pp 159–174
NAAS (2010) Degraded and wastelands of India–status of spatial distribution. National Academy of Agricultural Sciences, New Delhi
Nzuza P, Ramoelo A, Odindi J, Kahinda JMM, Lindeque L (2021) A triangulation approach for assessing and mapping land degradation in the Lepellane catchment of the greater Sekhukhune District, South Africa. S Afr Geogr J:1–25
Okpara UT, Stringer LC, Akhtar-Schuster M, Metternicht GI, Dallimer M, Requier-Desjardins M (2018) A social-ecological systems approach is necessary to achieve land degradation neutrality. Environ Sci Policy 89:59–66
Olsson L, Barbosa H, Bhadwal S, Cowie A, Delusca K, Flores-Renteria D, Hermans K, Jobbagy E, Kurz W, Li D, Sonwa DJ, Stringer L (2019) Land degradation. In: Climate change and land: an IPCC special report on climate change, desertification, land degradation, sustainable land management, food security, and greenhouse gas fluxes interrestrial ecosystems. Cambridge University Press. https://doi.org/10.1017/9781009157988.006

Orr BJ, Cowie AL, Castillo Sanchez VM, Chasek P, Crossman ND, Erlewein A, Walter S (2017) Scientific conceptual framework for land degradation neutrality: a report of the science-policy interface
IPBES (2018) Summary for policymakers of the assessment report on land degradation and restoration of the Intergovernmental SciencePolicy Platform on Biodiversity and Ecosystem Services. Zenodo. https://doi.org/10.5281/zenodo.3237411
Paramesh V, Arunachalam V, Nath AJ (2019) Enhancing ecosystem services and energy use efficiency under organic and conventional nutrient management system to a sustainable arecanut based cropping system. Energy 187:115902
Parrott N, Marsden T (eds) (2002) The real green revolution: organic and agroecological farming in the South. Greenpeace Environment Trust, London
Philippot L, Raaijmakers JM, Lemanceau P, van der Putten WH (2013) Going back to the roots: the microbial ecology of the rhizosphere. Nat Rev Microbiol 11:789–799
Ptichnikov AV, Martynyuk AA (2020) Adaptation of international indicators of land degradation neutrality for the assessment of forest ecosystems in arid conditions in Russia. Arid Ecosyst 10(2):127–134
Reddy GO, Kumar N, Singh SK (2018) Remote sensing and GIS in mapping and monitoring of land degradation. In: Geospatial technologies in land resources mapping, monitoring and management. Springer International Publishing, pp 401–424
SAC (2021) Desertification and land degradation atlas of India (Assessment and analysis of changes over 15 years based)
Sharma S, Kandpal V, Choudhury T, Gonzalez ED, Agarwal N (2023) Assessment of the implications of energy-efficient technologies on the environmental sustainability of rail operation. AIMS Environ Sci 10(5):709
Shaver TM, Peterson GA, Ahuja LR, Westfall DG, Sherrod LA, Dunn G (2002) Surface soil physical properties after twelve years of dryland no-till management. Soil Sci Soc Am J 66(4):1296–1303
Shively GE (1999) Risks and returns from soil conservation: evidence from low-income farms in the Philippines. Environ Monit Assess 62:55–69
Sims NC, Green C, Newnham G, England J, Held A, Wulder M, McKenzie N (2021) Good practice guidance. SDG Indic 15(1):2021-03
Singh M, Sinha B, Bisaria J, Thomas T, Srivastava P (2022) Understanding synergies and tradeoffs between forests, water, and climate change. Wiley Interdiscip Rev Water 9(6):e1614
Singh S, Giri K, Mishra G, Kumar M, Singh RK, Pandey S, Sharma R (2023) Pathways to achieve land degradation neutrality in India. Indian Council of Forestry Research and Education, Dehradun, p 3
Singh SN, Bisaria J, Sinha B, Patasaraiya MK, Sreerag PP (2024a) Developing a composite weighted indicator-based index for monitoring and evaluating climate-smart agriculture in India. Mitig Adapt Strateg Glob Chang 29(2):12
Singh S, Kumar A, Kumar A (2024b) Harmony in extraction: navigating mining, land degradation and ecological restoration for sustainable futures. In: Forests and climate change: biological perspectives on impact, adaptation, and mitigation strategies. Springer Nature Singapore, Singapore, pp 513–534
Sivakumar MV, Stefanski R (2007) Climate and land degradation—an overview. Springer Berlin Heidelberg, pp 105–135
Speranza CI, Adenle A, Boillat S (2019) Land degradation neutrality-potentials for its operationalisation at multi-levels in Nigeria. Environ Sci Policy 94:63–71
Tsymbarovich P, Kust G, Kumani M, Golosov V, Andreeva O (2020) Soil erosion: an important indicator for the assessment of land degradation neutrality in Russia. Int Soil Water Conserv Res 8(4):418–429
UNCCD (2012) Zero net land degradation, a sustainable development goal for Rio+20
UNCCD (2013) Refinement of the set of impact indicators on strategic objectives 1, 2 and 3. Recommendations of the ad hoc advisory group of technical experts. ICCD/COP(11)/CST/2

United Nations (2012) The future we want. Document A/CONF.216/L.1. 19 June. http://sustainabledevelopment.un.org/futurewewant.html

United Nations Development Programme (2017) Achieving land degradation neutrality for people and planet. https://www.undp.org/sites/g/files/zskgke326/files/publications/Achieving%20Land%20Degradation%20Neutrality.pdf

Uthappa AR, Bana OPS, Kumar M, Kanwal M (2015) Soil physico-bio-chemical properties as influenced by varying tree densities in poplar (Populusdeltoides Bartr. ex Marsh.) based agroforestry system. Indian J Agrofor 17(1):81–90

Von Maltitz GP, Gambiza J, Kellner K, Rambau T, Lindeque L, Kgope B (2019) Experiences from the South African land degradation neutrality target setting process. Environ Sci Pol 101:54–62

Westerberg V, Myint M (2016) An economic valuation of a large-scale rangeland restoration project through the Hima system in Jordan. ELD Initiative

Woodfine A (2009) The potential of sustainable land management practices for climate change mitigation and adaptation in sub Saharan Africa. Food and Agriculture Organization of the United Nations, Rome

Wuepper D, Borrelli P, Panagos P, Lauber T, Crowther T, Thomas A, Robinson DA (2021) A 'debt' based approach to land degradation as an indicator of global change. Glob Chang Biol 27(21):5407–5410

Yu J, Qu S, Li F, Wei D, Borjigidai A (2023) Plant nitrogen retention in alpine grasslands of the Tibetan Plateau under multi-level nitrogen addition. Sci Rep 13(1):877

Zhao Y, Peth S, Krümmelbein J, Horn R, Wang Z, Steffens M et al (2007) Spatial variability of soil properties affected by grazing intensity in Inner Mongolia grassland. Ecol Model 205(1–2):241–254

Global Efforts and Framework for Biodiversity Conservation

2

Harish Bahadur Chand, Aakash Kumar, Vatsala,
Padam Raj Joshi, Mukesh K. Gautam, and Sanjay Singh

Abstract

Biodiversity sustains life on Earth, providing essential ecosystem services. However, it faces alarming threats from habitat loss, climate change, and overexploitation of natural resources. This chapter explores the evolution of global biodiversity conservation efforts, examining key biodiversity frameworks like the convention on biological diversity and the Kunming-Montreal global biodiversity framework. It also highlights the importance of protected areas, indigenous knowledge, and community participation while addressing challenges for its implementation. The chapter finally provides way forward for strengthening the global biodiversity frameworks that include community engagement, cross-sectoral collaboration, research and education, use of advanced technologies and harmonizing conservation with livelihood. By prioritizing inclusive policies and

H. B. Chand
Forest Ecology and Climate Change Division, Forest Research Institute, Dehradun, India

A. Kumar
Himalayan Institute for Learning and Leadership (HILL), University of Petroleum and Energy Studies, Dehradun, India

Forest Research Institute (Deemed to be) University, Dehradun, India

Vatsala
Forest Research Institute (Deemed to be) University, Dehradun, India

P. R. Joshi
Chang'an University, Xi'an, Shaanxi, China

M. K. Gautam
Biology Department, Medgar Evers College, City University of New York, Brooklyn, NY, USA

S. Singh (✉)
Centre of Excellence, Sustainable Land Management, ICFRE, Dehradun, India

S. Singh, M. K. Gautam (eds.), *Biodiversity Conservation and Land Degradation Neutrality*, https://doi.org/10.1007/978-981-96-7210-3_2

equitable benefit-sharing, the chapter envisions a transformative path to harmonize biodiversity conservation with human well-being for a resilient and sustainable future.

Keywords

Ecosystem services · Global biodiversity framework · Indigenous knowledge · Community participation · Biodiversity conservation

2.1 Introduction

Biodiversity, encompassing the variety of plants, animals, and microorganisms on Earth, along with their genetic diversity and the diversity of ecosystems, is widely recognized as a priority for human society. Biodiversity denotes the quantity, variety, and population sizes of living species within their diverse physical environments (Wilson and Frances 1988). The extensive variety of species, genes, and ecosystems delivers vital ecosystem services, including food, water, medicine, and climate regulation, while also enhancing cultural, economic, and social values worldwide for human well-being (TEEB 2010; Singh et al. 2021; MEA 2005). Maintaining biodiversity is critically important for preserving the genetic diversity of species, that is, crucial for their resilience and adaptability to changing environment.

Human activities are seriously endangering biodiversity. Exploitation of natural resources beyond carrying capacity, ecosystem degradation, invasive species expansion, pollution, and climate change are major factors that are contributing to the global decline in biodiversity at an alarming rate (Ekardt et al. 2023). The rapid loss of species and ecosystems diminishes resilience (Vasiliev 2022), reducing the capacity of ecosystems to adapt to environmental changes. This poses a serious threat to the Earth's ecological integrity and threatens the services that biodiversity provides. Therefore, it is imperative to address these challenges through concerted global biodiversity conservation efforts and frameworks.

Numerous global frameworks and efforts have been established to address biodiversity loss. The convention on biological diversity (CBD) laid the foundation for uniting nations under a common goal to conserve biological diversity. Over the years, additional frameworks, such as the Aichi biodiversity targets and the Kunming-Montreal global biodiversity framework, have provided further strategic direction. This chapter explores the evolution of global biodiversity conservation efforts and the foundational frameworks over time. It also examines the significance of protected areas in conserving biodiversity, highlights the integration of indigenous knowledge, and emphasizes the importance of community participation in conservation initiatives. Additionally, the chapter delves into the challenges associated with implementing global biodiversity frameworks and offers insights into strategies for achieving more effective conservation outcomes.

2.2 Evolution of Global Biodiversity Conservation Effort

As humans recognized the significance of nature in meeting their needs and ensuring existence, worldwide conservation initiatives commenced. Initial conservation initiatives are associated with the establishment of sacred groves in India and the regulation of traditional hunting practices in Europe. These endeavors culminated in the establishment of the world's first National Park, Yellowstone, in the United States (1872), aimed at conserving its unique ecosystem and wildlife.

The concept of biodiversity conservation gained momentum during the Stockholm conference (1972), which framed environmental difficulties as global challenges necessitating a collective approach. This impetus resulted in the ratification of the CBD at the 1992 Earth Summit in Rio de Janeiro. The CBD marked a significant transformation by framing biodiversity conservation as a shared responsibility, associating it with sustainable development. Today, several international frameworks and conventions focus on addressing global challenges like habitat loss and climate change while fostering sustainable solutions for a biodiverse future.

2.2.1 International Framework and Convention for Biodiversity Conservation

The international framework for biodiversity conservation includes a series of agreements, conventions, and protocols designed to address the escalating loss of biodiversity and promote sustainable management of natural resources. These agreements focus collaboration among nations to integrate biodiversity into developmental policies, mitigate anthropogenic pressures, and achieve the targets of global sustainability agendas like the United Nations' Sustainable Development Goals (SDGs). Some of the important international conventions and frameworks for biodiversity conservation are shown in Fig. 2.1.

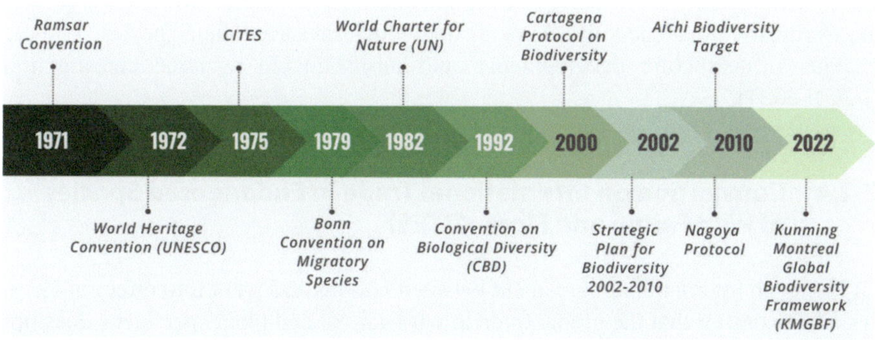

Fig. 2.1 Timeline series of International Convention and Frameworks of Biodiversity Conservation

2.2.2 Ramsar Convention

Ramsar convention is the only multilateral environmental agreement concentrating on a specific set of habitat types on wetlands, which was the first global biodiversity treaty of the modern period (Stroud et al. 2022). Wetlands are linked to large populations of fish, amphibians, birds, mammals, reptiles, and invertebrates. They are significant repositories of plant genetic material. Wetlands of international importance, sensible use of all wetlands, and international collaboration are the three pillars of this convention (Ramsar Convention 2000). Article 5 of this convention states that "Contracting Parties shall endeavor to coordinate present and future policies and regulations concerning the conservation of wetlands and their flora and fauna" (Ramsar 2016). A successful example of the Mediterranean wetlands initiative is MedWet initiative which was further followed by the establishment of the CREHO Ramsar Centre in Panama (Ramsar 2016).

2.2.3 United Nations Educational, Scientific and Cultural Organization (UNESCO)

Since the Convention's adoption in 1972, the idea of "sustainable development" has gained acceptance from the global community. A world heritage committee and a world heritage fund were established, that same year. Its aim was to identify, preserve, conserve, present, and transfer the cultural and natural legacy that has exceptional universal heritage to future generations (Organisation 2008).

On June 26, 2023, the secretariat of CITES and the UNESCO secretariat to the world heritage convention signed a historic agreement in a new attempt to guarantee the sustainable trade of wild animal and plant species found in regions of exceptional international significance listed on the UNESCO world heritage list (CITES 2023). UNESCO world heritage sites are home to an estimated half of the animal species listed in the CITES appendices (CITES 2023). For instance, the 2020 IUCN world heritage outlook revealed that around one-third of natural world heritage sites are seriously threatened by elements including invasive alien species, climate change, infrastructure developments, and unsustainable resource consumption (IUCN 2024).

2.2.4 Convention on International Trade in Endangered Species of Wild Fauna and Flora (CITES)

CITES is an international agreement between nations and went into effect in 1975. It seeks to ensure that the global trade in wild animal and plant specimens does not jeopardize the species' ability to thrive. It ensures that international commerce does not threaten the existence of about 35,000 species in the wild by regulating the global wildlife trade (CITES 2014). It depends on precise and targeted regulatory actions, such as trade restrictions and controls, which are put in place after a species' danger of extinction is evaluated and trade levels are subsequently tracked (CITES 2014; Challender et al. 2015).

2.2.5 Bonn Convention on Migratory Species (CMS)

The Bonn Convention is an environmental convention signed in 1979 under the guidance of the UNEP that provides a global platform for the conservation and sustainable use of migratory species and their habitats throughout their breeding and wintering range, and migratory corridors (flyways) including stop-over sites. Globally, 44% of CMS-listed species are experiencing population declines, with one in every five threatened with extinction. On average, habitat loss, degradation, and fragmentation affect three out of every four CMS-listed species (UNEP-WCMC 2024). Migratory species with unfavorable conservation status and other species that might greatly benefit from the negotiation of agreements are included in Appendix II of the CMS (Trouwborst 2012). There are various actions to conserve migratory species, and their habitats are (UNEP-WCMC 2024):

- Implementation of legally binding obligations under CMS
- Protecting and conserving major habitats for migratory species
- To promote ecological connectivity by removing migration barriers
- Mitigating light pollution
- Lessening overexploitation which includes mitigating incidental catch of non-target species

2.2.6 Convention on Biological Diversity (CBD)

CBD is an international legal agreement for "the conservation of biological diversity, the sustainable use of its components and the fair and equitable sharing of the benefits arising out of the utilization of genetic resources" that has been signed by 196 nations. It addresses biodiversity at the species, ecosystem, and gene levels. It covers biotechnology under the 2000 Cartagena protocol on biosafety and the 2010 Nagoya protocol on access and benefit sharing (UN 2024). Its goal is to encourage actions that will result in a sustainable future. It addresses all aspects of biodiversity, both directly and indirectly, as well as its role in development, economics, and politics, extending from science and education to agriculture, culture, and much more (UN 2024). It addresses biodiversity at the species, ecosystem, and gene levels.

2.2.7 Strategic Plan for Biodiversity (2002–2010)

To direct the CBD's future implementation globally, the parties to the convention on biological diversity (CBD) created this strategic plan in 2002, 10 years after it was made available for signature (CBD 2010). To ensure the continuation of its beneficial uses, the goal is to successfully halt the loss of biodiversity through the conservation and sustainable use of its constituent parts as well as fair and equal distribution of benefits resulting from the use of genetic resources. The CBD's work programs, national biodiversity policies and action plans, and other national, regional, and worldwide initiatives will all be used to carry out the strategic plan. Four goals and targets of the convention (CBD 2010) are follows:

Goal 1: Leadership role in international biodiversity issues.

Goal 2: Parties are better equipped to carry out the convention in terms of money, manpower, science, technology, and technology.

Goal 3: The convention's goals may be effectively implemented using national biodiversity strategies and action plans.

Goal 4: The convention and the value of biodiversity are now better understood, which has increased societal participation in its implementation.

2.2.8 Aichi Biodiversity Target (2010–2020)

Aichi biodiversity target was a decadal framework for action by all stakeholders to protect biodiversity and increase its benefits for people. Following the acceptance of the strategic plan for biodiversity 2011–2020 and its Aichi biodiversity targets and its progress over the "decade of biodiversity" (Maney et al. 2024). There were 5 goals and 20 targets of this strategic plan. The vision of this strategic plan was "Biodiversity is valued, restored, conserved and wisely utilized, maintaining ecosystem services, supporting a healthy planet and delivering benefits required for all people by 2050" (CBD 2020). Its goal is to stop biodiversity loss as quickly as possible so that by 2020 ecosystems are resilient and still provide vital services, protecting the planet's diversity of life and promoting human well-being and the eradication of poverty (CBD 2020). Following up on the Aichi targets, a new agreement was made in December 2022 under CBD, with targets set for 2030 and 2050. The most recent phase of agreements under the CBD is represented by the "Kunming-Montreal Global Biodiversity Framework," which includes a commitment to put in place a monitoring system to gauge advancement (CBD 2022).

2.2.9 Kunming Global Biodiversity Framework (GBF)

At a time when environmental degradation is threatening the survival of one million species and impacting the lives of billions of people, GBF seeks to halt and reverse it. This framework addresses the implementation gap left after the Aichi biodiversity targets, which were the previous global instrument for mainstreaming biodiversity conservation between 2010 and 2020 (IPBES 2019). Global targets to safeguard and sustainably use biodiversity by 2030 and beyond are part of the framework (UNEP 2022). To ensure that the world is on track to meet the 2050 vision for biodiversity by 2030, it requires transparent, responsible growth monitoring together with effective review mechanisms. Zero Draft's long-term goals for 2050 focused on ecosystem resilience and integrity, species conservation, genetic diversity, the benefits of nature to humans, meeting the goals of the Paris climate agreement, and equitable and fair distribution of the advantages of using genetic resources and traditional knowledge (Tsioumani 2020). Its 2030 mission is to take immediate action across society to put biodiversity on a road to recovery for the benefit of both people and the planet. Twenty action-oriented objectives for 2030 were also added, grouped into clusters on lowering biodiversity threats, addressing human needs via sustainable usage and benefit sharing, and providing tools and solutions for mainstreaming and implementation (Tsioumani 2020).

2.3 Global Biodiversity Initiatives and Programs

2.3.1 Coral Triangle Initiative on Coral Reefs, Fisheries, and Food Security (CTI-CFF)

The CTI, which spans 5.7 million km^2 and is biogeographically defined by high coral variety, is an intergovernmental agreement among six members governments (CT6): Malaysia, Indonesia, the Philippines, Papua New Guinea, Solomon Islands, and Timor Leste. The Coral Triangle region of Asia-Pacific, which is home to 76% of the world's coral and 37% of its reef fish species, is the worldwide epicenter of marine biodiversity (Foale et al. 2013). Approximately 120 million of the region's 370 million residents depend on the commodities and services provided by the marine environment for tourism, coastal protection, and fisheries output. Its bold and revolutionary vision is outlined in five regional goals: (i) the designation and effective management of priority seascapes; (ii) the full application of EAFM and other marine resources; (iii) the establishment and effective management of MPAs; (iv) the accomplishment of climate change adaptation measures; and (v) the improvement of threatened species status (Foale et al. 2013). CTI-CFF has two main objectives: (i) by 2025, coastal communities and coastal and marine ecosystems in the Coral Triangle region will be able to handle the effects of climate change and both natural and man-made threats; (ii) by 2030, food security, sustainable fisheries, and coastal livelihoods will have improved, making coastal communities and coastal and marine ecosystems in the CT region more resilient to the effects of climate change and man-made and natural threats (Foale et al. 2013).

2.3.2 Amazon Region Protected Areas (ARPA) Program

The primary biodiversity protection plan for the Amazon biome is represented by ARPA which is the largest conservation and sustainability project in the world focused on tropical forests (WWF 2017). The program was launched in 2002 and was formally initiated in 2003. By 2013, the initiative seeks to assist in the establishment and consolidation of 500,000 km^2 of protected land. Approximately 375, 000 km^2 of this total are designated as protected areas with creation assistance. Support will be provided for the consolidation and execution of the remaining 125,000 km^2. Its activities are carried out in three stages (2003–2010, 2010–2017, and 2014–2039) focuses on biological reserve, ecological station, and national park or state park (WWF 2017).

According to Soares-Filho et al. (2010), the areas are spread across all states in the Amazon region and safeguard 305,000 km^2, of which 217,000 km^2 are strictly preserved and 88,000 km^2 are reserves for sustainable usage. In regions not covered by the ARPA program, the deforestation was reduced at a rate comparable to that in stringent protection areas, sustainable use areas, and indigenous lands. The initiative promotes protected area establishment, consolidation, and upkeep (WWF 2017).

2.3.3 Roles of Protected Areas in Global Biodiversity Conservation

Protected areas (PAs) act as a central component for the conservation of biodiversity worldwide. The national parks, wildlife reserves, and marine protected areas (MPAs) are the important designated zones in efforts to reduce the loss of biodiversity caused by anthropogenic activities. To globally constrain the rate of biodiversity loss, important ecological, social, and policy interventions with innovative approaches are crucial.

- *Ecological Significance*

  Protected areas (PAs) are essential for conservation of ecosystems and maintenance of ecological integrity.

  PAs ensure the survival of endangered species and preserve genetic diversity through securing habitats. PAs like the Amazon rainforest national parks in Brazil and the Great Barrier Reef marine park in Australia are critical in conservation of biodiversity hotspots from threats such as deforestation, overfishing, and climate change. PAs also act as carbon sinks, which contributes to the regulation of climate by sequestering greenhouse gases (GHGs). For example, tropical forest reserves absorb a significant quantity of carbon dioxide (CO_2) which reduces the global carbon footprint (Mitchell Jr. 1991). A unique vertical in the ecological functioning is "climate change refugia" that is defined by Morelli et al. (2016) as "areas relatively buffered from contemporary climate change over time that enable persistence of valued physical, ecological, and socio-cultural resources" provide stable microhabitats where species can survive climate shifts. Alagador et al. (2014) underline the importance of PAs in buffering the climate change impacts on species distributions and ensuring regulations against environmental fluctuations.
- *Socio-Economic Benefits*

  Apart from ecological advantages, PAs contribute to socio-economic benefits by providing sustainable livelihoods and eco-tourism. Communities residing in the vicinity of PAs often get involved in activities like eco-friendly agriculture and nature-based tourism, which provide them with alternative sources of income as well as decreasing their dependency on the natural resources (Millennium Ecosystem Assessment 2005). For example, Nepal's Annapurna conservation area integrates local governance and conservation, demonstrating the role of participatory management in aligning human development with biodiversity goals (Schuett et al. 2016). Ecosystem services, like clean water, pollination, fresh air, and soil fertility, which are essential for global food security, are preserved by the protected areas. Coastal

PAs, like mangrove ecosystems, protect communities from storms and flooding, highlighting their role in disaster risks reduction.

- *Policy Framework and Global Commitments*

PAs are constitutive to international frameworks, such as Convention on Biological Diversity (CBD), which aims to diversify the global PA coverage under the post 2020 global biodiversity framework. Presently, 17% of the terrestrial and 8% of marine areas are under protection. The aim is to elevate this to 30% by the year 2030 (30 × 30 goal) (UNEP-WCMC 2020). Transboundary conservation efforts and initiatives can meet the interconnected nature of the ecosystems. The Kavango-Zambezi Transfrontier Conservation Area (KAZA TFCA) is a vast conservation area spanning five countries—Angola, Botswana, Namibia, Zambia, and Zimbabwe—in southern Africa. It houses an extraordinary diversity of ecosystems wildlife, plants, and indigenous communities. Through cross-border association, KAZA TFCA exemplifies the significance of in habitat connectivity and supporting species migration. Similarly, the concept of community-conserved areas (CCAs) emphasizes the local stewardship that promotes inclusiveness in the biodiversity conservation practices.

PAs also face challenges related to inappropriate funding, encroachment by local communities, and climate-induced stressors. Strengthening governance mechanisms and integrating them with indigenous knowledge into conservation and management plans are essential to overcome these challenges. Technological interventions like satellite monitoring and artificial intelligence-driven biodiversity assessments offer real-time insights into ecosystem health. Future conservation efforts should consider ensuring that benefits from PAs are equally shared with all the key stakeholders. The involvement of indigenous peoples and local communities in decision-making processes can increase the efficacy and legitimacy of the conservation measures.

2.4 Indigenous Knowledge and Community Participation in Conservation

The role of Indigenous knowledge and community participation in biodiversity conservation is an essential component that promotes sustainable ecosystems. IPLCs manage natural resources through traditional ecological knowledge (TEK). Their involvement boosts up the effectiveness of the conservation efforts as well as ensuring the equitable and culturally sensitive approaches.

Indigenous knowledge systems (IKS) relate to their cultural heritage and are adjusted to their local ecological conditions. IKS offers insights into behavior of species, ecosystem dynamics, and climate adaptation. The use of fire management by Aboriginal Australians has long maintained biodiversity in fire-prone ecosystems by preventing uncontrolled wildfires and promoting vegetation growth (Bird et al. 2008). Similarly, in the Amazon rainforest, indigenous communities practice agroforestry systems that integrate food production with forest preservation, enhancing soil fertility and biodiversity. These practices show how the traditional land-use strategies contribute to the sustainable management of natural resources and ecosystems and sequester carbon (Garnett et al. 2018).

2.5 Community Participation in Conservation

Community participation fulfills the gap between the top-down conservation strategies and efforts. It empowers the local population to become the custodians of biodiversity while providing a sense of responsibility and ownership. Community-conserved areas (CCAs) are one of the initiatives which showcase the potential of participatory conservation models. In the regions where policies restrict the access to lands of traditional significance, community engagement ensures that the efforts align with the need of the local people and provide long-term cooperation and conviction. Innovative policies like biocultural rights and payments for ecosystem services (PES) further encourage community involvement. For example, Ecuador's Socio Bosque Program compensates indigenous communities for conserving forests, blending economic incentives with traditional stewardship (de Koning et al. 2011). Community participation influences the collective efforts and knowledge of the local citizens, which not only enhances stewardship but also the socio-economic challenges. In South Asia, specifically India and Nepal, the best examples of participatory approaches to supplement formal conservation strategies include Joint Forest Management (Fig. 2.2a) and community forest (Fig. 2.2b), respectively.

2.6 Challenges to Implementing Global Biodiversity Frameworks

Challenges to implement global biodiversity framework is an ever-going discussion topic in the field of biodiversity conservation. Li et al. (2023) provided a landmark report for the list of challenges. The overall challenges can be summarized as:

(i) *Funding and fund disbursements*: inadequate financial support and unequal distribution of financial and technical resources for developing and least developed nations
(ii) *Development of quantifiable indicators*: challenge is to develop quantifiable indicators as Aichi Targets did not have measurable indicators for all global targets (Dudley and Stolton 2023)
(iii) *Coordination*: inadequate international coordination and collaboration on funding, research, and data sharing
(iv) *Knowledge sharing and data gap*: limited research and inadequate knowledge sharing among peers
(v) *Time constraints*: still in conceptual phase and time left is under 6 years (Li et al. 2023)

a

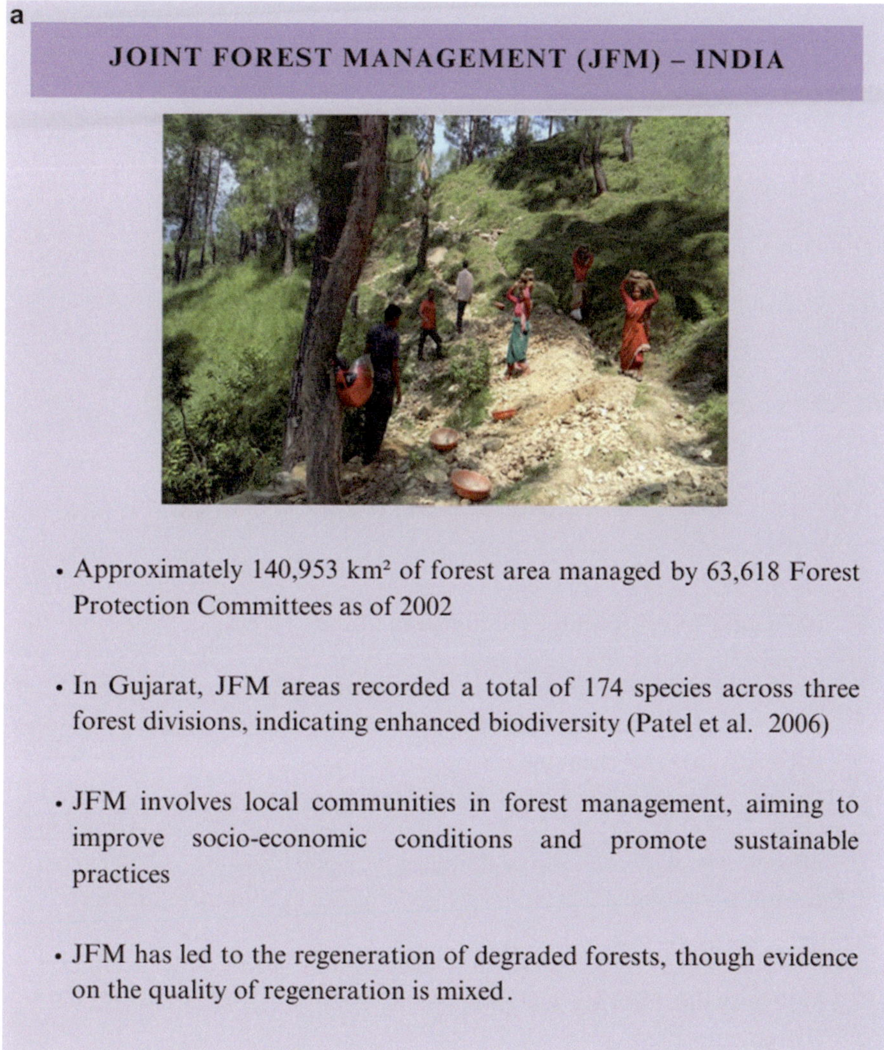

Fig. 2.2 (**a**) Joint Forest Management of India. (Source: Pooran Singh Rawal). (**b**) Community Forest of Nepal. (Source: Jenny Ohman)

Fig. 2.2 (continued)

(vi) *Social barrier*: conflicts between conservation efforts and the needs of local communities

(vii) *Inclusiveness in decision making*: insufficient participation of indigenous people and local communities in decision-making process

(viii) *Economic development conflict*: conservation approach versus agricultural expansion and infrastructure development

(ix) *Integrated policy approaches*: synergy between biodiversity conservation, agriculture, forestry, and industrial sector

(x) *Awareness and acceptance*: fostering environmental stewardship by public awareness of biodiversity conservation issues and making them realize the seriousness of the problem

2.7 Future Directions and Opportunities for Biodiversity Conservation

Efforts to conserve biodiversity are necessary and ongoing. There are successful endeavors in the area, but necessitate more substantial actions. The following are a few aspects that are necessary to strengthen the global biodiversity conservation efforts.

- *Community Engagement and Indigenous Knowledge*: biodiversity conservation efforts should prioritize involving a diverse group of stakeholders, including local and indigenous communities, underprivileged social groups, and women, in the decision-making process. Sharing indigenous knowledge and integrating it into conservation activities can significantly enhance the effectiveness and inclusivity of these efforts.
- *Research and Educational Activities*: need for more detailed research on restoration techniques and nature-based solutions is required. The research should focus on developing effective strategies for ecosystem restoration and harnessing nature's potential to address the decline in biodiversity. Alongside this, capacity building and training campaigns will be essential to equip stakeholders with the knowledge and skills necessary to implement the conservation efforts effectively.
- *Cross Sectoral and International Collaboration*: the success of the conservation projects heavily depends upon multilateral agreements (CBD, CITES, and other conventions) and requires strengthening collaboration between them. Economic development conflict is a major challenge; therefore engagement of industries and infrastructure development projects in corporate social responsibility and biodiversity-friendly approaches are a critical.
- *Technological Advancement*: use of advanced technology, such as RS and GIS, to monitor the efficiency of biodiversity conservation efforts. These could include monitoring critical habitats and ecosystems and land use changes. Additionally, the application of artificial intelligence and machine learning should be prioritized for assessing, analyzing, and monitoring biodiversity conservation projects.
- *Harmonizing Conservation with Livelihoods*: conservation strategies should be carefully designed to respect the socio-economic requirements of the community. Equitable benefit distribution and subsidiarity, with a focus on local rights and governance structures that enhance community well-being without undermining larger conservation goals, should be promoted.

2.8 Conclusion

Biodiversity is essential for maintaining ecological balance and human livelihoods. Even though international frameworks like the CBD and efforts like ARPA and CTI-CFF have led to progress, it is still hard to get fair and effective conservation results. Protected areas and community-driven approaches like Joint Forest Management and Community Forest highlight the potential for merging ecological and socio-economic concerns. Addressing challenges such as funding deficits, stakeholder participation, and conflicts with development objectives require advancements in research, enhanced community engagement, technological innovation, and interdisciplinary collaboration. Future efforts must reconcile biodiversity protection with livelihood demands, emphasizing education, research, and policy integration. Only by an integrated and adaptive strategy can biodiversity loss be mitigated, safeguarding a resilient planet for future generations.

References

Alagador D, Cerdeira JO, Araújo MB (2014) Shifting protected areas: scheduling spatial priorities under climate change. J Appl Ecol 51(3):703–713

Bird RB, Bird DW, Codding BF, Parker CH, Jones JH (2008) The "fire stick farming" hypothesis: Australian aboriginal foraging strategies, biodiversity, and anthropogenic fire mosaics. Proc Natl Acad Sci 105(39):14796–14801. https://doi.org/10.1073/pnas.0804757105

CBD (2010) Strategic plan 2002–2010. Retrieved from: https://www.cbd.int/sp/2010. Accessed on: 20 Nov 2024

CBD (2020) Aichi biodiversity targets. Retrieved from: https://www.cbd.int/sp/targets. Accessed on: 18 Sept 2024

CBD (2022) Kunming-Montreal global biodiversity framework. https://www.cbd.int/doc/c/e6d3/cd1d/daf663719a03902a9b116c34/cop-15-l-25-en.pdf. Accessed on: 18 Sept 2024

Challender DW, Harrop SR, MacMillan DC (2015) Understanding markets to conserve trade-threatened species in CITES. Biol Conserv 187:249–259

Chumbler C (2020) Community forestry is much more than just forests: Nepal's conservationists become local leaders. Exposure; USAID Publications, 16 June 2020. Retrieved from: https://usaidpubs.exposure.co/community-forestry-is-much-more-than-just-forests

CITES (2014) Appendices I, II and III. Retrieved from: https://cites.org/sites/default/files/eng/app/2014/E-Appendices-2014-09-14.pdf

CITES (2023) UNESCO World Heritage Convention and CITES unite to cooperate on conservation and sustainable wildlife trade in species occurring in World Heritage sites. https://cites.org/eng/news/UNESCO-World-Heritage-Convention-and-CITES-unite-to-cooperate-on-conservation-and-sustainable-wildlife-trade-in-species-occurring-in-World-Heritage-sites

Damodaran A, Engel S (2003) Joint Forest Management in India: assessment of performance and evaluation of impacts. ZEF Discussion Papers on Development Policy, No. 77, University of Bonn, Center for Development Research (ZEF), Bonn. https://doi.org/10.22004/ag.econ.18752

de Koning F, Aguiñaga M, Bravo M, Chiu M, Lascano M, Lozada T, Suarez L (2011) Bridging the gap between forest conservation and poverty alleviation: the Ecuadorian Socio Bosque program. Environ Sci Pol 14(5):531–542). Elsevier BV. https://doi.org/10.1016/j.envsci.2011.04.007

Dhungana N, Lee C-H, Khadka C, Adhikari S, Pudasaini N, Ghimire P (2024) Evaluating Community Forest User Groups (CFUGs)' performance in managing community forests: a case study in central Nepal. Sustainability 16(11):4471). MDPI AG. https://doi.org/10.3390/su16114471

Dudley N, Stolton S (2023) Best practice in delivering the 30 × 30 target, 2nd edn. The Nature Conservancy and Equilibrium Research, Cambridge

Ekardt F, Günther P, Hagemann K et al (2023) Legally binding and ambitious biodiversity protection under the CBD, the global biodiversity framework, and human rights law. Environ Sci Eur 35:8080. https://doi.org/10.1186/s12302-023-00786-5

Foale S, Adhuri D, Aliño P, Allison EH, Andrew N, Cohen P et al (2013) Food security and the coral triangle initiative. Mar Policy 38:174–183

Garnett ST, Burgess ND, Fa JE, Fernández-Llamazares Á, Molnár Z, Robinson CJ, Watson JEM, Zander KK, Austin B, Brondizio ES, Collier NF, Duncan T, Ellis E, Geyle H, Jackson MV, Jonas H, Malmer P, McGowan B, Sivongxay A, Leiper I (2018) A spatial overview of the global importance of Indigenous lands for conservation. Nat Sustain 1(7):369–374). Springer Science and Business Media LLC. https://doi.org/10.1038/s41893-018-0100-6

IPBES (2019) Summary for policymakers of the global assessment report on biodiversity and ecosystem services of the intergovernmental science-policy platform on biodiversity and ecosystem services. IPBES Secretariat, Bonn. https://doi.org/10.5281/zenodo.3553579

IUCN (2024) World Heritage sites: havens for biodiversity in a changing world. Retrieved from: https://iucn.org/blog/202405/world-heritage-sites-havens-biodiversity-changing-world. Accessed on: 20 Oct 2024

Li Q, Ge Y, Sayer JA (2023) Challenges to implementing the Kunming-Montreal global biodiversity framework. Land 12(12):2166. https://doi.org/10.3390/land12122166

Luintel H, Bluffstone RA, Scheller RM (2018) The effects of the Nepal community forestry program on biodiversity conservation and carbon storage. PLoS One 13(6):e0199526. https://doi.org/10.1371/journal.pone.0199526

Maney C, Guaras D, Harrison J, Guizar-Coutiño A, Harfoot MB, Hill SL, Sutherland W (2024) National commitments to Aichi Targets and their implications for monitoring the Kunming-Montreal global biodiversity framework. npj Biodivers 3(1):6

Millennium Ecosystem Assessment (2005) Ecosystems and human well-being: synthesis. Island Press, Washington, DC

Mitchell JM Jr (1991) Carbon dioxide and future climate. Weatherwise 44(4):17–23). Informa UK Limited. https://doi.org/10.1080/00431672.1991.9929375

Morelli TL, Daly C, Dobrowski SZ, Dulen DM, Ebersole JL, Jackson ST, Beissinger SR (2016) Managing climate change refugia for climate adaptation. PLoS One 11(8):e0159909

Organisation C (2008) Operational guidelines for the implementation of the World Heritage Convention. Paris, France

Patel R, Mali S, Tripathi JP, Kaushal V, Mudrakartha S (2006) Regeneration of teak forests under joint forest management in Gujarat. Int J Environ Sustain Dev 5(1):85–95

Ramsar (ed) (2016) An introduction to the Ramsar Convention on wetlands (previously The Ramsar Convention manual), 7th edn. Ramsar Convention Secretariat, Gland

Ramsar Convention (2000) Ramsar information paper no. 1. https://www.ramsar.org/sites/default/files/documents/library/info2007-01-e.pdf

Schuett MA, Dahal S, Nepal S (2016) Local perspectives on benefits of an integrated conservation and development project: the Annapurna conservation area in Nepal. Int J Biodivers Conserv 8(7):138–146

Singh S, Chand HB, Khatri PK, Kumar D, Kewat AK, Kumar A, Singh KP (2021) Phytosociology and regeneration status in different permanent preservation plots across different forest types in Madhya Pradesh, Central India. Grassroots J Nat Resour 4(2):179–198. The Grassroots Institute. https://doi.org/10.33002/nr2581.6853.040213

Soares-Filho B, Moutinho P, Nepstad D, Anderson A, Rodrigues H, Garcia R et al (2010) Role of Brazilian Amazon protected areas in climate change mitigation. Proc Natl Acad Sci 107(24):10821–10826

Stroud DA, Davidson NC, Finlayson CM, Gardner RC (2022) Development of the text of the Ramsar Convention: 1965–1971. Mar Freshw Res 73(10):1107–1126

TEEB (2010) The Economics of Ecosystems and Biodiversity: mainstreaming the economics of nature: a synthesis of the approach, conclusions and recommendations of TEEB. Retrieved from: https://www.teebweb.org/wpcontent/uploads/Study%20and%20Reports/Reports/Synthesis%20report/TEEB%20Synthesis%20Report%202010.pdf
Trouwborst A (2012) Transboundary wildlife conservation in a changing climate: adaptation of the Bonn Convention on Migratory Species and its daughter instruments to climate change. Diversity 4(3):258–300
Tsioumani E (2020) Convention on biological diversity: a review of the post-2020 global biodiversity framework working group negotiations. Environ Policy Law 50(1–2):55–59
UN (2024) Convention on biodiversity. United Nations. Retrieved from: https://www.un.org/en/observances/biological-diversity-day/convention
UNEP (2022) Kunming-Montreal global biodiversity framework. Retrieved from: https://www.unep.org/resources/kunming-montreal-global-biodiversity-framework. Accessed on: 19 Dec 2022
UNEP-WCMC (2020) Protected planet report 2020. UNEP-WCMC and IUCN: Cambridge UK; Gland, Switzerland. https://doi.org/10.34892/jg6t-xn70
UNEP-WCMC (2024) State of the world's migratory species. UNEP-WCMC, Cambridge
Vasiliev D (2022) The role of biodiversity in ecosystem resilience. IOP Conf Ser Earth Environ Sci 1072:012012
Wilson EO, Frances MP (eds) (1988) Biodiversity. National Academy Press, Washington, DC. 521 pp
WWF (2017) The impact of the ARPA Program on the management effectiveness of Amazon protected areas. Retrieved from: https://d3nehc6yl9qzo4.cloudfront.net/downloads/wwf_folder_ingles_paginas_2.pdf

Bridging Horizons: Global Biodiversity Framework and Land Degradation Neutrality

3

Vatsala, Aakash Kumar, Aishwarya Negi, Mukesh K. Gautam, Sanjay Singh, Abhishek Kumar, Moses Fayiah, and Harish Bahadur Chand

Abstract

The interconnected crises of biodiversity loss, land degradation, and climate change threaten planetary health, food security, and human well-being. These challenges, driven by unsustainable human activities and intensified by climate variability, form a feedback loop that exacerbates their impacts. This chapter explores the roles of the Kunming Global Biodiversity Framework (GBF) and Land Degradation Neutrality (LDN) principles in addressing these crises. It highlights their interlinkages and potential for integrated actions to mitigate biodiversity loss, combat land degradation, and achieve climate change mitigation

Vatsala
Forest Research Institute (Deemed to be) University, Dehradun, India

A. Kumar
Forest Research Institute (Deemed to be) University, Dehradun, India

Himalayan Institute for Learning and Leadership (HILL), University of Petroleum and Energy Studies (UPES), Dehradun, India

A. Negi · H. B. Chand (✉)
Forest Ecology and Climate Change Division, Forest Research Institute, Dehradun, India

M. K. Gautam
Biology Department, Medgar Evers College, City University of New York, USA

S. Singh
Centre of Excellence, Sustainable Land Management, ICFRE, Dehradun, India

A. Kumar
Forest Ecology and Climate Change Division, Forest Research Institute, Dehradun, India

Department of Biotechnology, Graphic Era (Deemed to be University), Dehradun, India

M. Fayiah
Department of Forestry, School of Natural Resources Management, Njala University, Njala, Sierra Leone

S. Singh, M. K. Gautam (eds.), *Biodiversity Conservation and Land Degradation Neutrality*, https://doi.org/10.1007/978-981-96-7210-3_3

goals. While the GBF sets ambitious targets to halt biodiversity loss, LDN focuses on sustainable land management to reverse degradation through flagship initiatives like the Great Green Wall and the Loess Plateau Restoration. The chapter demonstrates how these frameworks can transform degraded landscapes, stimulate biodiversity increase, and enhance climate change resilience. It identifies gaps in the LDN principles, particularly in its application and effectiveness, and provides recommendations to strengthen its impact. Recognizing the synergy between GBF and LDN is crucial in achieving sustainable development goals, biodiversity conservation, and enhanced ecosystem functions. The chapter advocates for unified policy frameworks, integrated approaches, and robust international collaboration to offset biodiversity loss and land degradation for a sustainable and prosperous future for all.

Keywords

Land degradation · Climate change · Global biodiversity framework · Land degradation neutrality · Sustainable development · Biodiversity conservation

3.1 Introduction

The problems of biodiversity loss, land degradation, and climate change are all connected to the health of the planet's ecosystems. These problems make up one of the most important issues of the twenty-first century. These phenomena are deeply interconnected, with biodiversity loss and climate change sharing common underlying causes and compounding each other's effects (Pörtner et al. 2023). Land degradation, characterized by the deterioration of land quality due to unsustainable practices, is a critical driver of this crisis, affecting nearly 29% of global land and the livelihood sustenance of approximately 3.2 billion people (Feng et al. 2022). Left unchecked, this degradation could impact up to 95% of the Earth's land areas within the next 30 years (UNEP 2021). The repercussions extend far beyond environmental degradation, jeopardizing food security, increasing greenhouse gas emissions, and adversely affecting the resilience of ecosystems to climate change (O'Grady et al. 2020).

The interplay between biodiversity loss, land degradation, and climate change creates a vicious cycle, where the degradation of one component exacerbates the others. For instance, land degradation contributes to greenhouse gas emissions through deforestation and degradation of carbon-rich ecosystems like peatlands, while climate change intensifies land degradation through rising temperatures and shifting precipitation patterns (O'Grady et al. 2020). These cascading effects directly threaten human well-being, with global crop yields projected to decline by up to 10%, and some regions facing losses of up to 50% (The Lancet Planetary Health 2018). Such outcomes could disrupt food security, trigger migration, and destabilize economies reliant on natural resources, particularly in vulnerable regions.

Amid these challenges, global initiatives like Land Degradation Neutrality (LDN) and the Kunming Global Biodiversity Framework (GBF) emerge as pivotal solutions. LDN aims to sustain and enhance land resources, focusing on preventing further degradation, restoring damaged ecosystems, and maintaining critical services such as water purification and carbon sequestration. Simultaneously, the Kunming Framework targets the halt and reversal of biodiversity loss by 2030, advocating for biodiversity conservation (Davies et al. 2015). These frameworks underscore the need for integrated approaches, where sustainable land management practices, such as agroforestry and reforestation, align with biodiversity conservation goals to enhance habitat quality and ecosystem resilience. This chapter delves into the synergies between LDN and the GBF, exploring how their integrated implementation can mitigate biodiversity loss, combat land degradation, and address climate goals.

3.2 Biodiversity Conservation

The biodiversity refers to the variety and variability of life forms across three different levels, i.e., species, genetic, and ecosystem. It is a cornerstone of ecological balance, underpinning numerous ecosystem services like nutrient cycling, pollination, and regulation of climate. Forest ecosystems, in particular, harbor immense biodiversity, sustaining terrestrial life while playing a vital role as carbon sinks and the achievement of sustainable development goals.

However, this intricate web of life is increasingly under threat. Biodiversity loss has emerged as one of the most pressing global challenges due to unsustainable anthropogenic activities. Land-use changes, deforestation, overexploitation, and pollution are primary drivers, with agriculture alone endangering 85% of threatened species globally (UNEP 2022). Since 1990, the world has lost approximately 420 million hectares of forest (FAO & UNEP 2020), with profound consequences for biodiversity and the ecosystems they support.

The effects of biodiversity loss extend far beyond ecological disruption. Forest degradation, habitat fragmentation, and species extinction destabilize ecosystems, reducing their resilience to climate change and other stressors. In biodiversity-rich regions like Southeast Asia, which saw a 14.5% loss in forest cover between 2000 and 2015 (Hughes 2017), this crisis is particularly acute. Such losses undermine ecological processes, weaken ecosystem services, and disrupt human livelihoods that depend on forest resources.

This alarming trend demands attention as the consequences of biodiversity loss cascade through ecosystems and societies. The decline in genetic diversity, for instance, limits species' adaptability, increasing their vulnerability to extinction. Furthermore, disruptions in ecological interactions—such as pollinator declines—can compromise food security and economic stability, highlighting the interdependence between biodiversity and human well-being.

3.3 Land Degradation

Land degradation is one of the most critical environmental challenges of our time, emanating from a combination of biophysical and socio-economic factors, often exacerbated by human activities. Unsustainable land use practices and urbanization significantly contribute to soil erosion, nutrient depletion, and biodiversity loss. These changes degrade ecosystems' ability to provide essential services like water regulation, carbon sequestration, and soil fertility, hence threatening food security and livelihoods, particularly in vulnerable regions. The annual loss of 50,000–70,000 km^2 of fertile land (Tokbergenova et al. 2018) underscores the urgency for addressing this issue, as regions like South Asia and Africa bear unequal burden of degradation caused by anthropogenic activities.

The consequences of land degradation extend far beyond ecological decline, intertwining with poverty, biodiversity loss, and climate change that create substantial barriers to sustainable development. In addition, soil erosion, a critical driver of degradation, reduces agricultural productivity and disrupts habitats, leading to species decline and the loss of ecosystem services. Additionally, degradation of land exacerbates natural disasters such as droughts, floods, dust storms, further displacing populations and straining resources (Kamboj et al. 2020). Also, land degradation has severe impact on nation's Gross Domestic Product (GDP) (Kumar et al. 2021). Addressing land degradation requires integrated approaches that combine sustainable land management practices, restoration efforts, and policy interventions to mitigate its cascading effects on ecosystems and human well-being.

3.4 Interconnections Between Biodiversity and Land Degradation

The interplay between biodiversity loss, land degradation, and climate change has long been recognized, with formal acknowledgment at the Earth Summit held in Rio de Janeiro in 1992. Today, international organizations such as the UNCCD and GEF emphasize that changes in one of these components inevitably trigger simultaneous shifts in others. The cumulative global impact of biodiversity loss, land degradation, and climate change is far greater than the sum of their individual effects (The Lancet Planetary Health 2018).

Land degradation undermines ecosystems' ability to provide ecosystem services and exacerbates biodiversity loss. As soil nutrients and moisture diminish, ecosystems struggle to maintain their functionality, triggering a feedback loop that accelerates these challenges.

Biodiverse ecosystems are inherently more resilient to land degradation. Diverse plant and microbial communities enhance soil structure, fertility, and nutrient cycling, while sustainable agricultural practices like cover cropping and organic farming preserve soil health and reduce erosion. These ecosystems, by promoting water infiltration and preventing surface runoff, help mitigate erosion and salinization. Wetlands and forests play an important role in hydrological functions, nutrient cycling, and carbon sequestering, thereby combating climate change—a significant driver of land degradation. Protecting and restoring these ecosystems is not only vital for climate change mitigation but also for improving overall land health, which creates favorable conditions for biodiversity conservation.

In regions like Southeast Asia, where rapid population growth and economic development exacerbate environmental degradation, biodiversity faces numerous threats from land-use changes, forest fires, and agricultural expansion. Shifting cultivation, particularly in tropical forests, highlights the link between land-use practices and land degradation. In the Chittagong Hills of Bangladesh, slash-and-burn cultivation has been shown to lead to significant soil erosion and nutrient loss, with soil erosion rates during cultivation up to six times higher than during fallow periods. However, the region's resilience allows ecosystems to recover during fallow periods, demonstrating both the detrimental effects of unsustainable land practices and the potential for recovery.

Deforestation further intensifies land degradation by reducing the soil's ability to retain water and diminishing its capacity to store carbon. Natural forests, rich in biodiversity, are especially vulnerable to these changes, with climate and weather shifts threatening the delicate balance of ecosystems. Similarly, semi-arid regions are experiencing altered grasslands and ecosystem dynamics, which negatively affect both wildlife and local human populations dependent on these resources.

Figure 3.1 shows that any degraded land can release the stored carbon which, along with Nitrous oxide, can lead to one of the factors of climate variability. This variability in climate affects marine, freshwater and terrestrial ecosystems negatively which, in other words, is known as biodiversity loss. The loss of biodiversity in terms of vegetation loss, aquatic diversity loss, and below-ground microbiota loss can significantly disturb biogeochemical cycles, and thermosteric expansion and ocean warming lead to land degradation. So, there is a circular loop between loss of biodiversity and land degradation. The change in one event can lead to another and vice versa. In turn, reducing land degradation through sustainable land management practices enhances biodiversity conservation, creating favorable conditions for species to thrive and ecosystems to function optimally.

3.5 The Kunming Global Biodiversity Framework

The urgent need to stop biodiversity loss and promote the equitable distribution of benefits led to the benefit of the GBF. It was adopted at CoP-15 Convention on Biological Diversity (CBD). This comprehensive framework resulted from 4 years of negotiations and aims to achieve global harmony with nature by 2050. It outlines an ambitious plan to achieve global congruence with nature by the year 2050, including 23 targets for 2030 and 4 goals for 2050 (CBD 2022). Goals include mainstreaming biodiversity across all sectors, incorporating conservation into working lands (Grumbine and Xu 2021), linking biodiversity with climate action (Pörtner et al. 2023), and focusing on the social and economic bias between the global north and south (Rammelt et al. 2023). The Kunming Biodiversity Framework addresses

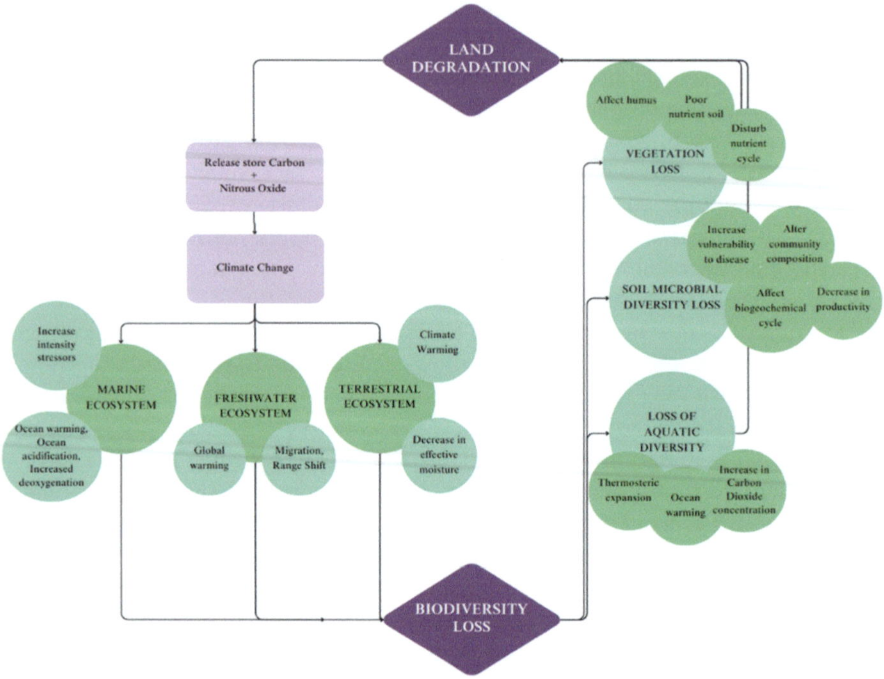

Fig. 3.1 A circular interloop between land degradation and biodiversity loss

these challenges by exploring possibilities for protection, restoration, and sustainable usage and mobilizing the essential resources to sustain being on Earth (Hughes and Grumbine 2023).

The framework emphasizes transparent monitoring and evaluation of progress, guiding the development of policies, guidelines, and national level biodiversity strategies. It fosters collaboration between the CBD and other international agreements to ensure effective implementation. By 2050, the framework aims to extend the area of natural ecosystems through restoration and protection, increase ecosystem functions, and preserve biodiversity. It stresses the importance of managing biodiversity sustainably and safeguarding traditional knowledge linked to genetic resources. Additionally, it promotes the fair and equitable sharing of benefits derived from genetic resources, especially with indigenous peoples and local communities, with a significant increase in these benefits by 2050 (CBD 2022).

3.5.1 2050 Goals for Biodiversity in Alignment with 2050 Vision

National commitments play a crucial role in driving effective and pertinent biodiversity actions at various scales. They bolster political ambition while fostering much-required changes across economic, social, political, and technological dimensions to meet the CBD obligations and align with 2050 vision (Xu et al. 2021).

These commitments seek to achieve substantial net improvements in biodiversity status by 2050 by stabilizing biodiversity loss over the next decade and promoting recovery over the next 20 years (CBD 2020). Known as minimum national targets, these global goals represent the fundamental requirements for parties to fulfill CBD obligations. They encompass policies, knowledge advancement, resource mobilization, information sharing, conservation, sustainable use, restoration, public engagement, and monitoring, all playing important roles in the success of this vision (Xu et al. 2021). Section G of the GBF outlines specific goals for 2050 in alignment with this vision, as represented in Table 3.1.

3.5.2 Target 30×30 of Global Biodiversity Framework

The "30×30" conservation target, outlined as draft target 3 of the GBF, calls for protecting 30% of the land and marine areas by 2030. This ambitious target builds on the previous Aichi Target 11, which aimed to protect land, coastal, and marine areas (IISD 2022). Key aspects of the target include:

Table 3.1 GBF goals in alignment with the 2050 vision

Goals	Description	Key focused area	Target by 2050
Goal A	Maintain, enhance, or restore the integrity, connectivity, and resilience of ecosystems, substantially increasing the natural ecosystem area.	Halt anthropogenic extinction of known threatened species. Increase the abundance of native wild species to healthy levels. Maintain genetic diversity within wild and domesticated species.	Increase the area of natural ecosystems. Reduce extinction rates. Safeguard genetic diversity.
Goal B	Sustainably use and manage biodiversity, valuing and enhancing nature's contributions to people, including ecosystem functions and services.	Restore declining ecosystems. Support sustainable development for current and future generations.	Ensure sustainable management of biodiversity. Enhance ecosystem functions and services.
Goal C	Share benefits from the utilization of genetic resources and traditional knowledge fairly and equitably, protecting traditional knowledge associated with these resources.	Fair sharing of monetary and non-monetary benefits. Protect the traditional knowledge of Indigenous people and local communities.	Increase equitable sharing of benefits. Protect traditional knowledge.
Goal D	Secure adequate means to implement, including financial resources, capacity-building, and technology transfer, to implement the framework effectively.	Close the biodiversity finance gap. Ensure equitable access to resources for all parties, especially developing countries.	Close the $700 billion annual biodiversity finance gap. Align financial flows with the framework.

(i) Focus on Nature: Target 3 emphasizes protecting regions important for ecosystem services, functions, and biodiversity.
(ii) Global Reach: Recognizing that biodiversity is unevenly distributed, the target urges parties and stakeholders to prioritize regions crucial for ecological services and functions, and ensure systems are ecologically representative, well-connected, and equitably governed. This includes careful consideration of physical locations and conservation methods.
(iii) Integrated Approach: It encompasses inland waters, terrestrial areas, and coastal and marine environments.
(iv) Indigenous Rights and Contributions: The GBF acknowledges the role of Indigenous peoples and local communities (IPLCs) in managing biodiversity and ecosystem services. It highlights the significance of their traditional territories in achieving Target 3.

Terrestrial and marine ecosystems absorb approximately 56% of anthropogenic CO_2 (COP-28 2023), underscoring the GBF's role in advancing the goal of people living in harmony with nature by 2050 (IUCN 2023). However, implementing the 30×30 target poses challenges. Estimates suggest that the cost of achieving all components of 30×30 ranges between US$103 and 178 billion annually (Waldron et al. 2020), or nearly 63 days' worth of global state subsidies to the oil sector (WEF 2018).

Challenges include:

(i) Biodiversity vs. Sustainable Use: Target 3 focuses primarily on biodiversity conservation rather than sustainable use (IUCN 2023).
(ii) Potential Diversion: There are concerns that the 30×30 target may divert attention from the underlying causes of biodiversity decline, such as unsustainable consumption (IISD 2022).
(iii) Limited Application: Sustainable use within Target 3 is often restricted, especially in more strictly protected areas, and may not fully address conservation management needs (IUCN 2023).
(iv) Effectiveness of Protected Areas: In some regions, existing protected area systems are not adequately designed or managed to address the decline in biodiversity (IUCN 2023).

To address these challenges and achieve the 30×30 goal, multiple strategies and approaches must be employed. Figure 3.2 outlines potential strategies for reaching 30% protection of global land and oceans by 2030.

3.6 Challenges of the Global Biodiversity Framework

GBF has been subject to various criticisms and identified challenges from critics, environmentalists, and scholars. Despite its ambitious targets and comprehensive approach, several issues hinder its effectiveness. One major criticism is the framework's broad and sometimes ambiguous targets, which can be challenging to implement and measure. Critics argue that while the GBF sets out clear objectives, the

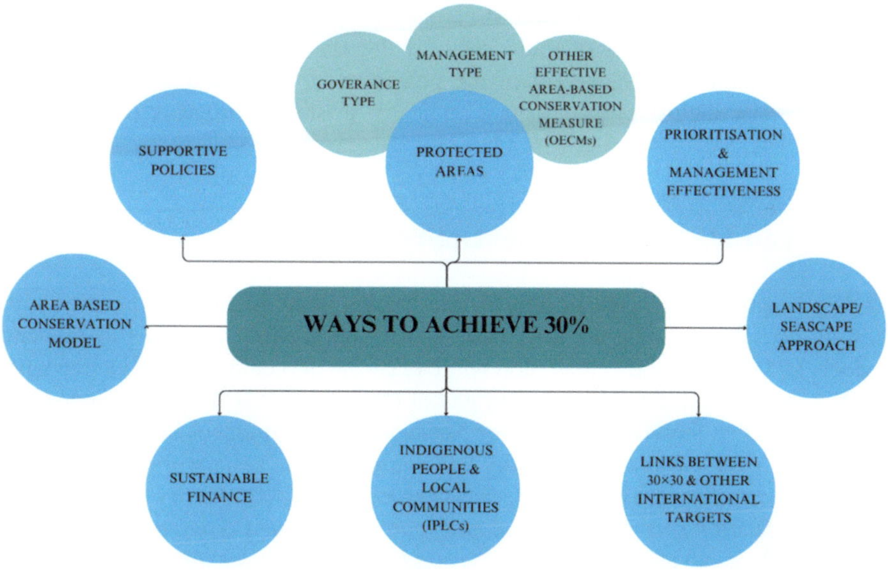

Fig. 3.2 Potential strategies for reaching 30% protection of global land and oceans by 2030

lack of specific, actionable steps makes it difficult for countries to develop concrete plans. For instance, Target 2 of the GBF aims to restore 20% of degraded ecosystems by 2030, but without clear guidelines on how to prioritize and implement restoration efforts, countries may struggle to achieve this goal (Visseren-Hamakers et al. 2021). The significant challenges of the global biodiversity framework include funding and resource allocation, along with several other challenges, which are as follows:

- The financial requirements for achieving the GBF targets are substantial, yet many countries, especially those in the global south, lack the necessary funding.
- The irregular distribution of resources leads to disparities in implementation, with wealthier nations being more capable of meeting their biodiversity conservation targets than poorer countries (UNEP 2020).
- Another issue is the integration of biodiversity conservation into national and local policies. Often, there is a disjoint between environmental goals and economic development priorities. This disjunction can result in policies that support monetary benefits in the short term but undermine the overall GBF's objectives.
- Lack of standardized indicators and metrics for tracking progress toward biodiversity targets, which complicates the assessment of conservation efforts and hinders meaningful comparisons across regions (Schmeller et al. 2015).
- The framework does not fully integrate climate change impacts or develop strategies for climate adaptation and mitigation, limiting its effectiveness in confronting the interwoven challenges of climate change and biodiversity loss (IPCC 2019).

- While indicators like the extent of natural ecosystems seem robust, they fail to account for ecosystem degradation, and the red list of ecosystems remains incomplete. The development of the percentage of land and seas covered by biodiversity-inclusive spatial plans indicator is still in progress (Hughes and Grumbine 2023).

3.7 Land Degradation Neutrality (LDN)

LDN was introduced to ensure stable or improving land quality and quantity, supporting ecosystem functions and food security by UNCCD. It aims to prevent, reduce, and reverse land degradation while ensuring no net loss of productive land. LDN was expanded from drylands to global land after the Rio+20 conference and was included in the 2030 Agenda for Sustainable Development in 2015. LDN is defined as "*a state where the quantity and quality of land resources needed to support ecosystem functions and services and to enhance food security, remain stable or increase within specified temporal and spatial scales and ecosystems*" (UNCCD 2022). While it can also be defined as "*a condition in which the quantity and quality of land resources required to sustain ecosystem functions and services and improve food security remain stable or increase at a given spatial and temporal scale and ecosystem scale*" (Clewell et al. 2004). It aims to sustain the foundation of land resources by employing diverse strategies to prevent, reduce, and reverse deterioration of the land, confirming no net loss of productive and healthy land. In addition to aiding climate change adaptation and mitigation, achieving LDN is essential for preserving species and biodiversity. Therefore, LDN should complement and support the United Nations Framework Convention on Climate Change (UNFCCC) in reaching its 2 °C goal and the Aichi biodiversity targets of the CBD (Davies et al. 2015). The principles of LDN consist of 19 guidelines that direct its implementation. These principles ensure that the LDN process successfully achieves its intended benefits while minimizing or preventing any potential negative or unintended consequences, which are as follows in Table 3.2.

3.8 Global Biodiversity Framework and Land Degradation

Biodiversity and land health are closely associated, as land degradation directly affects biodiversity, whereas biodiversity conservation can mitigate and rehabilitate the land. The GBF, instituted within the CBD framework, endeavors to mitigate these interdependencies by defining worldwide objectives for preserving biodiversity, sustainable land utilization, and rehabilitating ecologically degraded systems. One of the primary goals of the GBF is to achieve LDN by 2030.

Key sections of the GBF relevant to land degradation include:

(i) *Target 2*: Restoration of Degraded Ecosystems: This target aims to restore at least 30% of degraded ecosystems globally by 2030. Restoration activities

Table 3.2 Principles related to LDN

Principles underpinning the vision of LDN	Principles related to the frame of reference	Principles related to the mechanism for neutrality	Principles related to achieving neutrality
Maintain or enhance land-based natural capital	The LDN target equals the baseline	Counterbalance "like for like"	Leverage existing planning processes
Protect human rights and enhance human well-being	Neutrality is usually the minimum objective	Manage counter balancing at the same scale as land use planning	Base land use decisions on multi-variable assessments
Respect national sovereignty		Counterbalance anticipated losses in land-based natural capital with gains over the same timeframe, to achieve neutrality. Apply an integrated land use planning principle that embeds the neutrality mechanism in land use planning. Within a land type, counterbalancing cannot occur between protected areas and land managed for productive uses. Ensure that all stakeholders, public and private, pursue LDN responsibly by working in partnership with relevant levels.	Ensure that land management aligns with the capability of the land to minimize the risk of land degradation and help identify and prioritize appropriate land use practices. Balance economic, social, and environmental sustainability Apply the response hierarchy Quantify projected land degradation Apply a participatory process Apply good governance

Source: Orr et al. (2017)

focus on re-establishing native vegetation, improving soil health, and enhancing ecosystem services, which altogether reduces land degradation (CBD 2020).

(ii) *Target 3*: Sustainable Agriculture and Forestry: Promoting sustainable agricultural and forestry practices is essential for reducing land degradation. This target encourages the adoption of agro-ecological practices, sustainable forest management, and the reduction of chemical inputs, which shall contribute to healthier soils and ecosystems (CBD 2020).

(iii) *Target 8*: Sustainable Use of Biodiversity: This target emphasizes the sustainable use of biodiversity resources to avoid its exploitation which may lead to land degradation. It promotes practices that maintain ecosystem integrity and prevent overharvesting and habitat destruction (CBD 2020).

(iv) *Target 10*: Integrated Landscape and Seascape Management: Integrating biodiversity conservation into landscape and seascape management is crucial for maintaining ecosystem functions and reducing land degradation. This approach ensures that biodiversity conservation is considered in land-use planning and development projects (CBD 2020).

3.8.1 Interlinkages Between GBF and LDN

GBF and LDN share several objectives and present opportunities for synergies in their implementation which are essential for achieving sustainable development goals.

- Synergies: Both GBF and LDN share a common purpose to promote SLM (Sustainable Land Management) along with biodiversity conservation. Both share common goals of enhancing ecosystem services and restoring degraded ecosystems (Renaud et al. 2022).
- Shared Objectives: Both GBF and LDN target are for the management, conservation, and restoration of land resources in a sustainable manner (Streck 2023). Likewise, they also underline the significance of maintaining ecosystem health to support biodiversity and human well-being. Also, they focus on SLM practices to combat further degradation (Nilsson et al. 2022).
- Complementarities: GBF and LDN complement one another by integrating land management practices into biodiversity conservation and vice versa, which leads to a holistic and operative approach to biodiversity and land management (Nilsson et al. 2022).

3.8.2 Overlapping Indicators and Metrics

- Indicators: both frameworks use indicators of ecosystem health, such as vegetation cover, soil quality, and species diversity. The GBF system of headline and binary indicators can help monitor LDN progress at national and global levels (Streck 2023). These indicators help to follow progress toward achieving their respective goals (Renaud et al. 2022).
- Metrics: metrics like the extent of degraded land restored, the number of species protected, and the improvement in ecosystem services are common to both GBF and LDN. These metrics provide a basis for measuring success and identifying areas for improvement (Nilsson et al. 2022).

3.8.3 Integrated Approaches

- Synergies between biodiversity and climate can be enhanced by integrating LDN priorities and targets into National Biodiversity Strategies and Action Plans (NBSAPs) and National Adaptation Plans (NAPs) (IISD 2022).
- Integrated approaches are emphasized by incorporating indigenous landscapes into area-based conservation strategies, utilizing spatial planning to minimize land

alteration and promote sustainable practices (GBF Target 1 and GBF Target 3) (Streck 2023).

- Principles of both GBF and LDN support the integration of land management approaches that are used to address both biodiversity and land degradation issues. This includes practices like agroforestry, sustainable agriculture, and ecosystem restoration (Nilsson et al. 2022).
- Aligning national and local policies with the objectives of GBF and LDN can integrate the goals of both frameworks and create synergies and reduce conflicts (Nilsson et al. 2022).

3.8.4 Policy Coherence

- The GBF advocates coherence, complementarity, and collaboration among biodiversity-related conventions and agreements to effectively achieve its targets (Streck 2023).
- Collaborative governance structures can enhance policy coherence and ensure effective implementation of both frameworks, which might involve multiple stakeholders including government agencies, local communities, and NGOs (Nilsson et al. 2022).
- Effective cooperation between the GBF and the Paris Agreement at the political level is necessary for coordinating activities and enabling conditions like finance, capacity development, and transfer of technology (Streck 2023).

3.9 Gaps in Land Degradation Neutrality

To determine the phase of LDN, deterioration and repair must be graded in tandem. Surface vegetation and ecosystem changes are mostly caused by natural and human sources (Piao et al. 2020) which is one of the major contributing elements to land degradation. Climate parameters (such as temperature, precipitation, and wind) are the primary examples of natural factors. Population growth, consumption, urbanization, and ecological restoration initiatives are the primary examples of human causes (Guo et al. 2020). An examination of the factors influencing change for LDN is deficient. A uniform approach to LDN assessment does not exist. Further research is required to determine whether the driver factors for land degradation and restoration are reliable compared with those for LDN.

3.10 Success Stories on LDN

3.10.1 The Great Green Wall (GGW) Initiative

The Great Green Wall initiative in the Sahel region of Africa exemplifies a comprehensive strategy to combat land degradation and desertification while enhancing biodiversity. By restoring 100 million hectares of degraded land by 2030 (Reij et al. 2020), the GGW addresses both environmental and socio-economic challenges, aiming to create job opportunities and improve food security. Through extensive reforestation and sustainable land management techniques, the GGW seeks to halt the expansion of the Sahara Desert and its impacts on local communities (Gadzama 2017). This effort directly links to the GBF by promoting habitat restoration, enhancing biodiversity, and providing ecological corridors for plant and animal species. Furthermore, the initiative aligns with the LDN concept, focusing on improving soil health, land productivity, and ecosystem functions while reversing land degradation.

The GGW's efforts in combating desertification not only restore degraded lands but also play an important role in mitigating climate change. It makes a substantial contribution to the 2030 Sustainable Development Goals (SDGs) and the Paris Agreement by enhancing ecosystem resilience and sequestering carbon. GGW serves as a model for integrating sustainable land management practices with community-driven initiatives (Speranza et al. 2019), highlighting how reforestation and restoration activities can address land degradation and biodiversity loss. These efforts directly contribute to the LDN goal of maintaining or improving land productivity, while the restoration of habitats enhances biodiversity. Through this dual approach, the GGW initiative demonstrates how integrated environmental strategies can address complex challenges, furthering both land degradation neutrality and global biodiversity objectives.

3.10.2 The Loess Plateau in China

The restoration of the Loess plateau in China provides a compelling example of how land degradation can be reversed through large-scale ecological restoration and land management practices. The Grain for Green (GFG) and Gully Land Consolidation (GLC) projects have significantly improved the ecological quality of the plateau, restored vegetation cover, and reduced soil erosion (He et al. 2020). These efforts directly contribute to the LDN goal by improving soil health, reducing erosion, and enhancing land productivity. The plateau's transformation from a degraded and ecologically vulnerable site to a productive and sustainable land system demonstrates how effective land management can both prevent further degradation and restore ecosystems. The GLC project's success in creating level farmland in previously eroded areas aligns with the LDN framework's emphasis on sustainable land management techniques that balance ecological restoration with economic development.

Beyond land restoration, the Loess plateau projects have also contributed to biodiversity conservation, advancing the objectives of the GBF. By reforesting large areas and promoting sustainable agricultural practices, the projects have helped restore habitats for various species, enhancing biodiversity in the region. The improved land conditions have also facilitated the development of new farmland, which in turn has supported more intensive and higher-yield farming. This shift toward sustainable farming practices is crucial for supporting both local livelihoods and biodiversity. The Loess plateau restoration efforts exemplify the integration of biodiversity conservation with land restoration (Yu et al. 2020), illustrating how sustainable land management can contribute to both the LDN and GBF goals. This approach showcases the potential for global initiatives to address the challenges of land degradation and biodiversity loss through coordinated efforts that promote ecological balance and sustainable development.

3.11 Conclusion

The link between the GBF and LDN underscores the deep interconnection between biodiversity loss, land degradation, and climate change. Healthy ecosystems provide important ecosystem services such as carbon sequestration, soil stabilization, and water regulation, all of which are vital in mitigating climate change impacts. However, unsustainable land-use practices, deforestation, and urbanization exacerbate land degradation, leading to loss of biodiversity, reduced ecosystem functionality, and heightened greenhouse gas emissions. Climate change, in turn, intensifies these processes by accelerating soil erosion, desertification, and habitat disruption, creating a vicious cycle of ecological decline. Addressing this interconnected challenge requires integrated approaches that simultaneously target biodiversity conservation, land restoration, and climate resilience. Examples such as the GGW initiative and the Loess plateau project demonstrate how large-scale, community-based restoration efforts aligned with LDN principles can restore degraded lands, rejuvenate ecosystems, and safeguard biodiversity, benefiting both ecological health and local livelihoods.

The complementary goals of GBF and LDN provide a robust framework for addressing global environmental challenges holistically. Both initiatives emphasize sustainable land management, biodiversity conservation, and the restoration of ecosystem services, reinforcing the synergy between ecological restoration and human well-being. Aligning LDN priorities with national biodiversity strategy action plan and NAPs offers opportunities for coherent policy implementation that enhances resilience to environmental stress. Shared indicators, such as vegetation cover, soil quality, and species diversity, further strengthen monitoring and evaluation of progress under both frameworks. Coordinated governance, policy coherence, and collaborative action at national and international levels are critical to the success of GBF and LDN. Together, they exemplify the potential to combat desertification, restore biodiversity, and promote sustainable land management practices, creating a foundation for long-term environmental sustainability and resilience.

References

CBD (2022) Kunming-Montreal Global Biodiversity Framework. In: Fifteenth meeting of the Conference of the Parties to the Convention on Biological Diversity (Part Two) Decision 15/4

Clewell A, Aronson J, Winterhalder K (2004) The SER international primer on ecological restoration

Convention on Biological Diversity (2020) Zero draft of the post-2020 global biodiversity framework. CBD/WG2020/2/3, p 16
COP-28 (2023) The 30*30 biodiversity goal at COP28: bridging nature-based solutions and climate action, 2 Nov 2023. Retrieved from: https://www.cop28.com/en/thought-leadership/The-30x30-Biodiversity-Goal-at-COP28
Davies J, Gudka M, Laban P, Metternicht G, Alexander S, Hannam I, Welling L, Vasseur L, Siles J, Aguilar L, Poulsen L, Jones M, Nakanuku-Diggs L, Zeidler J, Hesselink F (2015) Land degradation neutrality: implications and opportunities for conservation. Nature based solutions to desertification, land degradation and drought. With contributions from Global Drylands Initiative, CEM, WCEL, WCPA, CEC 1 1 Contributors. www.iucn.org/drylands
FAO & UNEP (2020) The state of the world's forests 2020. Forests, biodiversity and people. FAO & UNEP, Rome. https://doi.org/10.4060/ca8642en
Feng S, Zhao W, Zhan T, Yan Y, Pereira P (2022) Land degradation neutrality: a review of progress and perspectives. Ecol Indic 144:109530
Gadzama NM (2017) Attenuation of the effects of desertification through sustainable development of Great Green Wall in the Sahel of Africa. World J Sci Technol Sustain Dev 14(4):279–289
Grumbine RE, Xu J (2021) Five steps to inject transformative change into the post-2020 global biodiversity framework. Bioscience 71(6):637–646
Guo B, Zhang J, Meng X, Xu T, Song Y (2020) Long-term spatio-temporal precipitation variations in China with precipitation surface interpolated by ANUSPLIN. Sci Rep 10(1):81
He M, Wang Y, Tong Y, Zhao Y, Qiang X, Song Y et al (2020) Evaluation of the environmental effects of intensive land consolidation: a field-based case study of the Chinese Loess Plateau. Land Use Policy 94:104523
Hughes AC (2017) Understanding the drivers of Southeast Asian biodiversity loss. Ecosphere 8(1):e01624. https://doi.org/10.1002/ecs2.1624
Hughes AC, Grumbine RE (2023) The Kunming-Montreal Global Biodiversity Framework: what it does and does not do, and how to improve it. Front Environ Sci 11:1281536
IISD (2022) The Global Biodiversity Framework's "30×30" target: catchy slogan or effective conservation goal? Retrieved from: https://www.iisd.org/articles/insight/global-biodiversity-framework-30x30-target. Accessed on: 6 Dec 2022
IPCC (2019) Special report on climate change and land. Intergovernmental Panel on Climate Change
IUCN (2023) 30*30. A guide to inclusive, equitable and effective implementation of target 3 of the Kunming-Montreal Global Biodiversity Framework, p 69. Retrieved from: https://www.iucn.org/sites/default/files/2023-09/30x30-target-framework.pdf
Kamboj N, Bisht A, Kamboj V, Pandey N, Bisht A (2020) Role of natural disasters in environmental degradation: an overview. In: Environmental degradation: causes and remediation strategies. Springer, Cham, pp 21–35. https://doi.org/10.26832/aesa-2020-edcrs-02
Kumar A, Singh S, Chand HB, Kewat AK, Singh KP (2021) Road towards restoring land degradation in India: a UNCCD perspective. Res J Chem Environ Sci 9(4):27–38
Nilsson M, Vijge MJ, Alva IL, Bornemann B, Fernando K, Hickmann T et al (2022) Interlinkages, integration and coherence. In: The political impact of the sustainable development goals: transforming governance through global goals? Cambridge University Press, New York, pp 92–115
O'Grady AP, Smith GS, Ascui F, Pinkard EA (2020) The rise and rise of natural capital: what role for forestry? Aust For 83(3):103–106
Orr BJ, Cowie AL, Castillo Sanchez VM, Chasek P, Crossman ND, Erlewein A, Louwagie G, Maron M, Metternicht GI, Minelli S, Tengberg AE, Walter S, Welton S (2017) Scientific conceptual framework for land degradation neutrality. A report of the Science-Policy Interface. United Nations Convention to Combat Desertification (UNCCD), Bonn
Piao S, Wang X, Park T, Chen C, Lian XU, He Y et al (2020) Characteristics, drivers and feedbacks of global greening. Nat Rev Earth Environ 1(1):14–27
Pörtner HO, Scholes RJ, Arneth A, Barnes DKA, Burrows MT, Diamond SE et al (2023) Overcoming the coupled climate and biodiversity crises and their societal impacts. Science 380(6642):eabl4881

Rammelt CF, Gupta J, Liverman D, Scholtens J, Ciobanu D, Abrams JF et al (2023) Impacts of meeting minimum access on critical earth systems amidst the Great Inequality. Nat Sustain 6(2):212–221

Reij C, Pasiecznik N, Mahamoudou S, Kassa H, Winterbottom R, Livingstone J (2020) Dryland restoration successes in the Sahel and Greater Horn of Africa show how to increase scale and impact. ETFRN News

Renaud FG, Zhou X, Bosher L, Barrett B, Huang S (2022) Synergies and trade-offs between sustainable development goals and targets: innovative approaches and new perspectives. Sustain Sci 17(4):1317–1322

Schmeller DS, Julliard R, Bellingham PJ, Böhm M, Brummitt N, Chiarucci A, Couvet D, Elmendorf S, Forsyth DM, Moreno JG, Gregory RD, Magnusson WE, Martin LJ, McGeoch MA, Mihoub J-B, Pereira HM, Proença V, van Swaay CAM, Yahara T, Belnap J (2015) Towards a global terrestrial species monitoring program. J Nat Conserv 25:51–57. https://doi.org/10.1016/j.jnc.2015.03.003

Speranza CI, Adenle A, Boillat S (2019) Land degradation neutrality-potentials for its operationalisation at multi-levels in Nigeria. Environ Sci Pol 94:63–71

Streck C (2023) Synergies between the Kunming-Montreal Global Biodiversity Framework and the Paris Agreement: the role of policy milestones, monitoring frameworks and safeguards. Clim Pol 23(6):800–811

The Lancet Planetary Health (2018) Land degradation: a solution is possible. Lancet Planet Health 2(5):e184). Elsevier BV. https://doi.org/10.1016/s2542-5196(18)30064-0

Tokbergenova A, Nyussupova G, Arslan M, Kiyassova SKL (2018) Causes and impacts of land degradation and desertification: case study from Kazakhstan. In: Vegetation of Central Asia and Environs. Springer International Publishing, Cham, pp 291–302. https://doi.org/10.1007/978-3-319-99728-5_11

UN Environment Programme (2022) Five drivers of the nature crisis. UNEP. https://www.unep.org/news-and-stories/story/five-drivers-nature-crisis

UNCCD (2022) Chronic land degradation: UN offers stark warnings and practical remedies in Global Land Outlook 2. Retrieved from: https://www.unccd.int/news-stories/press-releases/chronic-land-degradation-un-offers-stark-warnings-and-practical. Accessed on: 26 Apr 2022

UNEP (2020) State of finance for nature: tripling investments in nature-based solutions by 2030. United Nations Environment Programme

UNEP (2021) Land degradation. Global Environment Facility. https://www.unep.org/gef/focal-areas/land-degradation

Visseren-Hamakers IJ, Razzaque J, McElwee P, Turnhout E, Kelemen E, Rusch GM et al (2021) Transformative governance of biodiversity: insights for sustainable development. Curr Opin Environ Sustain 53:20–28

Waldron A, Adams V, Allan J, Arnell A, Asner G, Atkinson S et al (2020) Protecting 30% of the planet for nature: costs, benefits and economic implications

WEF (2018) Corruption is costing the global economy $3.6 trillion dollars every year. Retrieved from: https://www.weforum.org/agenda/2018/12/the-global-economyloses-3-6-trillion-to-corruption-each-year-says-u-n. Accessed on: 13 Dec 2018

Xu H, Cao Y, Yu D, Cao M, He Y, Gill M, Pereira HM (2021) Ensuring effective implementation of the post-2020 global biodiversity targets. Nat Ecol Evol 5(4):411–418

Yu Y, Zhao W, Martinez-Murillo JF, Pereira P (2020) Loess Plateau: from degradation to restoration. Sci Total Environ 738:140206

Biodiversity Conservation and Land Degradation Neutrality: Need of Synthesis in Policy and Practices

4

R. S. C. Jayaraj

Abstract

India is a signatory to conventions of biodiversity conservation as well as land degradation neutrality. The policy efforts and practices related to these conventions need synthesis at implementation level, as they are interrelated. In spite of government's efforts, there is a steady loss of biodiversity and also an increase in land degradation. The Kunming-Montreal Global Biodiversity Framework (KM-GBF) provides an opportunity for linking the efforts at biodiversity conservation with land degradation neutrality. There are seven global targets of the KM-GBF that have linkages with land degradation neutrality, and actions required to achieve those targets are elaborated in the chapter. The suggested actions include attention to tenurial issues in northeast India, where the biodiversity loss and forest degradation are high, introduction of land and water use planning in the northwest, improvement of dryland agriculture, enrichment and densification of open forests, protection of wetlands, elimination of invasives and a thrust on agroforestry.

Keywords

Conservation · Land degradation neutrality · Kunming-Montreal global biodiversity framework · North-east India

R. S. C. Jayaraj (✉)
UT Environmental Impact Assessment Authority, Andaman and Nicobar Islands, India

S. Singh, M. K. Gautam (eds.), *Biodiversity Conservation and Land Degradation Neutrality*, https://doi.org/10.1007/978-981-96-7210-3_4

4.1 Introduction

India is a signatory to the international conventions related to biodiversity conservation as well as land degradation neutrality and has formulated many policies and programmes for both the purposes. These actions have checked the pace of environmental degradation and promoted recovery of biodiversity to some extent but leave scope for improvement both at policy level and in practices. The chapter analyses the progress made so far and suggests need for a synthesis linking biodiversity conservation with reclamation of degraded landscapes.

4.2 Current Situation of Biodiversity Conservation

India is one of the 17 megadiverse countries, accounting for 7–8% of global biodiversity, and also housing 4 of the 36 global biodiversity hotspots. Various efforts have been taken to conserve biodiversity at landscape, species and genetic levels. At the landscape level the Protected Area Network of India extends over 175,169.42 km^2 which is about 5.32% of geographical area. This network includes 106 national parks, 573 wildlife sanctuaries, 115 conservation reserves and 220 community reserves (WII 2023). Included within these are 54 tiger reserves, 31 elephant reserves and 7 world heritage sites. For conservation of ecosystems, the country implements programmes, such as National Plan for Aquatic Ecosytems (NPCA) covering all lakes and wetlands, and National Mission for Sustaining the Himalayan Ecosystem (NM-SHE) focussing on the fragile nature of these mountain ranges and their rich biodiversity, to name a few. India is a member state of the United Nations Forum on Forests (UNFF) and has a mandate to work towards implementation of the UN Strategic Plan of Forests 2017–2030 (UNSPF) with a focus on sustainable forest management (SFM), enhancement of forest-based economic, social and environmental benefits, increasing the area of protected forests and the proportion of forest products from sustainably managed forests, mobilizing financial resources for SFM and strengthening of scientific and technical cooperation, promoting governance frameworks to implement SFM, and enhancing cooperation on forest-related issues within the UN system and across organizations, sectors and stakeholders.

At species level there are initiatives such as Project Tiger, Project Elephant, Project Hangul, Project Snow Leopard, Project Rhino, Project Dolphin, Project Crocodile, Project Cheetah, Project Sea Turtle, Vulture conservation, etc., focussing on key species but with benefit to the overall ecosystem. Project Tiger is considered one of the most successful conservation projects in the world. It was launched in 1973 when the tiger population was around 1200, and now the number has reached around 5000. Project Elephant launched in 1992 has helped in improving the habitats and migration corridors of the elephant population. In a similar way the other projects have nearly brought back the species from the brink of extinction. Besides these there are more than 200 zoological gardens in the country maintaining about 5 lakh animals for education and entertainment, and some of them are also engaged in captive breeding of endangered animals for subsequent release in the wild. Though there are 79 captive breeding programmes formulated success has been obtained only in a few as in the case of Mugger, Gharial, Saltwater Crocodile, Musk Deer, Great Indian Bustard, Lesser Florican, Gyps vulture, Black Softshell turtle, Greater One Horned Rhino, Western Tragopan, Indian Grey Wolf, Snow Leopard, Red Panda, Pygmy Hog, and Andaman wild pig, to name a few. It is reported that only 9% of the conservation breeding programmes have been successful in releasing animals into the wild (Ganesan 2016). As per the IUCN Red List of 2012 there are 132 species of plants and animals from India categorized as "Critically Endangered". Unlike the case of animals, there are not many species recovery programmes of plants. Out of 387 red-listed plants in India, 77 are critically endangered and are suitable candidates for recovery programmes, but greater attention is needed to prevent their extinction. A few successful recovery programmes are that of *Paphiopedilum druryi*, a slipper orchid, *Semecarpus kathalekanensis*, *Hubbardia heptaneuron*, and *Ceropegia fantastica* (Ravikanth et al. 2018).

For conservation of genetic resources, including wild relatives, varieties, clones and cultivars, India has specialized agencies conserving the germplasm of plants, animals, fishes and microbes in the form of national bureaus. There are separate bureaus for plants, animals, fishes, agriculturally important micro-organisms, and agricultural insect resources under the Indian Council of Agricultural Research. There is a National Bureau of Forest Genetic Resources under the Indian Council of Forestry Research and Education. India is rich in agrobiodiversity and is one of the 12 megadiversity regions of the world, and it is the place where about 25 crop species were domesticated. Among the flora of India are 168 major and minor crops and 730 of their wild relatives, falling under 817 taxa. In addition, there are 9500 species of ethnobotanical use, of which 7500 are used in ethnomedicine and 3900 are multipurpose/edible species (NAAS 1998; Jacob et al. 2020).

The collections are maintained in field gene banks, seed storage facilities, cryopreservation, in-vitro gene banks, on-farm conservation, ex-situ conservation areas and in-situ protection in natural habitats. There are around 57 National Active Germplasm sites conserving germplasm of horticultural and agricultural crops. The National Gene Bank holds the seeds, tissue culture materials and cryo-preserved samples of agricultural crops and field gene banks for trees and vines. Besides these, there are 122 major botanical gardens in the country conserving plant species for observation and research. There are 22 agrobiodiversity hotspots identified in the country (Nayar et al. 2009). There are 309 preservation plots, 287 in natural forests and 22 in plantations, to preserve representative areas of vegetation. There are over 13,000 sacred groves in India, held in reverence by communities. There are 14 national repositories notified by the National Biodiversity Authority where valuable germplasm is deposited for storage (Chandrakar et al. 2016). The diversity of domestic animals is also rich, with 61 breeds of cattle, 20 breeds of buffalo, 59 breeds of sheep, 8 breeds of camel, 6 breeds of horse, 3 breeds of pig and 18 breeds of poultry (Swaroop 2020).

In spite of all the efforts cited above, India is one of the regions experiencing the highest rate of biodiversity loss. Successive assessments of forest cover have shown a steady decline in dense forest areas with a concomitant increase in open forests, indicating progressive forest degradation, though the cover is shown to increase mainly due to increasing area under plantations (Aggarwal 2022; Rupavath 2022). India has seen the highest rate of deforestation between 1990 and 2020 with a surge between 2015 and 2020 (Jha 2023). According to Global Forest Watch, India lost 2.33 Mha of tree cover from 2000 to 2023, and the top regions responsible for about 60% of loss were Assam, Mizoram, Arunachal Pradesh, Nagaland and Manipur, all located in the northeastern region which is rich in biodiversity and natural forests. The loss of humid primary forests was 0.41 Mha, which is 18% of the total tree cover loss (GFW 2023). The loss of forests in the northeastern region has also been attributed to climate change, though this is only a secondary driver of forest loss, and human conversion of landscapes remains the main cause (Haughan et al. 2022). The genetic diversity of agricultural crops and domestic animals is threatened due to intensive cropping of selected varieties of crop or rearing and breeding of selected breeds of livestock and poultry, especially the exotic high-yielding varieties, to the detriment of the native breeds. Problem is compounded by soil degradation, fragmentation, excessive tillage, inappropriate crop rotation, water scarcity, post-harvest losses, natural disasters and climate change impacts (Jacob et al. 2020).

The ecological loss is not without an economic impact. The World Bank has noted that India was losing 5.7% of its GDP annually (based on 2009 figures) due to diminishing natural assets and reduction in flow of ecosystem services (World Bank 2013). India's natural capital stocks have reduced from 31% of national capital in 1990 to just 15% in 2014, as per the Inclusive Wealth Report of UNEP. The average annual loss of natural capital in India between 1990 and 2014 has remained at 0.4%, and between 2000 and 2019 at 1.6% of gross national income. This reduction in natural capital is combined with the expansion of produced capital, indicating that India has drawn down its natural capital stocks while investing heavily in the built infrastructure. The decline of natural capital has been significant in fossil fuels, fisheries and pastureland. Carbon loss damage due to fossil fuel use and deforestation has also contributed to the decline. The increase in wealth has not translated into improvement in human development, and India has suffered an overall loss of Human Development Index (HDI) due to inequality to the extent of 26.4%, the second highest after South Africa. The change in the wealth share of the top decile between 2000 and 2014 was more than 70% indicating very high inequality (UNEP 2018, 2023). While the average growth rate of GSDP during 2005–2015 was around 7–8% for almost all the states, the natural capital registered a decline in 11 states. While 13 states showed a marginal growth of 0–5%, just three states showed an increase of natural capital by more than 5%. This clearly indicates that this pace of economic growth is not sustainable in the long run (Pandey 2018).

4.3 Current Situation in Land Degradation

As per the latest estimates 97.85 million hectares (Mha) area of the country was undergoing land degradation during 2018–2019, which is 29.77% of the land area. A comparison of the data of the period from 2003–2005 to 2018–2019, in Table 4.1 shows a steady increase in the area under land degradation and desertification (SAC 2021). Deriving data from different sources a harmonized area statistic of degraded land/wastelands in India has been drawn recently, which shows that 120.72 Mha land stands degraded (Jinger et al. 2023), which is higher than the estimation of SAC.

Based on land use, 46% of degraded land is under agriculture (38% unirrigated and 8% irrigated), followed by forests and scrubland covering 36% of degraded land (22% forest and 14% scrubland). The unirrigated agricultural land and the forest together constitute 60% of the degraded land, mainly in the states and union territories of Rajasthan, Maharashtra, Gujarat, Ladakh, Karnataka, Jharkhand, Odisha, Madhya Pradesh and Telangana. The area under land degradation in these states and the breakup of the area due to various causes are given in Table 4.2. Though these areas have large degraded areas, they are not under accelerated degradation now, and in some of these states the trend is getting reversed due to afforestation. On the other hand, the states that have shown more than 25% degradation since 2003–2005 till 2018–2019 are Mizoram (188%), Punjab (80%), Arunachal Pradesh (47%), Assam (46%), Tripura (37%), Nagaland (29%) and Delhi (25%) largely due to man-made causes, viz., shifting cultivation, infrastructure development, groundwater depletion, water erosion and deforestation (SAC 2021).

Table 4.1 Progress of land area under degradation and desertification in India

Period of assessment	Area under land degradation (Mha)	Per cent land area	Area under desertification (Mha)	Per cent land area
2003–2005	94.53	28.76	81.48	24.78
2011–2013	96.40	29.32	82.64	25.13
2018–2019	97.85	29.77	83.69	25.45

Table 4.2 Area under land degradation due to various causes in major states (Mha)

State	Area under land degradation (2018–2019)	Vegetation degradation	Water erosion	Wind erosion	Settlement	Other reasons
Rajasthan	21.23	2.61	2.12	14.84	0.13	1.53
Maharashtra	14.30	5.14	8.2	0	0.37	0.59
Gujarat	10.24	2.32	3.83	1.17	0.21	2.71
Ladakh	7.11	1.12	0	1.73	0	4.26
Karnataka	6.95	1.69	5.01	0	0.11	0.14
Jharkhand	5.48	1.41	3.91	0	0.05	0.11
Odisha	5.35	0.74	4.43	0	0.04	0.14
Madhya Pradesh	3.85	2.53	1.12	0	0.10	0.10
Telangana	3.63	0.54	2.82	0	0.11	0.16

Of the total area of the country, 68% of the land area receives less than 1125 mm of rainfall and is thus drought-prone; tackling drought in these areas can contain desertification and land degradation. However, in the humid areas, water erosion remains the major reason for land degradation and needs to be controlled by appropriate vegetative measures in the catchments and the watercourses.

4.4 Biodiversity and Land Degradation Connection

Sustainable Development Goal 15—"Protect, restore and promote sustainable use of terrestrial ecosystems, sustainably manage forests, combat desertification, and halt and reverse land degradation and halt biodiversity loss"—is an overarching goal linking the objectives of biodiversity conservation and land degradation neutrality. The nine SDG targets under SDG 15 focus on sustainable forest management, combating desertification, biodiversity conservation, checking poaching and trafficking and control of invasive species. Though the overall SDG 15 score between 2010 and 2020 increased by more than 50% in India (Zhang et al. 2023), there has been significant loss of biodiversity and land degradation. The intimate connection between biodiversity loss and land degradation can be understood from the data reported by India to UN organizations. As can be seen from the UNCCD Data Dashboard, from 2015 to 2019, 30.51 Mha of India's total reported land got degraded, which is 9.45% of the land mass. The hotspots of degradation identified are 483 extending over 3.44 Mha, while the bright spots showing improvement were just 15 over 0.008 Mha. This extent of degradation is linked to the low Red List Index value of species survival at 0.68 (RLI close to 1 indicates low aggregate species extinction risk while RLI close to 0 indicate that more species are moving towards extinction). Between 2015 and 2019 India had moved from 0.69 to 0.68 in RLI and degradation had increased from 4.42% to 9.45%, in the successive assessments (UNCCD 2022, 2023).

4.5 Integration of Biodiversity Conservation with Land Degradation Neutrality

The latest measure towards conservation of biodiversity is the Kunming-Montreal global biodiversity framework (KM-GBF) which urges countries to conserve 30% of biodiversity on land and underwater by 2030. Recently, the Government of India has updated the national biodiversity strategy and action plan and adopted national biodiversity targets in alignment with the Kunming-Montreal global biodiversity framework (GOI 2024). The global targets, the national biodiversity targets and the action points linked to land degradation neutrality are tabulated in Table 4.3.

An impediment to successful implementation of the biodiversity action plan will be the lack of biodiversity finance policy/plan. As biodiversity financing at present is highly fragmented, involving multiple institutions with overlapping functions and

Table 4.3 Global and national biodiversity targets and their integration with land degradation neutrality

Global targets—No. and the target	National biodiversity targets—No. and the target	Action points under the national bodiversity strategy and action plan 2024–2030	Linkage with land degradation neutrality (LDN)	Action required to achieve LDN
1. Ensure that all areas are under participatory integrated biodiversity—inclusive spatial planning and/or effective management processes addressing land and sea use change—to bring the loss of areas of high biodiversity importance, including ecosystems of high ecological integrity, close to zero by 2030, while respecting the rights of indigenous peoples and local communities.	1. Biodiversity inclusive integrated land/sea use pl anning: ensure that all areas are under participatory and integrated biodiversity inclusive spatial planning and effective management processes addressing land- and sea-use change to bring the loss of areas of high biodiversity importance, including ecosystems of high ecological integrity, close to zero by 2030, while respecting the rights of local communities (LCs).	1. Prepare national and state level spatial plans using high resolution spatial data for delineating areas of high biodiversity importance and degraded ecosystems 2. Prepare forest working plans as per the National Working Plan Code to mainstream biodiversity conservation 3. Prepare management plans for all categories of protected areas and conservation plans for biodiversity heritage sites, coastal regulation zones, wetlands and other effective area-based conservation measures (OECM) sites 4. Promote integrated approaches to management of river basins, considering upstream and downstream inflows and withdrawals by season, pollution loads and natural regeneration capacities, in particular, for maintenance of in-stream ecological values 5. Adopt a comprehensive approach to integrated coastal management by addressing linkages between coastal areas, wetlands, and river systems in relevant policies, regulations and programmes	By the spatial planning process, the loss of land area of high biodiversity importance and ecosystems of high ecological integrity would be brought close to zero by 2030.	The states that have shown more than 25% land degradation since 2003–2005 till 2018–2019 are Mizoram (188%), Punjab (80%), Arunachal Pradesh (47%), Assam (46%), Tripura (37%), Nagaland (29%) and Delhi (25%). In the Northeastern states it is mostly due to deforestation for developmental projects, encroachments, soil erosion due to water, landslides and shifting cultivation. Appropriate land tenure systems need to be introduced to check the land degradation. In Delhi and Punjab, unsustainable infrastructure development and groundwater depletion need to be checked through land and water use planning.

(continued)

Table 4.3 (continued)

Global targets—No. and the target	National biodiversity targets—No. and the target	Action points under the national bodiversity strategy and action plan 2024–2030	Linkage with land degradation neutrality (LDN)	Action required to achieve LDN
2. Ensure that by 2030 at least 30% of areas of degraded terrestrial, inland water, and coastal and marine ecosystems are under effective restoration, in order to enhance biodiversity and ecosystem functions and services, ecological integrity and connectivity.	2. Ecosystems restoration: ensure that by 2030, at least the prioritized 30% of areas of degraded terrestrial, inland water, and marine and coastal ecosystems are under effective restoration, in order to enhance biodiversity and ecosystem functions and services, ecological integrity and connectivity.	1. Undertake mapping of degraded terrestrial, inland water and marine and coastal ecosystems for prioritizing areas for rehabilitation and restoration 2. Strengthen and augment the existing programmes, schemes and activities of the central and state governments relating to restoration of all prioritized degraded ecosytems 3. Undertake restoration of degraded ecosystems through multi-stakeholder partnerships to enhance ecological connectivity and restore ecosystem functions and services 4. Institutionalize a system of monitoring projects in all approved mining plans to ensure safe disposal of tailings and ecosystem rehabilitation following the principles of ecological succession 5. Strengthen capacity and make available techniques and technologies for the regeneration and restoration of ecosystems 6. Prepare and implement thematic action plans incorporating watershed management strategies to arrest and reverse desertification and expand green cover 7. Mitigate and manage the impacts on biodiversity due to river valley projects, power plants, industries, linear infrastructure and mines 8. Initiate actions to restore vital spatial and temporal connectivity (longitudinal, horizontal, vertical and temporal) within riverscapes and ensure environmental flow.	At least 30% of the degraded areas would be under effective restoration by 2030.	The degraded areas are mostly agricultural areas under dryland agriculture and forests with open canopy. Bringing the dryland agriculture under agroforestry systems and increasing the density of open forests by appropriate natural and artificial regeneration techniques is required.

3. Ensure and enable that by 2030 at least 30% of terrestrial, inland water, and of coastal and marine areas, especially areas of particular importance for biodiversity and ecosystem functions and services, are effectively conserved and managed through ecologically representative, well connected and equitably governed systems of protected areas and other effective area-based conservation measures, recognizing indigenous and traditional territories, where applicable, and integrated into wider landscapes, seascapes and the ocean, while ensuring that any sustainable use, where appropriate in such areas, is fully consistent with conservation outcomes, recognizing and respecting the rights of indigenous peoples and local communities including over their traditional territories.	3. Conserve biodiversity in land, water and sea: ensure and enable that by 2030, at least 30% of terrestrial, inland waters, and coastal and marine areas, especially areas of importance of biodiversity, ecosystem functions and services, are effectively conserved through ecologically representative, well-connected protected areas and other effective area-based conservation measures (OECMs). Also, integrate tribal areas wherever applicable into wider landscapes/ seascapes and ensure that sustainable use is legal and consistent with conservation outcomes while respecting the rights of local communities (LCs), including their traditional territories.	1. Promote the establishment of conservation reserves and community reserves to ensure adequate representation of all biogeographic zones in the country. 2. Establish a self-sustaining monitoring system for overseeing the management effectiveness of the PA network. 3. Implement site-specific ecodevelopment programmes in fringe areas of PAs to enhance livelihoods and provisioning of ecosystem services for local communities. 4. Strengthen and support research work on PAs, biosphere reserves and fragile ecosystems by involving research institutions and universities to develop baseline data on biological and management attributes and ecosystem functions and services. 5. Strengthen the protection of areas of high biological diversity, rich in threatened species, endemism and genetic resources, while providing alternative livelihoods and access to resources to local communities. 6. Develop partnerships with local communities for the enhancement of wildlife habitats and biodiversity conservation in conservation reserves and community reserves. 7. Ensure effective management of ecologically sensitive areas/eco-sensitive zones through implementation of a zonal master plan. 8. Integrate conservation and wise use of wetlands and river basins involving all stakeholders, in particular local communities, to ensure maintenance of hydrological regimes and conservation of biodiversity. 9. Identify hotspots of agrobiodiversity under different agroclimatic zones and promote on-farm conservation. 10. Accord priority attention to mitigate potential adverse impacts on designated UNESCO's natural world heritage sites and biodiversity heritage sites in view of their incomparable values that merit stricter standards than in otherwise comparable situations. 11. Update the database on sacred groves, ponds and landscapes and document bioresources and associated knowledge systems. 12. Promote expansion of the extent of OECMs in land, inland waters and marine areas, and develop a centralized national/state-level database on OECMs.	Bringing at least 30% of the land area under protected areas and other effective area-based conservation measures, would increase the area under protection, thus checking degradation in those areas.	The present area of notified forests, including the national parks and sanctuaries does not account for 30% of land area. There are non-forest areas and community forests, besides inland wetlands, coasts and marine areas which harbour rich biodiversity and provide ecosystem services, which can be identified and notified as forests or protected areas. The unclassed forests need to be classified as per the present human use and requirement for future, and ecological needs, either as notified forests or non-forest areas/ villages, so that there is certainty of tenure, and appropriate management. Most of the new areas proposed to be conserved, if located in tribal areas, may be made into community conserved areas, to be managed jointly by the community and the government, so that there is involvement of the local people in conservation.

(continued)

Table 4.3 (continued)

Global targets—No. and the target	National biodiversity targets—No. and the target	Action points under the national bodiversity strategy and action plan 2024–2030	Linkage with land degradation neutrality (LDN)	Action required to achieve LDN
6. Eliminate, minimize, reduce and/or mitigate the impacts of invasive alien species on biodiversity and ecosystem services by identifying and managing pathways of the introduction of alien species, preventing the introduction and establishment of priority invasive alien species, reducing the rates of introduction and establishment of other known or potential invasive alien species by at least 50%, by 2030, and eradicating or controlling invasive alien species, especially in priority areas such as islands.	6. Manage invasive alien species: eliminate, minimize, reduce and/or mitigate the impact of prioritized invasive alien species on biodiversity and ecosystem services by identifying and managing pathways for the introduction of alien species and eradicating or controlling invasive alien species, especially in priority sites, such as islands.	1. Strengthen domestic quarantine measures to monitor introduction pathways and contain the spread of invasive species from and to neighbouring areas and other countries. 2. Develop appropriate early warning and awareness systems against invasive alien species, and establish a national database on the extent and control of invasive alien species reported in India. 3. Promote regional cooperation in the adoption of uniform quarantine measures and containment of invasive alien species. 4. Assess invasion potential of introduced species and develop measures for their control and management. 5. Support capacity building for managing invasive alien species at different levels, especially of local bodies. 6. Provide priority funding to basic/applied research on managing invasive species.	One of the important drivers of forest degradation is the invasive alien species. Eliminating them will help restore the degraded forests. Most of the open forests are invaded by the invasive alien species, and their removal would help natural regeneration of the native species and increase the canopy density of these areas.	Special missions are required to eliminate the invasive alien species. Those on land need to be removed physically and either composted or compressed into pellets and used as fuel. If they can be pulped, they can go into paper production. The aquatic species can be collected and dumped on land where they cannot survive. They will add organic matter to the soil on land.

7. Reduce pollution risks and the negative impact of pollution from all sources, by 2030, to levels that are not harmful to biodiversity and ecosystem functions and services, considering cumulative effects, including: reducing excess nutrients lost to environment by at least half including through more efficient nutrient cycling and use; reducing the overall risk from pesticides and highly hazardous chemicals by at least half including through integrated pest management, based on science, taking into account food security and livelihoods, and also preventing, reducing, and working towards eliminating plastic pollution.	7. Reduce pollution risks and negative impact: reduce pollution risks and the negative impact of pollution from all sources by 2030 to levels that are not harmful to biodiversity and ecosystem functions and services, considering cumulative effects by (a) reducing excess nutrients lost to the environment, including through more efficient nutrient cycling and use: (b) reducing the overall risk from pesticides and highly hazardous chemicals by at least half, including through integrated pest management, based on science, taking into account food security and livelihoods; and (c) preventing, reducing, and working towards eliminating plastic pollution.	1. Develop and implement viable models of public-private partnerships that can benefit local bodies/ authorities/ local communities for collection, segregation, and setting up/operating secure landfills, incinerators, and deployment of other appropriate techniques for the treatment and disposal of municipal waste, toxic and hazardous wastes, both industrial and biomedical. 2. Survey and develop a national inventory of toxic and hazardous waste dumps and online monitoring system for the movement of hazardous wastes. Strengthen the capacity of institutions responsible for monitoring and enforcement in respect of toxic and hazardous wastes. 3. Strengthen the legal instruments and response measures for addressing emergencies arising out of transportation, handling and disposal of hazardous wastes as part of the chemical accident regime. 4. Minimize and eliminate activities leading to biodiversity loss due to point and non-point sources of pollution and promote the development of clean technologies 5. Strengthen the monitoring and enforcement of emission standards for both point and non-point sources 6. Treat and manage domestic and industrial effluents so as to minimize adverse impacts on terrestrial and aquatic (freshwater and marine) biological resources 7. Promote biodegradable and recyclable substitutes for non-biodegradable materials and develop and implement strategies for their recycling, reuse, and final environmentally benign disposal, including through the promotion of relevant technologies and the use of incentive-based instruments. 8. Develop and implement a strategy for strengthening regulation and addressing the impacts of ship-breaking activities on human health and coastal and marine bioresources. 9. Support assessments of e-waste and plastic pollution, including micro- and nanoplastics, in terrestrial, inland waters, and coastal and marine ecosystems and develop innovative approaches 10. Promote and support the approach of India's Blue Flag certification to maintaining clean beaches 11. Develop strategies for creating awareness and enhancing capacity of all stakeholders, especially youth and schoolchildren, about the adverse effects of pollution and disseminate possible mitigating action 12. Encourage utilization of non-hazardous industrial waste (e.g. fly ash) to reduce pollution 13. Raise awareness of pollution risks and the negative impact of pollution from all sources on biodiversity and human health	One of the drivers of land degradation is pollution from pesticides and hazardous chemicals and also excessive use of chemical fertilizers. Their reduction by half through integrated nutrient management and integrated pest management will reduce land degradation. Eliminating plastic pollution will reduce the land degraded by landfills and dumps of municipal solid wastes.	The degradation of agricultural land will be reduced by introduction of appropriate nutrient management and pest management measures.

(continued)

Table 4.3 (continued)

Global targets—No. and the target	National biodiversity targets—No. and the target	Action points under the national bodiversity strategy and action plan 2024–2030	Linkage with land degradation neutrality (LDN)	Action required to achieve LDN
8. Minimize the impact of climate change and ocean acidification on biodiversity and increase its resilience through mitigation, adaptation, and disaster risk reduction actions, including through nature-based solutions and/or ecosystem-based approaches, while minimizing negative and fostering positive impacts of climate action on biodiversity.	8. Minimize the impact of climate change	1. Identify and delineate the key ecologically rich areas of the country vulnerable to climate change at national, state and local levels and incorporate ecosystem-based approaches and relevant measures in various programmes, including watershed management, coastal and marine areas planning and regulation, agricultural technologies and practices, forest management, and health programmes 2. Explicitly consider the vulnerability of coastal areas and their biodiversity to climate change and sea level rise in coastal management plans, as well as infrastructure planning and construction norms for taking mitigation and adaptation measures 3. Develop ecological criteria for identifying the species and ecosystems at greater risk from climate change and their priority habitats. 4. Strengthen efforts for partial substitution of fossil fuels with biofuels and other alternative energy resources 5. Create awareness among the people for the adoption of the country's programme on lifestyle for environment (mission life) 6. Develop the capacity of diverse stakeholders for integrating climate risks into the conservation of biodiversity 7. Promote research on various aspects of climate change and its impact on biodiversity 8. Promote ecosystem-based approaches for climate change mitigation and adaptation. 9. Promote approaches for integration of biodiversity conservation and ecosystem management into state action plans on climate change and the Mahatma Gandhi National Rural Employment Guarantee Scheme (MGNREGS). 10. Develop/promote methodologies for assessment of climate co-benefits of biodiversity conservation and promote conservation actions that maximizes co-benefits 11. Promote measures and develop stakeholders' capacity for integration of ecosystem-based disaster risk management into national, state and district level disaster management plans and strategies 12. Strengthen capacity to assess and address the interlinkages of biodiversity, climate change and human health, including through research, education, knowledge and communication tools, and enhance international cooperation	The national action plan on climate change includes missions on water, forestry, agriculture and sustainable habitat, all of which are related to sustainable land use.	The missions are suffering from financial constraints which need to be addressed.

10. Ensure that the areas under agriculture, aquaculture, fisheries and forestry are managed sustainably, in particular through the sustainable use of biodiversity, including through a substantial increase in the application of biodiversity-friendly practices, such as sustainable intensification, agroecological and other innovative approaches contributing to the resilience and long-term efficiency and productivity of these production systems and to food security, conserving and restoring biodiversity and maintaining nature's contributions to people, including ecosystem functions and services.	10. Sustainable management of agriculture, animal husbandry, fisheries, aquaculture and forest areas	1. Encourage adoption of science-based and traditional sustainable land use practices through research and development, knowledge sharing, pilot-scale demonstrations and large-scale dissemination, including farmers' training and, wherever necessary, access to institutional finance. 2. Promote sustainable alternatives to shifting cultivation where it is no longer ecologically viable, ensuring that the culture and social fabric of the local people are not disrupted 3. Encourage agroforestry, organic farming, agroecological approaches, environmentally sustainable cropping patterns, traditional crop varieties and the adoption of efficient irrigation techniques 4. Integrate wetland conservation, including management of village ponds and tanks, springs and streams into sectoral development plans for poverty alleviation and livelihood improvement, and link efforts for conservation and sustainable use of wetlands with the ongoing rural infrastructure development and employment generation programmes 5. Enhance restoration and strengthen sustainable management of mangroves to ensure the protection of coastal belts and conservation of flora and fauna 6. Encourage integrated pest management practices, use of organic manures, bio-pesticides and biofertilizers and discourage use of chemical fertilizers, fungicides, herbicides, pesticides and insecticides 7. Promote natural and sustainable farming of locally adapted and traditional crop varieties through traditional practices (draught animals) and appropriate incentives and direct access to markets duly supported by appropriate certification systems. 8. Strengthen effective management of forests for multiple benefits while ensuring a sustained flow of ecosystem services. 9. Provide support for effective regulation/enforcement of guidelines/rules/laws for sustainable inland and marine fisheries 10. Promote climate-smart agriculture and water efficient crops 11. Promote best animal husbandry practices for semi-domesticated and indigenous livestock 12. Support integrated and sustainable management for pastoral/ grazing areas, especially for alpine pastures 13. Support individual experts, farmers, and NGOs helping in reviving traditional practices of farming, maintaining indigenous breeds of livestock and germplasm conservation 14. Raise awareness of the interlinkages between biodiversity and health for nutrition, food security, livelihoods and food system resilience	Since all these are land-based production systems, improving their productivity while simultaneously maintaining their biodiversity and ecosystem functions reduces land degradation.	Agroforestry systems combine various production systems. Increasing the area under agroforestry will improve biodiversity and check land degradation.

no systematic tracking, there is a need for evolving mechanisms to map key sources and quantum of funds (Pandey et al. 2020).

A proposal has been mooted for a national mission on biodiversity and human well-being linking the requirements of biodiversity conservation with agriculture, health, bio-economy and capacity building (Bawa et al. 2021). A recent analysis prioritizing India's landscapes for biodiversity, ecosystem services and human well-being has found that only about 15% of the priority areas identified are included within the present protected area network, and several landscapes of high importance were omitted from all previous global-scale assessments. The study has examined the spatial congruence of aspirational districts identified by the NITI Aayog with the priority areas for conservation and recommended management zoning through state-driven and participatory approaches, to meet the twin objectives of biodiversity conservation and sustainable development. Since substantially large biodiversity exists in unprotected human-use landscapes, the study has suggested invoking other effective area-based conservation measures (OECMs), viz., granting community forest rights, and declaring areas as critical wildlife habitats under the Forest Rights Act. In areas of overlap of aspirational districts with priority sites for biodiversity, the negative effects of infrastructure development arc proposed to be mitigated through economic incentives, such as payment for ecosystem services (PES), promotion of agroforestry, demarcation of conservation and/or community reserves where sustainable use of resources by communities is permissible. The study has also suggested inclusion of biodiversity and ecosystem service considerations into the programmes for reversal of land degradation, Green India Mission and renewable energy projects (Srivathsa et al. 2023).

Bonn Challenge is a global effort to bring 150 Mha of degraded and deforested land into restoration by 2020, later extended to cover 350 Mha by 2030. India has committed to restore 26 Mha of degraded and deforested land by 2030 (UNCCD 2019). Between 2011 when the Bonn Challenge was launched and 2016–2017 India had restored 9.8 Mha, mainly through the National Afforestation Programme and Green India Mission. National Bamboo Mission, National Green Highway Mission, National Mission for a Clean Ganga, Nagar Van Yojana and various other schemes have also contributed to these restoration efforts (Borah et al. 2018). India's nationally determined contribution (NDC) aims to create an additional carbon sink of 2.5–3 billion tonnes of CO^2eq through additional forest and tree cover by 2030.

NITI Aayog has prepared a technical report on Greening and Restoration of Wastelands with Agroforestry (GROW) which has identified 20.74 Mha (6.31% of total geographical area) as highly suitable for agroforestry and 16.23 Mha (4.94% of TGA) as moderately suitable (NITI Aayog 2024). These "wastelands" if taken up for agroforestry can help in conserving forestry species on-farm and also in protecting the agrobiodiversity, provided the selection of wastelands for agroforestry and choice of species are appropriate. There is a danger of grasslands and scrub forest getting classified as "wastelands", though they are ecosystems in their own right, with unique flora and fauna required to be maintained as such.

Recently, the Government of India has launched the Green Credit Scheme, under which the forest departments would identify degraded land parcels—including open forest and scrubland, wasteland and catchment areas—which shall be used for tree plantation and increase of green cover, generating credits that can be used by industries and other funding agencies to compensate for loss of forests in any of their developmental projects. The scheme has been faulted as an endeavour to create a land bank for easy and quick diversion of forest lands, thus facilitating "ease of doing business" for user agencies and in clash with the Forest (Conservation) Act. Technically also, it is found wanting as the so-called scrublands and wastelands support herbs, shrubs and grasses which check runoff, maintain soil moisture, preserve seeds for forest regeneration and provide habitats for wildlife, which would be lost if planted with trees; further, the rationale behind assigning credits based on the number of trees, which vary in ecological impact and functionality, is also questioned (Koshy 2024; Nitnaware 2024). There is also a proposal to roll out a domestic Carbon Credit Trading Scheme by 2026, which will also involve raising plantations. These market-based mechanisms for afforestation efforts need to be effective on the ground and equitable, providing adequate safeguards for forest-dependent people, meeting the socio-ecological requirements. A recent analysis has suggested design principles to make such market-based mechanisms successful in the forestry sector. It has proposed four design principles, viz., comprehensive governance regimes, strengthened oversight in forest carbon accounting, lower transaction costs and modified engagement terms with local communities, and encouraged locally relevant flexibilities in temporal and spatial commitments (Roy and Bhan 2024).

4.6 The Way Forward

The major land uses where degradation is noticed are agriculture in the arid and semi-arid regions and forests with open canopy, that is, having canopy density between 10% and 40%. Paying attention to these two areas, aimed at improvement of dryland agriculture and densification of the open forests, can check degradation of land to a great extent and also enrich the biodiversity. The following suggestions are worth consideration, to achieve the twin objectives of biodiversity conservation and checking land degradation, while simultaneously increasing the agricultural output and ensuring the environmental services of the forests.

1. The open forests can be made denser by allowing for natural regeneration of the local species, offering effective protection from grazing and fire. Wherever required the regeneration can be supplemented by planting nursery-grown seedlings of the native species.
2. Dryland agriculture needs a boost by deployment of soil and water conservation measures and raising of crops suitable for arid land, thus reducing the requirement of water and also conserving whatever is available. The oilseeds are generally suitable for cultivation under arid conditions, and raising them would increase production and help in reducing the imports of edible oil. The agrobiodiversity of the oilseeds can also be conserved on-farm.
3. Medicinal plants suitable for arid lands can be cultivated to meet the requirements of the indigenous medical systems and the ethnomedicinal practices, thus ensuring primary health care of the population.

4. Development of water bodies in the arid lands by suitable rainwater harvesting practices and cascading systems of water storage can help in improving agriculture in the area nearby and also provide scope for fish cultivation.
5. The wetlands and the catchments need to be protected and restored for their contribution to the hydrology. The wetland management may include avoidance of water pollution, shoreline protection, desilting, weed control and fisheries development, which will improve aquatic biodiversity and also contribute to livelihoods.
6. Production forestry, especially of bamboo, pulpwood, and timber, needs a thrust increasing the productivity through breeding and deployment of high-yielding varieties in order to reduce the dependence on imports. Increased production outside the forests will reduce the pressure on forests and help in conserving biodiversity.
7. Green Highways Mission is an important activity undertaken to compensate for the loss of vegetation due to expanding network of roads. The plantations along the roadside can include trees and shrubs native to the locality, with flower production throughout the year by ensuring an appropriate mix of plants so that the pollinators are benefitted.
8. Implementation of agroforestry in the agricultural drylands identified by NITI Aayog can help in containing land degradation, and by appropriate choice of species, the species that are facing genetic erosion or overexploitation in their native habitats can be conserved on-farm. This is especially true of high-value agroforestry where the tree component can be a species of high value, viz., sandalwood, agarwood, red sanders, rosewood, etc. This will also contribute to the conservation and management of CITES listed species, which have an export market and potential to improve livelihoods.
9. Since urbanization is one of the major factors contributing to land degradation, urban greening needs a thrust in the form of parks, rooftop gardens and roadside plantations, where biodiversity can also be conserved.
10. The market-based mechanisms to fund afforestation efforts should take into consideration the governance regimes, proper oversight and accounting, transaction costs and involvement of local communities with flexible temporal and spatial commitments. They should address the concerns of the experts that these schemes may be just means of facilitating ease of doing business and make them technically sound and feasible.

References

Aggarwal M (2022) The state of India's forests: losing forests, gaining plantations. Mongabay Series: Climate Connections, The Indian Forest Story, 24 Jan 2022

Bawa KS, Sengupta A, Chavan V, Chellam R, Ganesan R, Krishnaswamy J, Mathur VB, Nawn N, Olsson SB, Pandit N, Quader S, Rajagopal P, Ramakrishnan U, Ravikanth G, Sankaran M, Shankar D, Seidler R, Shaankar RU, Vanak AT (2021) Securing biodiversity, securing our future: a national mission on biodiversity and human well-being for India. Biol Conserv 253:108867. https://doi.org/10.1016/j.biocon.2020.108867

Borah B, Bhattacharjee A, Ishwar NM (2018) Bonn Challenge and India: progress on restoration efforts across states and landscapes. International Union for Conservation of Nature and Natural Resources and Ministry of Environment, Forest and Climate Change, Government of India, New Delhi

Chandrakar AK, Yadav KK, Kumar V, Gupta N (2016) Scenario of biodiversity conservation in India: an overview. Photon eBooks, p 30

Ganesan R (2016) Caught in captivity. Down To Earth

GFW (2023) India deforestation rates and statistics. http://globalforestwatch.org/dashboards/country/IND

GOI (2024) India's updated National Biodiversity Strategy and Action Plan in alignment with Kunming-Montreal Global Biodiversity Framework: a roadmap for conservation of India's biodiversity: 2024-2030. Ministry of Environment, Forest and Climate Change and National Biodiversity Authority, Government of India, New Delhi

Haughan AE, Pettorelli N, Potts SG, Senapathi D (2022) The role of climate in past forest loss in an ecologically important region of South Asia. Glob Chang Biol 28:3883–3901

Jacob CT, Parida A, Krishna Kumar NK (2020) Conservation of India's agrobiodiversity towards increasing food, nutritional and livelihood security. Curr Sci 119(4):607–612

Jha S (2023) India lost 668,400 ha forests in 5 years, 2nd highest globally: report. Down To Earth, Mar 2023

Jinger D, Kaushal R, Kumar R, Paramesh V, Verma A, Shukla M, Chavan SB, Kakade V, Dobhal S, Uthappa AR, Roy T, Singhal V, Madegowda M, Kumar D, Khatri P, Dinesh D, Singh G, Singh AK, Nath AN, Joshi N, Joshi E, Kumawat S (2023) Degraded land rehabilitation through agroforestry in India: achievements, current understanding, and future prospectives. Front Ecol Evol 11:1088796. https://doi.org/10.3389/fevo.2023.1088796

Koshy J (2024) Can green credits benefit India's forests? The Hindu, 26 Apr 2024

NAAS (1998) Conservation, management and use of agrobiodiversity. Policy Paper 4. National Academy of Agricultural Sciences, India

Nayar MP, Singh AK, Nair KN (2009) Agrobiodiversity hotspots in India: conservation and benefit sharing, vol II. PPV&FR Authority, New Delhi

NITI Aayog (2024) Technical report on Greening and Restoration of Wastelands with Agroforestry (G.R.O.W). National Institution for Transforming India, New Delhi

Nitnaware H (2024) Green Credit Rules will negatively impact forest ecology, say experts. Down To Earth, 27 Feb 2024

Pandey K (2018) India loses natural resources to economic growth: report. Down To Earth, 9 Oct 2018

Pandey R, Gupta M, Sachdeva P, Singh A, Sugand S (2020) Biodiversity conservation in India: mapping key sources and quantum of funds. NIPFP Working Paper No. 311. National Institute of Public Finance and Policy, New Delhi

Ravikanth G, Jagadish MR, Vasudeva R, Uma Shaanker R, Aravind NA (2018) Recovery of critically endangered plant species in India: need for a comprehensive approach. Curr Sci 114(3):504–511

Roy A, Bhan M (2024) Forest carbon market-based mechanisms in India: learnings from global design principles and domestic barriers to implementation. Ecol Indic 158:111331

Rupavath P (2022) Indian forests around the size of Nagaland, thinned down in two years. Mongabay Series: The Indian Forest Story, 27 Jan 2022

SAC (2021) Desertification and land degradation atlas of India (assessment and analysis of changes over 15 years based on remote sensing). Space Applications Centre, Ahmedabad

Srivathsa A, Vasudev D, Nair T, Chakrabarti S, Chanchani P, de Fries R, Deomurari A, Dutta S, Ghose D, Goswami VR, Nayak R, Neelakantan A, Thatte P, Vaidyanathan S, Verma M, Krishnaswamy J, Sankaran M, Ramakrishnan U (2023) Prioritizing India's landscapes for biodiversity, ecosystem services and human well-being. Nat Sustain 6:568–577. https://doi.org/10.1038/s41893-023-01063-2

Swaroop D (2020) Livestock biodiversity in India: a review. Int J Agric Ext Soc Dev 3(2):26–29

UNCCD (2019) Land climate: key message from the IPCC special report on climate change and land by the United Nations Convention to Combat Desertification

UNCCD (2022) UNCCD Data Dashboard. United Nations Convention to Combat Desertification. data.unccd.int

UNCCD (2023) Performance review and assessment of implementation process. Seventh reporting process. Report from India. United Nations Convention to Combat Desertification

United Nations Environment Programme (2018) Inclusive Wealth Report 2018: Measuring Sustainability and wellbeing. https://www.unep.org/resources/inclusive-wealth-report-2018

United Nations Environment Programme (2023) Inclusive Wealth Report 2023: Measuring Sustainability and Equity. https://wedocs.unep.org/20.500.11822/43131

WII (2023) National Wildlife Database. Wildlife Institute of India, Dehradun. Updated as on 27 Nov 2023

World Bank (2013) India: Diagnostic assessment of select environmental challenges. An analysis of physical and monetary losses of environmental health and natural resources: Volume I. Report No. 70004-IN

Zhang Y, Li Y, Liu J (2023) Global decadal assessment of life below water and on land. iScience 26:106420. https://doi.org/10.1016/j.isci.2023.106420

An Overview of Assessment Methods and Factors Involved for Sustainable Land Management and Land Degradation Neutrality

5

Juhi Agrawal, Anil Kumar, Devesh Dubey, Aakash Kumar, and Abhishek Kumar

Abstract

Sustainable Land Management (SLM) and Land Degradation Neutrality (LDN) are development components to be confederated through the mare belief to sustain the environment globally. The study demonstrates methodologies to be deployed for assessment along with the cutting-edge parameters for land degradation; however, it emphasizes the prominence of such factors in land degradation. Conserving biodiversity, generating jobs, and achieving the sustainable development goals are some of the outcomes when integrated with sustainable land-management practices. Methodologies include indicators and questionnaire-based approaches for soil health and productivity assessment and their application in regional and global contexts. Rajasthan case study demonstrates new framework techniques, such as the Pressure-State-Response model and Google Earth Engine for monitoring land degradation coupled with combating it. It puts a specific emphasis on such seminal ecosystem-based indicators as soil organic carbon, vegetation index, and climatic indices which ultimately culminate in the realization of LDN. The chapter critiques challenges in integrating land-use practices, economies, and topographical influences, creating a systemic analysis of sustainable land management strategies.

J. Agrawal · A. Kumar
University of Petroleum and Energy Studies, Dehradun, Uttarakhand, India

D. Dubey
Arahas Technologies, Gurugram, Haryana, India

A. Kumar (✉)
Forest Research Institute (Deemed to be) University, Dehradun, India

A. Kumar
Department of Biotechnology, Graphic Era (Deemed to be University), Dehradun, Uttarakhand, India

S. Singh, M. K. Gautam (eds.), *Biodiversity Conservation and Land Degradation Neutrality*, https://doi.org/10.1007/978-981-96-7210-3_5

Keywords

Sustainable land management · Land degradation neutrality · Environmental sustainability

5.1 Introduction

Land resources play an important role in providing food, materials, and often-overlooked ecosystem services that support provisioning services (Critchley et al. 2021). However, global demands are growing because of increasing population and affluence, resulting in the deterioration of land health and productivity. Urbanization is also claiming prime agricultural land. This elevated competition over available land poses risks of instability in society and politics, making issues like migration, poverty, food insecurity, and conflict. Effectively addressing degradation of land is essential (Barbier and Hochard 2014; Liniger et al. 2017). Managing land degradation not only enhances the ability to adapt to and mitigate climate change but also has significant role in biodiversity conservation, ensuring food security, and creating long-term employment opportunities. The concept of Land Degradation Neutrality (LDN) initially induced an enhanced reaction from policymakers to land degradation (Reichhuber et al. 2019). LDN is now considered as part of Sustainable Development Goal 15; also enhancing the ability to LDN is one among United Nations Convention to Combat Desertification's (UNCCD's) main objectives (UNCCD 2015; Kust et al. 2017).

According to UNCCD (2017), LDN is the condition wherein the amount of land needed to sustain ecosystem-based services to enhance the safety of food remains stable or grows within certain temporal and geographical ecosystems and scales (Smyth and Dumanski 1993). This idea was presented to focus attention on the specific goal of "no net loss," which is to protect the global productive and wholesome land (Dumanski 1997). The strategy includes actions to stop or lessen the deterioration of land and to reverse any existing degradation, ensuring that gains outweigh losses. Countries are urged to support LDN freely formally on a nationwide scale (Cf 2015; Anderson et al. 2006). The LDN aim is an international objective (Akhtar-Schuster et al. 2017). The UNCCD Science-Policy Interface (SPI) was tasked the creation of a scientific "Conceptual Framework for LDN" by the Conference of the Parties (COP) to the UNCCD. This framework aims to establish a solid scientific basis for comprehending and carrying out LDN, providing advice for its pursuit and monitoring developments toward the LDN objective (Orr et al. 2017; Sanz et al. 2017). However, the UNCCD's application of the LDN conceptual framework is restricted to dry lands for different kinds of land, uses of land, and ecosystem services. LDN is important for realizing multiple Sustainable Development Goals (SDGs) pertaining to sustainable resource utilization, environmental preservation, and food security (Liniger et al. 2011; Forest Europe 2020). The evolution of sustainable land management (SLM) is intricately linked to the principles of LDN. It is believed that implementation of effective land management practices, often termed as "best practices," can help prevent or alleviate land degradation and potentially reverse it, all while preserving ecological functioning and productivity (Chotte et al. 2019). However, performing an analysis concludes that not all individual practices and technologies align with the goal of achieving land degradation neutrality (Critchley et al. 2021). On the other hand, sustainable approaches to land management does not necessarily certify that land degradation neutrality has been achieved.

Sustainable Land Management (SLM) is an integrated strategy designed to stop the degradation of land resources while maintaining ecosystem processes and land productivity (Barbier and Hochard 2014; Al Sayah et al. 2019). The UN Conference on Environment and Development (UNCED) in 1992 is credited with the concept's emergence between 1980 and 1990. The Framework for assessing SLM established five fundamental assumptions, characterizing SLM as a combination of methods, tools, and techniques that combine environmental concerns with socio-economic theories (Al Sayah et al. 2021). The objective is to preserve or enhance output and services, lower risks, conserve natural resources, stop soil and water deterioration, and maintain economic viability and societal acceptance (Stavi and Lal 2015; Liniger et al. 2019). The recent emphasis on SLM is closely tied to the LDN approaches aligned with Sustainable Development agenda of Goal-15 of 2023 (Gichenje et al. 2019).

This objective centers on safeguarding, rejuvenating, and encouraging the sustainable usage of the earth's ecosystems. It includes reversing land degradation, managing forests, and preventing desertification. The consensus is that SLM technologies and practices can prevent, reduce, and mitigate the negative impacts of unwise land use, thereby helping to reverse land degradation (Zolotokrylin 2019). The Food and Agriculture Organization of the United Nations (FAO) regards SLM as a strategy focused on mitigating land degradation, rehabilitating degraded areas, and optimizing the utilization of land resources for the benefit of both present and future generations. The Economics of Land Degradation Initiative emphasizes how important the economic component of SLM is. SLM is a term used to describe a grouping of methods, procedures, and technologies that are utilized for managing land resources at the local level. Numerous studies assessing the economic value of ecosystem services demonstrate the financial benefits of investing in SLM practices. The most recognized definition of SLM presents a strong emphasis on using land resources for the generation of products, including plants, water, soil, animals, with the intention of fulfilling the shifting needs for human resources. It preserves the environmental benefits of these resources while ensuring their long-term productivity.

The concept of SLM has gained widespread acceptance but currently faces challenges in determining the sustainability of specific land management practices and technologies (Abel et al. 2021). Existing methods often rely on environmental metrics, such as rates of soil erosion or water quality, which vary depending on natural area and the type of land management and may not always be definitive results (Agbeshie et al. 2022). There are no widely recognized standards for evaluating the sustainability of effects and the efficacy of implemented measures. For example,

Mario et al. (2022) suggest that implementing SLM practices can reduce the harmful impacts of droughts on plantations, farms, pastures, and forests. Further research by Al Sayah et al. (2021) and Feng et al. (2022) exemplify how SLM practices contribute to the preservation and increase of soil organic carbon stocks. These stocks are important indicators for evaluating ecosystem health and are significant considerations in selecting SLM approaches.

The effectiveness of SLM practices and technologies can be assessed by aiming to achieve Land Degradation Neutrality (LDN) in a specific area, as suggested by Chotte et al. (2019). Another study by Kust et al. (2017) emphasizes that sustainable land management is a crucial mechanism for achieving LDN. Critchley et al. (2021) underscore the ability to select optimal; in contrast, FAO presents the "LDN index" as a straightforward and useful tool that shows a region or farm's successful land policy and decreased chances of land degradation (Barbier and Hochard 2014). Through integration of SLM and LDN approaches, the authors propose a definition of land degradation for use in Russia, encompassing all categories of lands and farmlands (Al Sayah et al. 2019). It defines land degradation as "a broad variety of natural and man-made factors that combine to reduce an area's potential economically and/or naturally, as well as the ecosystem services it offers and its ability to withstand adverse effects" (Al Sayah et al. 2021). Although there are direct recommendations and seemingly clear working hypotheses, there has been a lack of algorithms developed to assess sustainable land management using LDN techniques (Stavi and Lal 2015; Liniger et al. 2019). Not every practice deemed "good" for land management helps achieve LDN, according to a thorough examination of several land management approaches done by Gichenje et al. (2019), while cases of neutrality are not always linked to specific land management models.

5.2 Classification of Sustainable Land Management (SLM) Models

The enhanced classification of SLM models allows for a precise identification of scenarios in which attaining Land Degradation Neutrality (LDN) should be viewed as SLM and situations where the adopting a set of effective practices will result in LDN achievement. To achieve this, the algorithms incorporate the concepts of passive and active land management practices (Zolotokrylin 2019). Passive practices denote situations where no specific actions are taken to preserve natural or anthropogenic ecosystems, which, in certain instances (e.g., protected areas or fallow lands classified under different categories), can still be considered as land management, possibly contributing to land restoration or non-deterioration. However, their distinctive characteristic lies in the absence (or discontinuation) of direct human action at a specific moment, while active practices always entail modifying the ecosystem in some way by outside forces.

In certain instances, such as protected areas or fallow lands categorized differently, these scenarios may be regarded as land management, sometimes geared towards restoration, or preventing further deterioration. A key distinguishing factor

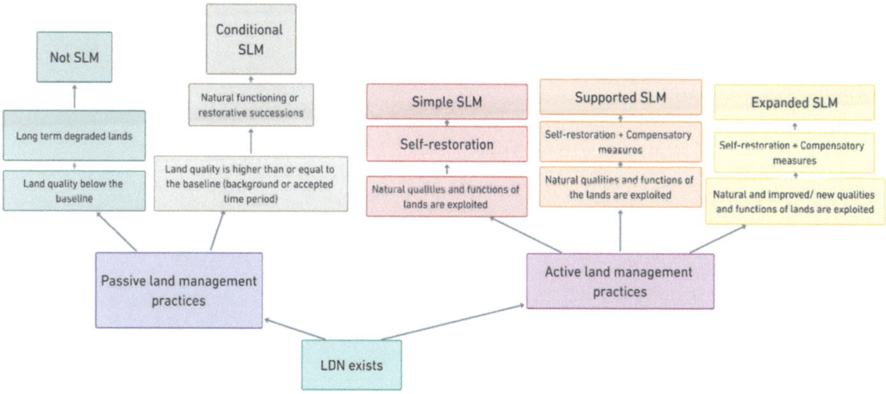

Fig. 5.1 LDN and SLM are related. (i) SLM recognition algorithm for the formation of LDNs

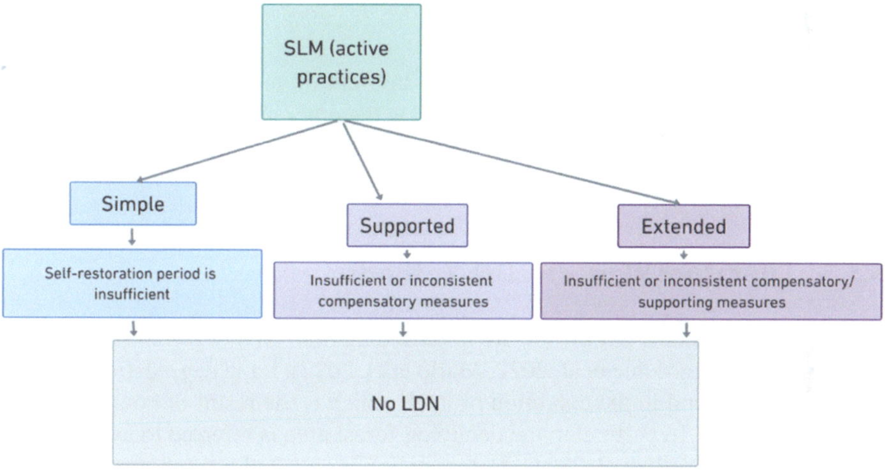

Fig. 5.2 Algorithmic inversion to determine the feasibility of implementing LDN for SLM (active practices) models

is the cessation or absence of direct human intervention at a specific moment. Active practices, on the other hand, invariably involve some external influence or alteration to the ecosystem (Lobkovsky et al. 2022). As shown in Fig. 5.1, in order of achieving LDN, only models of SLM that incorporate self-restoration, adequate compensatory and supportive measures, and proactive land management techniques may be identified using formal indicators.

Conditionally sustainable models, such as old-growth forest or long-term fallows, are those in which natural processes actively help to achieve LDN standards based on formal indicators, even when there are passive approaches. Figure 5.2 describes that even with the use of a series of efficient active techniques, LDN remains unattainable if compensatory measures are inadequate, poorly coordinated,

Fig. 5.3 Algorithmic inversion to determine the feasibility of implementing LDN for SLM (passive practices) models

or if these practices neglect the necessary duration of self-restoration for natural systems (Kust et al. 2022; Abel et al. 2021). On the other hand, certain passive practices (as mentioned in Fig. 5.3) can successfully achieve LDN, whether the trends are perceived as positive or negative.

5.3 Literature Work on LDN Methods

People and ecosystems worldwide are impacted by both land degradation (LD) and climate change (Agbeshie et al. 2022; Mario et al. 2022). Land degradation refers to an unfavorable trend in the condition of land, which is the result of both direct and secondary factors. In particular, the declining forest area is referred to as forestland degradation (Al Sayah et al. 2021). Deforestation is an act of turning forestland into non-forest land that can exacerbate land degradation (Feng et al. 2022; Kaffas et al. 2022).

The processes of land degradation, categorized as direct drivers by the Millennium Ecosystem Assessment (2005), directly lead to land degradation. Indirect factors like changing land tenure or cash crop prices can lead to shifts in land management and usage, also considered as factors in land degradation (Mikhailova et al. 2024; Pereira et al. 2020). To mitigate it, the LDN can be employed, involving halting, slowing, and potentially reversing land degradation. Land degradation arises from a combination of physical, chemical, and human activities (Zhao et al. 2021). Biological productivity and ecological integrity encompass the structural and functional attributes of the ecosystem. Human value, providing benefits from the land, may erode due to initiation of processes at a larger-than-usual scale. Pressures leading to degradation of specific land system components include irrigation, salinization of groundwater sources, and water erosion under agricultural practices (soil, water, and biota). Various pressures and degradation processes, such as overgrazing, landscape drainage, warming, and climatic variables like rainfall changes, contribute to depletion of soil organic matter (Allen et al. 2020; Bobushev and Sultanaliev 2020).

5.4 Land Degradation Neutrality (LDN) Assessment Methods

The evaluation indicators and questionnaires are the main tools used to achieve neutrality for land degradation. Considering that land degradation is a complex problem with socioeconomic and environmental implications, dimensions, relying on a single indicator, may not offer a comprehensive assessment of the land's health or condition (Manna et al. 2024). Land degradation has led to the emergence of various interpretations of different assessment techniques. Three indicators have been identified by the UNCCD to produce a systematic assessment for LDN in line with sustainable development goals (SDGs): land cover, crop produce, and soil organic carbon. Indicators designed to assess the land's capacity to deliver ecosystem services were formally approved in November 2017 by the Inter-Agency and Expert Group on Sustainable Development Indicators (IAEG SDG). Additional indicators have been introduced by certain research to improve the evaluation of land degradation.

Moreover, the questionnaire method serves as a valuable supplement to the indicator method. Research conducted in a paper (UNCCD 2017) indicates that combining the questionnaire method with the indicator method enhances the assessment of land degradation.

5.4.1 Indicator Method

Enhancing the three assessment indicators for Land Degradation Neutrality (LDN) that the UNCCD designated has been the focus of several studies. Some critics argue that the UNCCD's LDN indicator, which categorizes land as either degraded or non-degraded, is overly simplistic. Various studies have proposed combining LDN evaluation incorporating the causes of land degradation, such as drought, landslides, and wildfires, resulting in the identification of 12 categories of land degradation (Moonrut et al. 2021). The creation of local indicators is encouraged by the UNCCD to improve LDN assessment. Global-scale LDN assessments have incorporated additional indicators, such as timber supplies, forest cover, biodiversity levels, wind erosion, and water erosion (Al Sayah et al. 2021; Li et al. 2021b; Liniger et al. 2019). Regional-scale LDN assessments have introduced new indicators based on different land degradation characteristics, such as total soil loss, soil loss rate, gully erosion (Ding et al. 2020), and land contamination and atmospheric humidity.

While productivity, soil organic carbon, and land cover are vital LDN indicators, the multitude of data regarding these indicators raises questions about consistency in LDN values calculated using different products. This poses uncertainty regarding the most valuable products for evaluating the degradation of land. For instance, salinization, wind, water, and freeze-thaw cycles are the main factors in land degradation in China. Even though these LDN-adopted indicators can show shifts in the state of the land, comprehending how particular soil degradation processes lead to changes in these indicators requires further study.

Existing research methods primarily focus on assessing land degradation, but LDN involves both degradation and restoration. Therefore, methods considering both aspects are needed to monitor LDN progress. Additionally, grading degradation and restoration can help identify the status of LDN.

5.4.2 Questionnaire Method

Several research endeavors have augmented LDN outcomes by incorporating participatory mapping, field surveys, and farmer interviews (Li et al. 2021b, with the aim of fostering increased involvement of stakeholders in LDN assessments. The location and degree of degradation in South Africa were ascertained using a mapping questionnaire (QM) and the World Overview of Conservation Approaches and Technologies (WOCAT). The findings showed differences between land degradation regions determined by this approach and the ones derived using markers like soil organic carbon, vegetation productivity, and land cover (Crossland et al. 2018). This underscores the necessity to develop more flexible methods for LDN assessment. Additionally, prior findings have shown that combining questionnaire surveys with data from remote sensing can improve accuracy of land degradation assessments (Liniger et al. 2019).

5.5 Case Study of LDN in Rajasthan

This study was conducted in Rajasthan, India, situated between 69° 30′ to 78° 17′ E longitude and 23° 30′ to 30° 12′ N latitude. Rajasthan is a significant state located in western India, spanning an area of 342,239 square kilometers. It comprises 10.4% of the country's land area and is home to 5.67% of its population. As per the census data of 2021, Rajasthan has a population of approximately 79 million, with 75.13% residing in rural areas and dependent on rainfall for agriculture, etc. In Rajasthan, the summer temperatures fluctuate between 26 and 46 °C; during winter, the temperature typically varies between 0 and 28 °C. The state encounters an average yearly precipitation of 574 mm, falling short of the national average of 1100 mm. Rainfall patterns exhibit substantial diversity throughout the area. For instance, Jaisalmer district records less than 100 mm of rainfall, while Banswara and Jhalawar districts receive over 900 mm. The southwest monsoon constitutes roughly 91% of the total annual rainfall (Nzuza et al. 2022).

5.5.1 Data and Methodology

The LDN indicators outlined within the framework of the UNCCD conceptualize principles like mapping, non-continuous and continuous monitoring, and recovery of degraded land. These principles are to be implemented at global and regional levels (Raj and Sharma 2022). The aim of this convention is to minimize the effects of drought in countries severely affected by these issues and combat desertification, especially in Africa. This is to be achieved through required measures by every nation, reinforced by international partnership and cooperation arrangements, within a comprehensive strategy aligned with the principles of Agenda 21. The aim is to contribute to achieving sustainable development in affected areas (Akhtar-Schuster et al. 2017).

5.5.2 Pressure-State-Response (PSR) Model

The PSR model is employed to structure the data utilized in developing indicators, enhancing their accessibility, and augmenting their value. Additionally, the PSR framework can facilitate the integration of individual monitoring programs, identify redundancies and gaps, foster the creation of new indicators, and enhance the utilization of this data for policy and program development. The World Bank has recently revised the PSR framework and tailored it for the development of land quality indicators (LQI) and environmental indicators. The PSR framework serves as a tool to structure and organize information for LQIs, aiding in selecting the most relevant indicators that effectively illustrate how stakeholders manage their land and the resulting implications. This investigation categorizes indicators into three groups: state (climatic and topographical variables), pressure (socio-economic variables), and response (ecological variables). These categories depict human pressure and its effects on the condition of the lands (Kust et al. 2017).

5.5.3 Google Earth Engine (GEE)

GEE serves as a geospatial processing engine tailored for large-scale environmental monitoring and analysis, operating on a cloud server infrastructure. To expedite computational processes, the GEE platform leverages Google's extensive computational resources, enabling concurrent processing of geospatial data. The Application Programming Interfaces (APIs) provided by GEE, including a comprehensive development environment for Python, and JavaScript are available on GitHub, facilitating seamless communication with the Earth Engine servers (Raj et al. 2023).

5.5.4 Land Cover Type

Land cover and land use play crucial roles in influencing both land neutrality and degradation. This study utilized the MODIS Land Cover Type Product (MCD12Q1) to generate global maps at yearly intervals of land cover and a spatial resolution of 500 meters spanning from 2001 to 2020. The MCD12Q1 product was created by employing supervised categorization techniques on MODIS reflectance data (Sharma et al. 2021). Across the study area, 17 Land Use and Land Cover (LULC) classes were consolidated into categories including tree-covered areas (comprising protected forests and open forest lands), grasslands (herb lands, and encompassing shrub), crops, wetlands (including water bodies), artificial land, and other land types. In this analysis, we examine the trends in LULC types between 2001 and 2020.

5.5.5 Land Productivity of Rajasthan

Any land's productivity can potentially be appraised via taking into consideration variables like soil type, land cover, and vegetation diversity, and these each may have an unpredictable effect on Net Primary Productivity (NPP). The distribution of NPP for the three primary land cover types was examined in this study: regions covered in trees (including protected forests and open forest lands), grassland (encompassing shrub and herb lands), and crops, spanning from 2001 to 2020 within the study. The estimation of NPP at a global scale was conducted using the MODIS MOD17A3 NPP product, accessible through the GEE editor's data catalogue. This product offers a spatial resolution of 500 m and a temporal resolution of 1 year (Solomun et al. 2018). NPP is calculated below:

$$\mathrm{NPP} = \mathrm{GPP} - R_{\mathrm{m}} - R_{\mathrm{g}}$$

In this context, GPP refers to gross primary productivity, R_{g} denotes the energy expenditure required for synthesizing organic compounds through the process of photosynthesis, and R_{m} denotes the annual plant maintenance respiration.

In another regard, the type of land cover, the yearly NDVI composite from the baseline to the goal years, and the soil types of the land types are used to determine productive performance categories, such as deterioration and improvement. Specifically, land is classified as productive if the NDVI exceeds the 90th percentile frequency for that land type, while land with an NDVI below the 50th percentile is considered degraded from 2000 to 2020.

5.5.6 Fluctuations in Climatic Conditions

One crucial indication that is used in the evaluation of land degradation is the climate. Some of the crucial variables that have asymmetries in dry settings include

Table 5.1 Methods employed in researching the assessment of LDN

Technique	Scale	Indicator	References
Indicator method	Global	Water erosion	Nazuza et al. (2021)
		Wind erosion	Wuepper et al. (2021)
		Levels of biodiversity, forest cover, and timber stocks	Chappell et al. (2019)
	Region	Contaminated land and eroding ravines	Ptichnikov and Martynyuk (2020)
		Soil loss rate, total soil loss	Speranza et al. (2019)
		Atmospheric humidity	Tsymbarovich et al. (2020)
Questionnaire method	Global	Field surveys, farmer interviews, and participatory mapping	Kuderina et al. (2020)
		Mapping questionnaire	Crossland et al. (2018)
		Expert empirical data	Von Maltitz et al. (2019)
		Triangulation	Garcia et al. (2019)

soil moisture, rainfall, aridity index, and the potential evapotranspiration. The monthly datasets for the last 20 years (2001–2020) that were evaluated for this study are displayed in Table 5.1 and have been compiled into an annual average.

5.5.7 Soil Moisture

Soil moisture serves as a vital component for vegetation growth and serves as an indicator of land productivity. It is closely related to soil organic matter, and this relationship is particularly influenced by topography in humid environments, directly impacting land erosion (Rajendram and Patel 2021).

In Fig. 5.4, monthly mean soil moistures dataset spanning from 2000 to 2020 were utilized. The datasets were obtained from TerraClimate data accessed via Climate Engine (Rao et al. 2023). The Climate Engine computes soil moisture by utilizing the climatic water balance extracted from a global terrestrial surface model.

5.5.8 Yearly Pattern of Various Metrics for LD

Trend analysis presents a visual representation illustrating the average quantitative condition of a specific region. In this research, an examination was conducted on the annual linear trends of various factors including NPP, NDVI, aridity index, soil moisture, PET (potential evapotranspiration), and rainfall spanning from 2000 to 2020 in western India. The mean values for NDVI stand at 0.26 and NPP stand at 0.13. Notably, their peaks are observed at 0.30 in 2020 and 0.20 in both 2011 and 2020, while their lowest points were recorded at 0.003 and 0.20 in 2002. Both NDVI and NPP exhibit an upward trend, with R2 values of 0.39 and 0.81, respectively. On average, rainfall and PET measure 38.01 mm and 141.2 mm, respectively, with

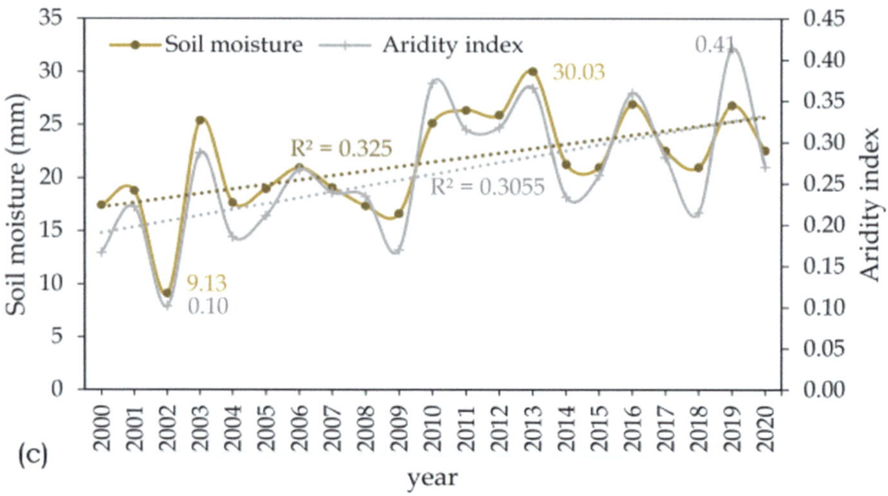

Fig. 5.4 Annual linear trend of aridity index and soil moisture from year 2000 to 2020 in Rajasthan

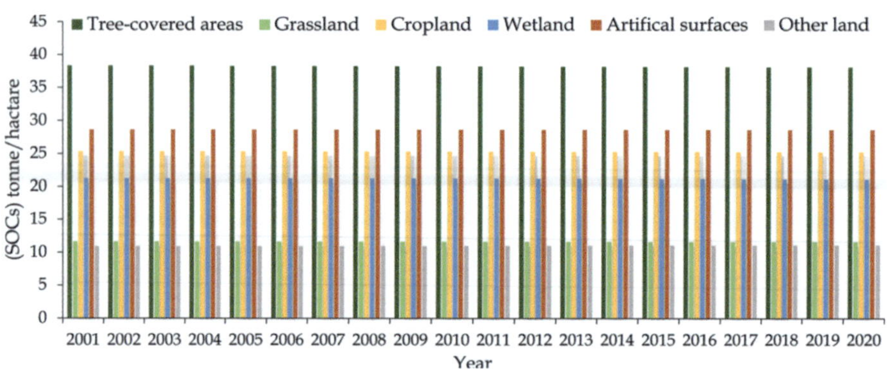

Fig. 5.5 SOC stock of various land cover types from 2001 to 2020

peaks at 57.11 mm in 2019 and 157.40 mm in 2000 for rainfall, and 15.33 mm and 133.29 mm in 2002 and 2019, respectively, for PET. Their respective R2 values are 0.38 and 0.58. The average values for aridity index and soil moisture are 21.48 mm and 0.26 mm, with peaks at 30.03 mm and 0.41 mm in 2013 and 2019, respectively, and lows at 9.13 mm and 0.10 mm in 2002.

5.5.9 Land Class Soil Organic Carbon

Figure 5.5 illustrates the soilorganic carbon (SOC) percentage across different land cover class from 2001 to 2020. These classes encompass croplands, grasslands, tree-covered area, wetlands, and artificial surfaces. SOC values for tree-covered area and other land class fall inside the range of 38.11 to 38.32 t/ha and 10.95 to 11.17 t/ha, respectively. Consequently, tree-covered areas, croplands, and wetland areas, which serve as the primary reservoirs of SOC in this region, collectively account for approximately 84% of the total.

5.6 Factors Contributing to the Consequential Status of Land Degradation

Land degradation poses significant challenges to the health and quality of soil in the region. Surface soil has high importance in supporting plant development, containing a high concentration of organic matter and approximately 50% of essential minerals like potassium, calcium, and phosphate. Various factors contribute to land degradation, including change in soil organic composition, vegetation dynamics, population growth, natural disasters such as droughts and famines, climate change, mineral mining, land encroachment, overgrazing, dust storms, and urbanization (Rao et al. 2023).

The study area experiences arid climatic conditions, and semi-arid, with annual rainfall ranging from 350 to 400 mm. Temperature change depends on topographical features, ranging from 20 to 45 °C. Over the past 20 years, there has been a notable increment in population, rising from 162 to 262 people per km^2 (a 67% increase), along with an improvement in literacy rates from 52% to 67%. While natural calamities including droughts and floods contribute modestly to LD, land use and land cover change (LULC) emerge as significant drivers. Climatic and soil-related factors, including soil alkalinization and salinization, also exert considerable influence, although topographical effects are relatively minor. Net primary productivity and soil carbon stock are influenced by topographical and climatic aspects but play a relatively modest role in LD.

Overall, the study indicates a moderate increase in LD status in between 2001 and 2020, reflecting the complex interplay of various environmental and anthropogenic factors impacting soil health and quality in the region.

5.7 Conclusion

Land degradation (LD) is a gradual and complex phenomenon influenced by various factors, both abiotic and biotic, with its mitigation also subject to these same factors. This chapter on Land Degradation Neutrality (LDN) introduces a fresh perspective that underscores the interconnectedness of these aspects and their significance in LD challenges faced by western India. The chapter highlights that this paradigm relies on the Pressure-State-Response (PSR) model, utilizing UNCCD factors based on remotely sensed data. These factors encompass vegetation indices (NDVI and NPP), socioeconomic indicators (literacy rate and population density), Land Use and Land Cover (LULC), topographical features (slope and DEM), and climatic variables (aridity index, rainfall, PET) at various resolutions.

Consequently, this chapter presents an analysis of soil organic matter degradation, land degradation and transitions, performance, productivity state, and the annual trends of various indicators from 2001 to 2020. Specifically, it examines changes in land degradation (LD), productivity, susceptibility, neutrality response, and the pressure-state response for land degradation. It also covers soil organic carbon and related aspects in western India. In Rajasthan, Net Primary Productivity (NPP) increased by 50% and the Normalized Difference Vegetation Index (NDVI) increased by 220%, while changes in Land Use and Land Cover (LULC) classes such as water bodies, forest regions, croplands, and artificial lands exhibited varying rates over the past two decades. Annual trend analysis revealed correlations between different variables, with varying degrees of positive and negative relationships. LD susceptibility and associated factors suggest that a substantial portion of land in western India, particularly in Rajasthan, remains stable, while a small percentage experiences high degradation conditions. Overall, factors such as LULC type, forest cover loss, and changes in soil organic carbon stocks are influenced by increasing soil moisture and rainfall from 2001 to 2020. However, addressing Land Degradation Neutrality (LDN) requires further development of effective techniques to control and globally mitigate land degradation processes. This includes assessing the effectiveness of offsetting mechanisms and their viability in maintaining the equilibrium of terrestrial ecosystem qualities, functions, and services.

References

Abel C, Horion S, Tagesson T, De Keersmaecker W, Seddon AW, Abdi AM, Fensholt R (2021) The human–environment nexus and vegetation–rainfall sensitivity in tropical drylands. Nat Sustain 4:25–32

Agbeshie AA, Abugre S, Atta-Darkwa T, Awuah R (2022) A review of the effects of forest fire on soil properties. J For Res 33(5):1419–1441. https://doi.org/10.1007/s11676-022-01475-4. Springer Science and Business Media LLC

Akhtar-Schuster M, Stringer LC, Erlewein A, Metternicht G, Minelli S, Safriel U, Sommer S (2017) Unpacking the concept of land degradation neutrality and addressing its operation through the Rio Conventions. J Environ Manag 195:4–15. https://doi.org/10.1016/j.jenvman.2016.09.044. Elsevier BV

Al Sayah MJ, Abdallah C, Khouri M, Nedjai R, Darwich T (2019) Application of the LDN concept for quantification of the impact of land use and land cover changes on Mediterranean watersheds – Al Awali basin – Lebanon as a case study. In: CATENA, vol 176. Elsevier BV, pp 264–278. https://doi.org/10.1016/j.catena.2019.01.023

Al Sayah MJ, Abdallah C, Der Sarkissian R, Abboud M (2021) A framework for investigating the land degradation neutrality – disaster risk reduction nexus at the sub-national scales. J Arid Environ 195:104635. Elsevier BV. https://doi.org/10.1016/j.jaridenv.2021.104635

Allen C, Metternicht G, Verburg P, Akhtar-Schuster M, da Cunha MI, Santivañez MS (2020) Delivering an enabling environment and multiple benefits for land degradation neutrality: stakeholder perceptions and progress. Environ Sci Pol 114:109–118

Anderson J, Blackie MJ, Burcroff R, Ceesay MM, Esikuri EE, Fernandes ECM, Gillison AN, Sanginga EN, Smaling E, Styger ED (2006) Sustainable land management: challenges, opportunities, and trade-offs (English). Agriculture and rural development. World Bank Group, Washington, DC. http://documents.worldbank.org/curated/en/973741468135939915/Sustainable-land-management-challenges-opportunities-and-trade-offs

Barbier E, Hochard J (2014) Land degradation, less favored lands and the rural poor: a spatial and economic analysis a report for the economics of land degradation initiative. http://www.edwardbbarbier.com/Projects/ELD/Economics_of_Land_Degradation_Initiative_Barbier_and_Hochard_abridged.pdf

Bobushev TS, Sultanaliev KE (2020) Evaluation and adaptation of the land degradtion neutrality approach to land classification resources in the Kyrgyz Republic. Arid Ecosyst 10:123–126

Ct ODDS (2015) Transforming our world: the 2030 agenda for sustainable development. United Nations, New York

Chappell A, Webb NP, Leys JF, Waters CM, Orgill S, Eyres MJ (2019) Minimising soil organic carbon erosion by wind is critical for land degradation neutrality. Environ Sci Pol 93:43–52

Chotte JL, Aynekulu E, Cowie A, Campbell E, Vlek P, Lal R et al (2019) Realising the carbon benefits of sustainable land management practices: guidelines for estimation of soil organic carbon in the context of land degradation neutrality planning and monitoring, A report of the Science-Policy Interface. United Nations Convention to Combat Desertification (UNCCD), Bonn, Germany

Critchley W, Harari N, Mekdaschi-Studer R (2021) Restoring life to the land: the role of sustainable land management in ecosystem restoration. UNCCD and WOCAT, Bern

Crossland M, Winowiecki LA, Pagella T, Hadgu K, Sinclair F (2018) Implications of variation in local perception of degradation and restoration processes for implementing land degradation neutrality. Environ Dev 28:42–54

Ding Z, Peng J, Qiu S, Zhao Y (2020) Nearly half of global vegetated area experienced inconsistent vegetation growth in terms of greenness, cover, and productivity. Earths Future 8(10):e2020EF001618

Dumanski J (1997) Criteria and indicators for land quality and sustainable land management. ITC J 3(4):216–222

Feng S, Zhao W, Zhan T, Yan Y, Pereira P (2022) Land degradation neutrality: a review of progress and perspectives. Ecol Indic 144:109530. https://doi.org/10.1016/j.ecolind.2022.109530. Elsevier BV

Forest Europe (2020) State of Europe's Forests (2020) Ministerial conference on the protection of forests in Europe (FOREST EUROPE), Bratislava, Slovakia. Online available at: https://foresteurope.org/wp-content/uploads/2016/08/SoEF_2020.pdf

Garcia CL, Teich I, Gonzalez-Roglich M, Kindgard AF, Ravelo AC, Liniger H (2019) Land degradation assessment in the Argentinean Puna: comparing expert knowledge with satellite-derived information. Environ Sci Pol 91:70–80

Gichenje H, Munoz-Rojas J, Pinto-Correia T (2019) Opportunities and limitations for achieving land degradation-neutrality through the current land-use policy framework in Kenya. Land 8(8):115

Kaffas K, Papaioannou G, Varlas G, Al Sayah MJ, Papadopoulos A, Dimitriou E et al (2022) Forecasting soil erosion and sediment yields during flash floods: the disastrous case of Mandra, Greece, 2017. Earth Surf Process Landf 47(7):1744–1760

Kuderina TM, Suslova SB, Lunin VN, Kudikov AV (2020) Atmospheric moisture as a factor of land degradation neutrality in forest-steppe landscapes. Arid Ecosyst 10(2):156–160

Kust G, Andreeva O, Cowie A (2017) Land degradation neutrality: concept development, practical applications and assessment. J Environ Manag 195:16–24. https://doi.org/10.1016/j.jenvman.2016.10.043. Elsevier BV

Kust, G., Andreeva, O., Lobkovskiy, V., & Annagylyjova, J. (2022). Experience in Developing Land Degradation Neutrality Concept in the Russian Federation. Authorea, Inc.New York, USA. https://doi.org/10.22541/au.164285470.03310502/v1

Li XS, Lu Q, Jia XX (2021b) Harnessing big earth data to facilitate land degradation neutrality goals—practices and prospects. Bull Chin Acad Sci 36:896–903

Liniger HP, Mekdaschi Studer R, Hauert C, Gurtner M (2011) Sustainable land management in practice – guidelines and best practices for Sub-Saharan Africa. TerrAfrica, World Overview of Conservation Approaches and Technologies (WOCAT) and Food and Agriculture Organization of the United Nations (FAO), Rome, Italy. Online available at: https://wocat.net/documents/82/SLM_in_Practice_E_Final_low.pdf

Liniger HP, Mekdaschi Studer R, Moll P, Zander U (2017) Making sense of research for sustainable land management. Centre for Development and Environment (CDE), University of Bern/ HelmholtzCentre for Environmental Research GmbH – UFZ, Bern/Leipzig

Liniger H, Harari N, van Lynden G, Fleiner R, de Leeuw J, Bai ZG, Critchley W (2019) Achieving land degradation neutrality: the role of SLM knowledge in evidence-based decision-making. Environ Sci Pol 94:123–134

Lobkovsky VA, Andreeva OV, Kust GS (2022) Integration of international and national systems for monitoring and assessing land degradation in Russia. Izv Akad Nauk, Ser Geogr 1:1–18

Manna P, Agrillo A, Bancheri M, Di Leginio M, Ferraro G, Langella G et al (2024) A geospatial decision support system for supporting the assessment of land degradation in Europe. Land 13(1):89

Mario J, Al-Sayah, Sarkissian RD, Abdallah C (2022) Review of recent events in Lebanon (2011–2020) and their effect on land degradation, United Nations Human Settlements Programme (UN-Habitat), Nairobi, Kenya

Mikhailova EA, Zurqani HA, Lin L, Hao Z, Post CJ, Schlautman MA, Shepherd GB (2024) Possible integration of soil information into land degradation analysis for the United Nations (UN) Land Degradation Neutrality (LDN) concept: a case study of the contiguous United States of America (USA). Soil Syst 8(1):27

Millennium Ecosystem Assessment (2005) Ecosystems and human well-being: synthesis. Island Press, Washington, DC

Moonrut N, Takrattanasaran N, Khamkajorn T, Chaikaew P (2021) Integrated remote sensing and GIS approaches for land degradation neutrality (LDN) assessment in the agricultural area. IOP Conf Ser: Earth Environ Sci 626(1):012025. 2nd International Conference on Advances in Civil and Ecological Engineering Research

Nzuza P, Ramoelo A, Odindi J, Mwenge Kahinda J-M, Lindeque L (2022) A triangulation approach for assessing and mapping land degradation in the Lepellane catchment of the greater Sekhukhune District, South Africa. In S Afr Geogr J 104(4):514–538. https://doi.org/10.1080/03736245.2021.2000481

Orr BJ, Cowie AL, Castillo Sanchez VM, Chasek P, Crossman ND, Erlewein A, Louwagie G, Maron M, Metternicht GI, Minelli S, Tengberg AE, Walter S, Welton S (2017) Scientific conceptual framework for land degradation neutrality. A report of the Science-Policy Interface. United Nations Convention to Combat Desertification (UNCCD), Bonn

Pereira P, Barcelo D, Panagos P (2020) Soil and water threats in a changing environment. Environ Res 186:109501

Ptichnikov AV, Martynyuk AA (2020) Adaptation of international indicators of land degradation neutrality for the assessment of forest ecosystems in arid conditions in Russia. Arid Ecosyst 10(2):127–134

Raj A, Sharma LK (2022) Assessment of land-use dynamics of the Aravalli range (India) using integrated geospatial and CART approach. Earth Sci Inf 15(1):497–522. https://doi.org/10.1007/s12145-021-00753-9

Raj A, Sharma LK, Verma R, Naik R, Nathawat MS, Francaviglia R (2023) Land degradation neutrality (LDN) in Rajasthan, Western India: a combined approach of pressure-state-response model and MODIS data products. Environ Monit Assess 195(12):1468

Rajendram K, Patel N (2021) Geospatial application in desertification monitoring Rajasthan, India. Int J Res Sci Innov 08(03):136–143. https://doi.org/10.51244/ijrsi.2021.8309

Rao MCS, Udayana SK, Swami DV (2023) Achieving land degradation neutrality to combat the impacts of climate change. In: Naorem A, Machiwal D (eds) Enhancing resilience of dryland agriculture under changing climate. Springer. https://doi.org/10.1007/978-981-19-9159-2_5

Reichhuber A, Gerber N, Mirzabaev A, Svoboda M, Lopez A, Graw V, Stefanski R, Davies J, Vuković A, García M, Fiati C, Jia X (2019) The Land-Drought Nexus: enhancing the role of land-based interventions in drought mitigation and risk management. A report of the Science-Policy Interface. United Nations Convention to Combat Desertification (UNCCD). Bonn, Germany

Sanz MJ, De Vente J, Chotte JL, Bernoux M, Kust G, Ruiz I et al (2017) Sustainable land management contribution to successful land-based climate change adaptation and mitigation: a report of the science-policy interface. United Nations Convention to Combat Desertification (UNCCD), Bonn

Sharma LK, Raj A, Somawat K (2021) Spatio-temporal assessment of environmentally sensitive areas (ESA) in The Thar Desert India, to combat desertifcation under UNCCD framework. J Arid Environ 194:104609. https://doi.org/10.1016/j.jaridenv.2021.104609

Smyth AJ, Dumanski J (1993) FESLM: an international framework for evaluating sustainable land management: a discussion paper. World Soil Resources Report 73. Food and Agriculture Organization, Rome, 74 p

Solomun MK, Barger N, Cerda A, Keesstra S, Marković M (2018) Assessing land condition as a first step to achieving land degradation neutrality: a case study of the Republic of Srpska. Environ Sci Pol 90:19–27. https://doi.org/10.1016/j.envsci.2018.09.014

Speranza CI, Adenle A, Boillat S (2019) Land degradation neutrality: potentials for its operationalisation at multi-levels in Nigeria. Environ Sci Pol 94:63–71

Stavi I, Lal R (2015) Achieving zero net land degradation: challenges and opportunities. J Arid Environ 112:44–51

Tsymbarovich P, Kust G, Kumani M, Golosov V, Andreeva O (2020) Soil erosion: an important indicator for the assessment of land degradation neutrality in Russia. Int Soil Water Conserv Res 8(4):418–429

UNCCD (2015) Integration of the Sustainable Development Goals and targets into the implementation of the United Nations Convention to Combat Desertification and the report of the Intergovernmental Working Group on Land Degradation Neutrality. https://www.unccd.int/sites/default/files/sessions/documents/ICCD_COP12_4/4eng.pdf

UNCCD (2017) China final national report of the voluntary land degradation neutrality (LDN) target setting programme. https://knowledge.unccd.int/sites/default/files/ldn_targets/China_LDN%20TSP%20Country%20Report.pdf

Von Maltitz GP, Gambiza J, Kellner K, Rambau T, Lindeque L, Kgope B (2019) Experiences from the South African land degradation neutrality target setting process. In Environmental Science & Policy (Vol. 101, pp. 54–62). Elsevier BV. https://doi.org/10.1016/j.envsci.2019.07.003

Wuepper D, Borrelli P, Panagos P, Lauber T, Crowther T, Thomas A, Robinson DA (2021) A 'debt' based approach to land degradation as an indicator of global change. Glob Chang Biol 27(21):5407–5410

Zhao M, Geruo A, Zhang JE, Velicogna I, Liang CZ, Li ZY (2021) Ecological restoration impact on total terrestrial water storage. Nat Sustain 4(1):56–U85

Zolotokrylin AN (2019) Global warming, desertification/degradation, and droughts in arid regions. Izvestiya Rossiiskoi Akademii Nauk Seriya Geograficheskaya (1):3–13. https://doi.org/10.31857/s2587-5566201913-13

Pixel to Preservation: Harnessing Remote Sensing and GIS for Biodiversity Conservation

6

Divesh Pangtey, Manoj Kumar, and Hitendra Padalia

Abstract

Remote sensing (RS) together with Geographical Information System (GIS) offers valuable insights, making them essential tools for mapping habitats, species distributions, and ecological interactions. Advanced remote sensing techniques, such as LiDAR and hyperspectral imaging, enhance the ability to analyze ecosystem dynamics, identify critical wildlife corridors, and evaluate the impact of environmental change. Pixel level observations captured in remote sensing images are extensively used in conservation planning, land-use management, and ecological modelling, facilitating informed decision-making for sustainable ecosystem preservation. Satellite-based remote sensing offers comprehensive global coverage, enabling continuous monitoring of Earth's surface with high temporal resolution. This chapter highlights how pixel-based observations and RS-GIS can bridge the gap between data and decision-making, playing a critical role in protecting our planet. Tools such as Marxan, Zonation, and CAPTAIN exemplify how spatial optimization enhances protected area design and biodiversity conservation. The chapter also provides an insight on the use of emerging innovations like machine learning and data fusion for conservation planning. The chapter identifies key barriers, including data gaps in biodiversity-rich regions, the subjectivity of conservation target weighting, and the pressing need for standardized methodologies to ensure consistency and reliability in conservation planning, and advocates for global collaboration, advanced analytics, and adaptive strategies for effectively addressing these challenges.

D. Pangtey (✉) · M. Kumar
Centre of Excellence on Sustainable Land Management (CoE-SLM), Indian Council of Forestry Research & Education, Dehradun, India

H. Padalia
Forestry and Ecology Department, Indian Institute of Remote Sensing (IIRS), Indian Space Research Organisation (ISRO), Dehradun, India

S. Singh, M. K. Gautam (eds.), *Biodiversity Conservation and Land Degradation Neutrality*, https://doi.org/10.1007/978-981-96-7210-3_6

Keywords

Ecosystem · Zonation · Habitat mapping · Habitat fragmentation · Landscape · Land use · Land cover

6.1 Introduction

A pixel (picture element) in a remote sensing image represents the smallest unit of an image that corresponds to a specific area on the Earth's surface. Each pixel carries information about reflectance, emissivity, or radiance at a given wavelength, depending on the sensor used in mapping. Different land cover types (e.g., forests, wetlands, grasslands) exhibit distinct spectral signatures at the pixel level. Pixel-based supervised and unsupervised classification techniques help delineate ecosystems and habitat connectivity. Pixels serve as the building blocks of biodiversity assessment by enabling spatial mapping, habitat classification, and environmental monitoring. Advancements in remote sensing, AI, and GIS modelling further enhance the ability to extract valuable biodiversity insights at multiple scales (Kala and Kumar 2021).

Biodiversity is a cornerstone of the Earth's ecosystems, playing an essential role in maintaining ecological balance and supporting human well-being. Despite its significance, biodiversity faces unprecedented threats, making conservation a pressing global concern (Haq et al. 2021a, 2024; Kumar et al. 2019c). The accelerated rate of species extinction and ecosystem degradation underscores this urgency. According to Ceballos et al. (2010), one in every 10,000 species is lost annually, and the current rate of extinction is estimated to be 10,000 times faster than it was just a century ago (IPBES 2019). The IUCN Red List, a global benchmark for assessing species' conservation status, currently includes over 157,100 species, with more than 44,000 at risk of extinction (IUCN 2024). This crisis is aggravated by a multitude of global challenges, including habitat destruction, climate change, pollution, and the spread of invasive species. Over the years, several key international frameworks and initiatives have been established to conserve biodiversity and combat biodiversity loss, including the Convention on Biological Diversity (CBD), the Aichi Biodiversity Targets, United Nations Sustainable Development Goals (SDGs), and Ramsar Convention, while Convention on International Trade in Endangered Species of Wild Fauna and Flora (CITES) and the International Union for Conservation of Nature (IUCN) are two pivotal organizations in biodiversity conservation. Additionally, initiatives such as IPBES (Intergovernmental Science-Policy Platform on Biodiversity and Ecosystem Services) synthesize scientific knowledge to inform policy decisions. And more recent Kunming-Montreal Global Biodiversity Framework (CBD 2022) aims to protect 30% of the Earth's land and sea areas by 2030. These initiatives have promoted international cooperation, policy development, and the exchange of scientific knowledge, to conserve biodiversity.

Despite concerted efforts, global conservation targets remain largely unmet, and biodiversity continues to decline (Capellesso et al. 2021). The persistent global decline in biodiversity highlights the need for continuous refinement of these technological tools and approaches. Ensuring their effective implementation remains crucial for addressing urgent conservation challenges. The fundamental step in conservation framework is the biogeographical identification of areas important for conservation, commonly referred to as "spatial prioritization" (Kukkala and Moilanen 2013). The utility of spatial prioritization is in balancing or trading-off multiple factors that determine the importance of an area for conservation interventions (Singh et al. 2020).

Post the pioneering work of Myers (1988), considerable progress has been done for strategizing conservation resources (Brooks et al. 2006.). Several methods to support decision-making on efficiently allocating conservation interventions have been developed under the framework referred to as "systematic conservation planning" (Kukkala and Moilanen 2013). Given the constraints of limited resources, conservation prioritization involves a strategic trade-off among multiple factors that determine the significance of an area in terms of biodiversity (Haq et al. 2024). These factors include species richness, endemism, ecological uniqueness, and vulnerability to threats such as habitat destruction and climate change. By systematically evaluating these criteria, assessing spatial configurations helps answer key scientific questions related to conservation prioritization, ensuring that efforts are focused on areas that maximize both ecological and socio-economic benefits (Haq et al. 2021b, 2022, 2024). This approach ensures that limited resources are allocated efficiently to safeguard habitats and species that are most critical for maintaining ecological balance and providing ecosystem services. Globally, 36 biodiversity hotspots have been recognized for their high levels of endemic species and severe threats from habitat loss (Fig. 6.1). However, addressing biodiversity decline requires more than hotspot identification. Spatial prioritization, rooted in biogeographical analysis, employs quantitative data and spatial analysis to identify critical areas for conservation investment, thereby providing a robust foundation for targeted conservation efforts.

Spatial prioritization aims to target areas of high ecological importance, such as biodiversity hotspots, and protect species and ecosystems that are most vulnerable or irreplaceable. Effective spatial prioritization requires accurate, large-scale data on species distributions, habitat conditions, and ecological processes, since something that traditional field-based methods alone cannot provide. The integration of RS-GIS has emerged as a transformative approach in biodiversity conservation. It provides unparalleled insights into habitat conditions, species distributions, and ecosystem changes, offering comprehensive, large-scale observations across various spatial and temporal dimensions. It enables the identification of critical areas for conservation, monitoring the effectiveness of protected area networks, and supports the planning of ecological corridors that maintain connectivity in fragmented landscapes. In recent years, the integration of remote sensing with geospatial analysis tools and machine learning has further enhanced its potential for informing conservation strategies. Brum et al. (2017) demonstrated multidimensional biodiversity conservation strategies by mapping priority areas based on taxonomic, phylogenetic, and trait diversity, revealing limited overlap with existing protected areas. Ward et al. (2020) underscores the critical need to enhance connectivity within protected area networks for effective biodiversity conservation under climate change.

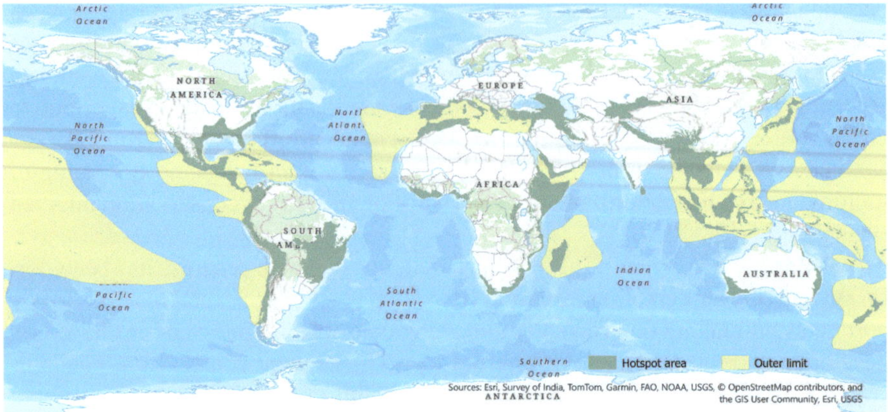

Fig. 6.1 Conservation International defined 36 recognized biodiversity hotspots that contain at least 1500 endemic species of vascular plants and have lost at least 70% of its primary native vegetation (www.conservation.org)

The integration of RS and GIS also facilitates community engagement and policy-making by offering clear, visual representations of environmental degradation or recovery, which can mobilize action from local communities to international governments. These technologies provide vital data for monitoring protected areas and tracking compliance with international biodiversity agreements. This chapter explores how RS and GIS are essential for biodiversity conservation and tools that transform raw spatial information into actionable insights. It delves into the principles, approaches, and methods of prioritization, illustrates how these tools guide conservation efforts. By linking pixels to preservation, remote sensing not only enhances our understanding of biodiversity patterns but also empowers decision-makers to effectively protect nature's most vital assets.

6.2 Biodiversity and Remote Sensing

Before proceeding, it is essential to grasp the context of biodiversity. Biodiversity often has different meaning when looking from a different perspectives. It can just be the list of species in a certain taxon or set of taxa for a systematist. For a community ecologist, it is more like the variety and distribution of species or plant types, whereas to a wildlife manager, managing biodiversity may mean interspersing habitats to maximize edge effects, thereby building populations of popular game species. Hence, it makes sense to describe biodiversity before moving ahead. Biodiversity is typically defined as the variety of genes, species, and ecosystems that constitute life on Earth and the diversity of interactions and processes at different levels of biological organization (Noss 1990; Laush et al. 2016). Biodiversity comprises multiple levels of diversity from genetic to species and ecosystems with three primary attributes composition, structure, and function that is nested hierarchy incorporating elements of each attribute (Noss 1990; Reddy et al. 2021). Biodiversity is inherently scale-dependent, as its measurement and significance vary across spatial and temporal scales. At a local scale, biodiversity may focus on species richness or functional diversity within a specific habitat, such as a wetland or forest fragment (Olokeogun and Kumar 2020). At a regional scale, the emphasis often shifts to beta diversity, which measures species turnover or differences between ecosystems across landscapes. At the global scale, biodiversity encompasses large-scale patterns, such as global biodiversity hotspots or biome-level ecosystem diversity. Temporal scales further influence biodiversity, capturing changes over time due to natural processes or anthropogenic impacts such as deforestation and climate change. Remote sensing technologies offer an effective means of measuring biodiversity across various scales using proxies derived from spectral data, structural metrics, and spatial patterns. While remote sensing is well-suited for large-scale biodiversity assessments, particularly at the ecosystem and habitat levels, its effectiveness decreases at the species level, where indirect approaches are more prevalent.

Remote sensing involves the acquisition of information about Earth's surface without direct contact, providing a broad perspective on environmental changes, offering critical data on land cover, vegetation health, climate patterns, and habitat fragmentation. Satellites, aircrafts, and unmanned aerial vehicles (UAVs) are equipped with sophisticated sensors that collect data across the electromagnetic spectrum, measuring the energy reflected or emitted by Earth's surface. Each surface object reflects or emits radiation differently at different wavelengths forming a unique spectral signature (Sabins Jr and Ellis 2020). By analyzing these spectral signatures, RS can differentiate between types of vegetation, identify specific species, detect stress in plants, and monitor changes in land cover and ecosystems (Singh et al. 2021). Although remote sensing still has several challenges, it is excellent for capturing information that can directly or indirectly reflect several vegetation parameters (Kumar et al. 2021b).

The optical remote sensing data in conjunction with field measurements have been widely used for the estimation of forest biophysical parameters (e.g., forest density, LAI, biomass, etc.). However, as far as forest height, structure, and aboveground biomass is concerned, optical remote sensing data have limited scope for regional and global scale studies. Radio Detection and Ranging (RADAR) and Light Detection and Ranging (LiDAR) remote sensing techniques are particularly useful for capturing 3-D information on forests enabling precise monitoring of biomass stock and changes. The spatial resolution is also crucial, as it provides high-resolution imagery that captures fine details of the Earth's surface, enabling the mapping of habitats and the assessment of habitat fragmentation. Temporal resolution offers repeated coverage of the same area over time, which is vital for tracking seasonal variations, phenological changes in vegetation, deforestation rates, and the impact of climate change on biodiversity.

6.3 The Role of Remote Sensing in Biodiversity Mapping and Conservation

Remote sensing provides synoptic view of earth cover at regular time intervals and hence is an ideal tool for biodiversity assessment and monitoring at various levels (GEO BON, Innes and Koch 1998). GIS has emerged as a handy tool to carry out criteria-based spatial modelling. RS measures the energy that is reflected and emitted from the Earth's surface. Because the properties of materials commonly found on the surface are known, RS provides insights into the surface composition. There are also biodiversity-relevant situations, that may not be directly observable with RS but may be correlated with what can be observed. This allows remotely sensed observations to act as a "proxy" for surface activities if sufficient surface measurements are available to establish the link. In the last decade, studies have been published that use remote sensing to assess and monitor biodiversity (Kacic and Kuenzer 2022). This information supports monitoring at scales relevant to research and decision-making, underpins various biodiversity and ecosystem function indicators, and envisions a future where data collection is seamlessly integrated with models and infrastructure for effective biodiversity assessment (Kumar et al. 2019a).

Remote sensing techniques are invaluable for mapping land cover and identifying critical habitats (Lavania et al. 2024). RS helps in the assessment of habitat quality, distribution, and changes over time, which are crucial for conservation planning (Tyagi and Kumar 2022). The species distribution modelling (SDM) can predict the potential habitats of species across large areas (Singh et al. 2023a, b). These models are essential for identifying biodiversity hotspots, prioritizing areas for conservation, and assessing the impacts of climate change on vegetation (Kumar et al. 2019b). Satellite imagery provides a temporal dimension to conservation, enabling the monitoring of changes in ecosystems over time. This capability is vital for detecting deforestation, desertification, and other processes that threaten biodiversity (Olokeogun and Kumar 2022). The use of time-series analysis helps in understanding trends and developing strategies to mitigate adverse impacts (Chafiq et al. 2023).

While GIS is crucial for managing, analyzing, and visualizing spatial data. When combined with remote sensing, GIS enhances our ability to interpret complex datasets and apply them to real-world conservation challenges. GIS enables the integration of various spatial datasets, such as land use, topography, and climate data, to identify priority areas for conservation. Techniques like overlay analysis, buffer analysis, and suitability modelling were used to delineate critical habitats, migration corridors, and conservation areas. By analyzing the spatial patterns of different layers, GIS can identify various levels of threat under varying topographic conditions (Joshi et al. 2021; Kumar et al. 2021a; Pokhriyal et al. 2020). Protecting identified threatened areas is crucial for maintaining global biodiversity. GIS-based modelling allows conservationists to simulate the outcomes of different management strategies, such as protected area expansion or habitat restoration. These models provide a decision support system, helping policymakers evaluate the potential impacts of various actions on biodiversity. For example, Sharma et al. (2020) used satellite imagery and spatial modelling to identify high-value biodiversity corridors in the Eastern Himalayas, ensuring landscape connectivity for wildlife migration. Some specialized tools and techniques widely used in biodiversity assessment and conservation, integrating RS-GIS, are explained ahead.

6.3.1 Spectral Variation Hypothesis

Estimating biodiversity using field data presents several challenges, primarily due to high time and financial costs, along with difficulties in developing standardized and reproducible data collection methods. To simplify and streamline this process, remote sensing-based diversity assessment serves as a viable alternative, where spectral heterogeneity at the pixel level is used as a proxy for biodiversity (Rocchini et al. 2010). According to the Spectral Variation Hypothesis (SVH), higher spatial variability in spectral diversity observed through remote sensing imagery is associated with a greater number of ecological niches or species in a given area (Palmer et al. 1999; Torresani et al. 2020). Thus, spectral heterogeneity, which represents the diversity of pixel values in remote sensing data, is closely linked to biodiversity (Rocchini et al. 2017). SVH has shown promising results across various ecosystems, including grasslands (Lopes et al. 2017), conifer forests (Torresani et al. 2018), and tropical forests (Pangtey et al. 2023). However, while SVH is a powerful approach, there are several factors to consider when applying it.

The effectiveness of SVH is strongly influenced by the spatial resolution of the remote sensing data, size of field plots used for validation, and choice of spectral heterogeneity (SH) index also affects SVH outcomes, as different indices capture varying aspects of spectral diversity (Rocchini et al. 2017). Additionally, the selection of spectral bands or spectral indices used to derive SH significantly impacts biodiversity estimation, making it crucial to use appropriate combinations for accurate assessments (Féret and Asner 2014; Torresani et al. 2019). Another key consideration is that SVH performs best when applied within a single feature class, as it tends to overestimate biodiversity when used across multiple land cover types with high spectral differences.

Several heterogeneity indices have been applied in ecological research, with the Shannon Diversity Index (Shannon 1948) being one of the most widely used. Originally developed for ecological studies, this index can be readily adapted to remote sensing data, where it is calculated from reflectance values (or digital numbers) in remotely sensed images. Additionally, among the various heterogeneity indices, Rao's Q index has emerged as one of the most promising index for biodiversity assessments using remote sensing (Rocchini et al. 2017). This index has demonstrated superior performance in capturing ecological variability and has produced excellent results in biodiversity estimation. By carefully considering these factors, SVH can serve as a powerful tool for biodiversity assessment, providing cost-effective and scalable insights into ecosystem diversity through remote sensing technology.

6.3.2 Spatial Prioritization

Biodiversity is not evenly distributed throughout the world; some regions host a large number of endemic species, while some do not. Protecting these regions is crucial for preserving the diversity of life on Earth. Spatial prioritization focuses on strategically allocating resources to areas where they can yield the highest ecological impact, thereby addressing critical challenges like habitat loss, species extinction, and ecosystem degradation. This approach typically identifies regions of high biodiversity, endemic species, and essential ecosystem services, such as global biodiversity hotspots. These hotspots, like the Amazon rainforest, Coral Triangle, and Himalayas, are prioritized because of their ecological significance and vulnerability to human activities. Modern tools and technologies have revolutionized spatial prioritization by enabling data-driven, quantitative assessments of species distribution, habitat conditions, and ecosystem connectivity. By combining ecological and socio-economic data, these tools support conservation planning that balances biodiversity preservation with human needs, ensuring the sustainable use of natural resources.

While the initial focus has been on maximizing the protection of species while minimizing costs, additional parameters can sometimes be modelled, such as species rarity and threat, total protected area, and evolutionary diversity. The field was initially focused on the conservation of nature for itself, without human interference, but gradually incorporated the bidirectional links to people, recognizing our ubiquitous influence on nature and the multifaceted contributions we derive from it, including the sustainable use of species. At the global scale, prioritization is often based on metrics like species richness, endemism, or ecosystem vulnerability. Spatial prioritization tools like Marxan and Zonation use optimization algorithms to identify areas that meet biodiversity goals with minimal socio-economic costs. For instance, Weeks et al. (2014) employed Marxan to design marine protected areas in the Coral Triangle, balancing biodiversity preservation with fishing community needs, while Harlio et al. (2019) effectively used Zonation to address multi-objective conservation goals in agricultural landscapes by minimizing habitat quality loss, maximizing connectivity, and incorporating landscape heterogeneity in southwest Finland. The general mechanism of prioritization in show in Fig. 6.2.

6.3.2.1 Zonation

Zonation is a spatial conservation planning tool designed to prioritize areas for biodiversity conservation by optimizing the representation of multiple species or ecosystems. Unlike other tools that identify discrete conservation units, Zonation produces a hierarchical ranking of all areas based on their ecological value, ensuring that high-priority regions are identified even under varying resource constraints. It considers factors such as species richness, rarity, and habitat quality and integrates spatial connectivity to maintain ecological processes like migration and dispersal. Gordon et al. (2009) applied Zonation in Greater Melbourne to prioritize conservation for 30 threatened fauna species, guide urban growth boundaries, and assess conservation areas. Geange et al. (2017) used Zonation in New Zealand to design Marine Protected Areas, balancing biodiversity representation and economic

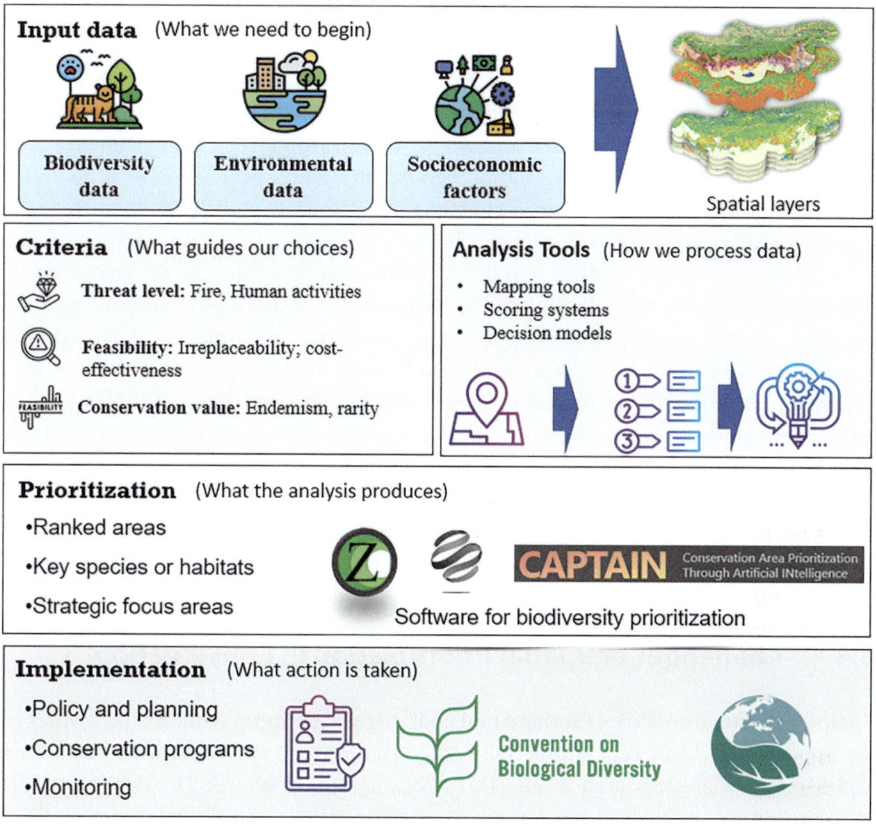

Fig. 6.2 General framework for biodiversity prioritization (Silvestro et al. 2022)

objectives. Zonation's flexibility allows users to incorporate diverse conservation goals, making it particularly effective for large-scale planning across heterogeneous landscapes.

6.3.2.2 Marxan

Marxan (Ball et al. 2009) is a widely used conservation planning tool that employs simulated annealing optimization to design cost-effective protected area networks. It aims to meet specific biodiversity conservation targets while minimizing socio-economic costs, such as land acquisition or resource use restrictions. Marxan has been applied globally in both terrestrial and marine contexts, such as designing marine protected areas in the Coral Triangle (Weeks et al. 2014) and prioritizing avian biodiversity hotspots in the Rocky Mountains of North America (Pearce et al. 2008). The tool balances trade-offs between ecological objectives and human livelihoods, ensuring equitable and efficient conservation. While effective for static policy optimization, Marxan has limitations in addressing dynamic changes like climate impacts and evolving anthropogenic pressures, although recent extensions have sought to incorporate these aspects.

6.3.2.3 CAPTAIN

Silvestro et al. (2022) developed a novel AI-driven framework for systematic conservation planning that optimizes biodiversity protection policies over time called CAPTAIN (Conservation Area Prioritization Through Artificial INtelligence). Unlike traditional tools, CAPTAIN employs a reinforcement learning (RL) approach to dynamically balance exploration (data gathering) and exploitation (action implementation), allowing for adaptive decision-making in response to evolving ecological and anthropogenic conditions. This platform integrates species-specific sensitivities to threats such as habitat disturbance, climate change, and economic pressure, providing a more nuanced and realistic prioritization. CAPTAIN can model both static and dynamic conservation policies, enabling the design of flexible strategies that address immediate needs and long-term sustainability. By simulating biodiversity dynamics and incorporating empirical data, CAPTAIN offers an innovative solution for navigating trade-offs in resource allocation, ensuring robust conservation outcomes in a rapidly changing world (Fig. 6.2).

6.4 Challenges of Spatial Prioritization in Conservation

Spatial prioritization faces fundamental challenges stemming from the principles of the approach and inherent subjectivity of conservation target weighting (Di Minin and Moilanen 2012; Marsh et al. 2007; Possingham et al. 2002). At its core, the process requires balancing ecological, socio-economic, and political considerations to identify areas of highest conservation value. However, decisions about what to prioritize such as species richness, endemism, or ecological uniqueness often involve subjective judgments. This subjectivity is further complicated by the lack of standardized criteria or frameworks, leading to inconsistencies in how conservation priorities are determined across regions (De Grammont and Cuarón 2006; Miller et al. 2006). Additionally, many prioritization efforts rely on incomplete or surrogate datasets, particularly in biodiversity-rich but data-poor regions like the tropics. These limitations in data availability restrict the ability to comprehensively evaluate biodiversity patterns and often result in skewed or partial conservation strategies. For example, species with rare distributions or high economic value may disproportionately influence outcomes, potentially overshadowing other critical conservation needs. The lack of a clear justification for assigning specific weights or targets undermines the transparency and effectiveness of conservation planning.

Spatial prioritization for biodiversity conservation faces significant challenges, primarily due to data limitations and uneven data distribution across regions (Arponen 2012). Underrepresented areas, such as tropical regions, which host some of the world's most biodiverse ecosystems, often lack sufficient data on species distributions, ecological processes, and environmental threats. Moreover, the temporal availability of data is equally problematic, as historical records are either sparse or non-existent, making it difficult to analyze long-term biodiversity trends. High-resolution data, although increasingly accessible through remote sensing technologies, remain costly and resource-intensive, further limiting their use in conservation projects in low-income countries. Ground-truthing, a critical step for validating spatial data, is similarly constrained by logistical and financial challenges, hindering the accuracy of spatial prioritization outcomes.

Additionally, spatial prioritization must account for the spatial characteristics of species distributions and their correlations, which are often poorly understood and underexplored. Factors such as species rarity, distribution nestedness, and spatial overlap significantly influence prioritization outcomes but are rarely systematically analyzed (Wright and Reeves 1992). Trade-offs become increasingly complex when prioritizing conservation actions across multiple species and locations, as the spatial dynamics of individual species interact with overall biodiversity patterns. These complexities are further magnified by the need to adapt prioritization frameworks to address dynamic threats such as climate change and land-use changes, ensuring that conservation efforts remain effective in evolving ecological and socio-economic contexts.

6.5 Conclusion

Remote sensing offers transformative opportunities for monitoring and conserving biodiversity on a global scale. By delivering detailed, extensive, and cost-effective data on habitat conditions, species distributions, and ecosystem dynamics, it complements traditional field-based methods and significantly enhances the ability to understand and manage biodiversity. As environmental challenges intensify, continued innovation in remote sensing technologies, combined with collaboration among researchers, conservationists, and policymakers, will be critical for developing more effective conservation strategies. Standardized protocols for data collection and analysis are essential to facilitate cross-regional comparisons, integrate findings across studies, and build a cohesive body of knowledge to accelerate the creation of robust, scalable algorithms. Looking ahead, remote sensing will remain an indispensable tool for biodiversity research and conservation. By incorporating advancements in machine learning, data fusion, and emerging analytical techniques, researchers can overcome these limitations and unlock the full potential of remote sensing. These innovations will not only refine how we monitor biodiversity trends and assess ecosystem health but also empower conservationists to identify threats and implement timely, evidence-based interventions. As technology evolves, pixel level observations retrieved through remote sensing will continue to revolutionize biodiversity assessment and conservation, ensuring the preservation of the Earth's diverse ecosystems and species resilience for future generations.

References

Arponen A (2012) Prioritizing species for conservation planning. Biodivers Conserv 21:875–893

Ball IR, Possingham HP, Watts M (2009) Marxan and relatives: software for spatial conservation prioritisation. Spatial conservation prioritisation: Quantitative Methods and Computational Tools 14:185–196

Brooks TM, Mittermeier RA, Da Fonseca GA, Gerlach J, Hoffmann M, Lamoreux JF, Mittermeier CG, Pilgrim JD, Rodrigues AS (2006) Global biodiversity conservation priorities. Science 313(5783):58–61

Brum FT, Graham CH, Costa GC, Hedges SB, Penone C, Radeloff VC, Rondinini C, Loyola R, Davidson AD (2017) Global priorities for conservation across multiple dimensions of mammalian diversity. Proc Natl Acad Sci 114(29):7641–7646

Carlson KM, Asner GP, Hughes RF, Ostertag R, Martin RE (2007) Hyperspectral remote sensing of canopy biodiversity in Hawaiian lowland rainforests. Ecosystems 10:536–549

CBD (2022) Kunming-Montreal global biodiversity framework. In Fifteenth meeting of the Conference of the Parties to the Convention on Biological Diversity (Part Two) Decision 15/4

Capellesso ES, Cequinel A, Marques R, Sausen T, Bayer C, Marques MCM (2021) Co-benefits in biodiversity conservation and carbon stock during forest regeneration in a preserved tropical landscape. For Ecol Manag 492:119222

Ceballos G, García A, Ehrlich PR (2010) The sixth extinction crisis: loss of animal populations and species. J Cosmol 8(1821):31

Chafiq T, Hmamou M, Ouhammou I, Azmi R, Kumar M (2023) Modelling change detection for unveiling urban transitions: using machine learning algorithms and Sentinel-2 data in Larache City, Morocco. Model Earth Syst Environ 10:1–15

De Grammont PC, Cuarón AD (2006) An evaluation of threatened species categorization systems used on the American continent. Conserv Biol 20(1):14–27

Di Minin E, Moilanen A (2012) Empirical evidence for reduced protection levels across biodiversity features from target-based conservation planning. Biol Conserv 153:187–191

Féret JB, Asner GP (2014) Mapping tropical forest canopy diversity using high-fidelity imaging spectroscopy. Ecol Appl 24(6):1289–1296

Geange SW, Leathwick J, Linwood M, Curtis H, Duffy C, Funnell G, Cooper S (2017) Integrating conservation and economic objectives in MPA network planning: a case study from New Zealand. Biol Conserv 210(4):136–144. https://doi.org/10.1016/j.biocon.2017.04.011

Gordon A, Simondson D, White M, Moilanen A, Bekessy SA (2009) Integrating conservation planning and landuse planning in urban landscapes. Landsc Urban Plan 91(4):183–194. https://doi.org/10.1016/j.landurbplan.2008.12.011

Haq SM, Calixto ES, Kumar M (2021a) Assessing biodiversity and productivity over a small-scale gradient in the protected forests of Indian Western Himalayas. J Sustain For 40:675–694

Haq SM, Yaqoob U, Calixto ES, Kumar M, Rahman IU, Hashem A, Abd Allah EF, Alakeel MA, Alqarawi AA, Abdalla M (2021b) Long-term impact of transhumance pastoralism and associated disturbances in high-altitude forests of Indian Western Himalaya. Sustainability 13:12497

Haq MS, Calixto ES, Rashid I, Hussain Malik A, Kumar M, Ahmad Khuroo A (2022) Anthropogenic pressure and tree carbon loss in the temperate forests of Kashmir Himalaya. Bot Lett 169:400–412

Haq SM, Waheed M, Darwish M, Siddiqui MH, Goursi UH, Kumar M, Song L, Bussmann RW (2024) Biodiversity and carbon stocks of the understory vegetation as indicators for forest health in the Zabarwan Mountain Range, Indian Western Himalaya. Ecol Indic 159:111685

Harlio A, Kuussaari M, Heikkinen RK, Arponen A (2019) Incorporating landscape heterogeneity into multi-objective spatial planning improves biodiversity conservation of semi-natural grasslands. J Nat Conserv 49:37–44

Innes JL, Koch B (1998) Forest biodiversity and its assessment by remote sensing. Glob Ecol Biogeogr Let 7(6):397–419

IPBES (2019) Global assessment report on biodiversity and ecosystem services of the Intergovernmental Science-Policy Platform on Biodiversity and Ecosystem Services. In: Brondizio ES, Settele J, Díaz S, Ngo HT (eds) IPBES secretariat, Bonn, 1148 pages. https://doi.org/10.5281/zenodo.3831673

IUCN (2024) The IUCN red list of threatened species. Version 2023-1. ISSN 2307-8235. https://www.iucnredlist.org

Joshi S, Garg JK, Kaur A, Kumar M (2021) Assessment of wildfire landslide risk using spatial analytics and deep learning techniques for Rudraprayag Forest Division, Uttarakhand. Indian For 147:824–833

Kacic P, Kuenzer C (2022) Forest biodiversity monitoring based on remotely sensed spectral diversity—a review. Remote Sens 14(21):5363

Kala AK, Kumar M (2021) Role of geospatial technologies in natural resource management. In: Climate impacts on sustainable natural resource management, pp 19–34. https://doi.org/10.1002/9781119793403.ch2

Kukkala AS, Moilanen A (2013) Core concepts of spatial prioritisation in systematic conservation planning. Biol Rev 88(2):443–464

Kumar M, Padalia H, Nandy S, Singh H, Khaiter P, Kalra N (2019a) Does spatial heterogeneity of landscape explain the process of plant invasion? A case study of Hyptis suaveolens from Indian Western Himalaya. Environ Monit Assess 191:794. https://doi.org/10.1007/s10661-019-7682-y

Kumar M, Savita, Singh H, Pandey R, Singh MP, Ravindranath NH, Kalra N (2019b) Assessing vulnerability of forest ecosystem in the Indian Western Himalayan region using trends of net primary productivity. Biodivers Conserv 28:2163–2182

Kumar M, Singh MP, Singh H, Dhakate PM, Ravindranath NH (2019c) Forest working plan for the sustainable management of forest and biodiversity in India. J Sustain For:1–22. https://doi.org/10.1080/10549811.2019.1632212

Kumar M, Kalra N, Singh H, Sharma S, Rawat PS, Singh RK, Gupta AK, Kumar P, Ravindranath NH (2021a) Indicator-based vulnerability assessment of forest ecosystem in the Indian Western Himalayas: an analytical hierarchy process integrated approach. Ecol Indic 125:107568

Kumar M, Phukon SN, Paygude AC, Tyagi K, Singh H (2021b) Mapping Phenological Functional Types (PhFT) in the Indian Eastern Himalayas using machine learning algorithm in Google Earth Engine. Comput Geosci 158:104982

Laush et al (2016) Linking earth observation and taxonomic, structural and functional biodiversity: local to ecosystem perspectives

Lavania P, Singh RK, Kumar P, Savad K, Gupta G, Dobriyal M, Pandey AK, Kumar M, Singh S (2024) Unveiling the green guardians: MAPPING AND IDENTIFICATION of Azadirachta indica trees with semantic segmentation deep learning neural network technique. Egypt J Remote Sens Space Sci 27:491–500

Lopes M, Fauvel M, Ouin A, Girard S (2017) Spectro-temporal heterogeneity measures from dense high spatial resolution satellite image time series: application to grassland species diversity estimation. Remote Sens 9(10):993

Marsh H, Dennis A, Hines H, Kutt A, McDONALD K, Weber E, Williams S, Winter J (2007) Optimizing allocation of management resources for wildlife. Conserv Biol 21(2):387–399

Miller RM, Rodríguez JP, Aniskowicz-Fowler T, Bambaradeniya C, Boles R, Eaton MA, Gärdenfors U, Keller V, Molur S, Walker S, Pollock C (2006) Extinction risk and conservation priorities. Science 313(5786):441–441

Myers N (1988) Threatened biotas: "hot spots" in tropical forests. Environmentalist 8(3):187–208

Noss RF (1990) Indicators for monitoring biodiversity: a hierarchical approach. Conserv Biol 4(4):355–364

Olokeogun OS, Kumar M (2020) An indicator based approach for assessing the vulnerability of riparian ecosystem under the influence of urbanization in the Indian Himalayan city, Dehradun. Ecol Indic. https://doi.org/10.1016/j.ecolind.2020.106796

Olokeogun OS, Kumar M (2022) Indicator-based vulnerability assessment of riparian zones in Nigeria's Ibadan region due to urban settlement pressure. Environ Chall 7:100501

Palmer MW, Wohlgemuth T, Earls P, Arévalo JR, Thompson SD (1999) Opportunities for long-term ecological research at the Tallgrass Prairie Preserve, Oklahoma. In: Proceedings of the ILTER regional workshop: cooperation in long term ecological research in Central and Eastern Europe, Budapest, Hungary, vol 22. Oregon State University, Corvallis

Pangtey D, Padalia H, Bodh R, Rai ID, Nandy S (2023) Application of remote sensing-based spectral variability hypothesis to improve tree diversity estimation of seasonal tropical forest considering phenological variations. Geocarto Int 38(1):2178525

Pearce JL, Kirk DA, Lane CP, Mahr MH, Walmsley J, Casey D, Muir JE, Hannon S, Hansen A, Jones K (2008) Prioritizing avian conservation areas for the Yellowstone to Yukon Region of North America. Biol Conserv 141(4):908–924. https://doi.org/10.1016/j.biocon.2007.12.033

Pokhriyal P, Rehman S, Krishna GA, Rajiv P, Manoj K (2020) Assessing forest cover vulnerability in Uttarakhand, India using analytical hierarchy process. Model Earth Syst Environ. https://doi.org/10.1007/s40808-019-00710-y

Possingham HP, Lindenmayer DB, Tuck GN (2002) Decision theory for population viability analysis. Population Viability Analysis 470–489

Qiu L, Jing L, Hu B, Li H, Tang Y (2020) A new individual tree crown delineation method for high resolution multispectral imagery. Remote Sens 12(3):585

Reddy CS, Kurian A, Srivastava G, Singhal J, Varghese AO, Padalia H, Ayyappan N, Rajashekar G, Jha CS, Rao PVN (2021) Remote sensing enabled essential biodiversity variables for biodiversity assessment and monitoring: technological advancement and potentials. Biodivers Conserv 30:1–14

Rocchini D, Balkenhol N, Carter GA, Foody GM, Gillespie TW, He KS, Kark S, Levin N, Lucas K, Luoto M (2010) Remotely sensed spectral heterogeneity as a proxy of species diversity: recent advances and open challenges. Ecol Inform 5:318–329

Rocchini D, Marcantonio M, Ricotta C (2017) Measuring Rao's Q diversity index from remote sensing: an open source solution. Ecol Indic 72:234–238

Sabins FF Jr, Ellis JM (2020) Remote sensing: principles, interpretation, and applications. Waveland Press, Long Grove

Shannon CE (1948) A mathematical theory of communication. Bell Syst Tech 27:379–423

Sharma P, Chettri N, Uddin K, Wangchuk K, Joshi R, Tandin T, Pandey A, Gaira KS, Basnet K, Wangdi S, Dorji T (2020) Mapping human–wildlife conflict hotspots in a transboundary landscape, Eastern Himalaya. Glob Ecol Conserv 24:e01284

Silvestro D, Goria S, Sterner T, Antonelli A (2022) Improving biodiversity protection through artificial intelligence. Nat Sustain 5(5):415–424

Singh H, Kumar N, Kumar M, Singh R (2020) Modelling habitat suitability of western tragopan (Tragopan melanocephalus) a range-restricted vulnerable bird species of the Himalayan region, in response to climate change. Clim Risk Manag 29:100241

Singh RK, Singh P, Drews M, Kumar P, Singh H, Gupta AK, Govil H, Kaur A, Kumar M (2021) A machine learning-based classification of LANDSAT images to map land use and land cover of India. Remote Sens Appl Soc Environ 24:100624

Singh H, Kumar N, Singh R, Kumar M (2023a) Assessing the climate change impact on the habitat suitability of the range-restricted bird species (Catreus wallichii) in the Indian Himalayan ecosystem. Environ Sci Pollut Res 30:121224–121235

Singh A, Wagner B, Kasel S, Baker PJ, Nitschke CR (2023b) Canopy composition and spatial configuration influences beta diversity in temperate regrowth forests of Southeastern Australia. Drones 7(3):155

Torresani M, Rocchini D, Zebisch M, Sonnenschein R, Tonon G (2018) Testing the spectral variation hypothesis by using the RAO-Q index to estimate forest biodiversity: effect of spatial resolution. In IGARSS 2018-2018 IEEE International Geoscience and Remote Sensing Symposium (pp 1183–1186). IEEE

Torresani M, Rocchini D, Sonnenschein R, Zebisch M, Marcantonio M, Ricotta C, Tonon G (2019) Estimating tree species diversity from space in an alpine conifer forest: the Rao's Q diversity index meets the spectral variation hypothesis. Eco Inform 52:26–34

Torresani M, Rocchini D, Sonnenschein R, Zebisch M, Hauffe HC, Heym M, Pretzsch H, Tonon G (2020) Height variation hypothesis: a new approach for estimating forest species diversity with CHM LiDAR data. Ecol Indic 117:106520

Tyagi K, Kumar M (2022) The resilience of Indian Western Himalayan forests to regime shift: are they reaching towards no return point? Ecol Inform 69:101644

Ward M, Saura S, Williams B, Ramírez-Delgado JP, Arafeh-Dalmau N, Allan JR, Venter O, Dubois G, Watson JE (2020) Just ten percent of the global terrestrial protected area network is structurally connected via intact land. Nat Commun 11(1):4563

Weeks R, Aliño PM, Atkinson S, Beldia P, Binson A, Campos WL, Djohani R, Green AL, Hamilton R, Horigue V, Jumin R (2014) Developing marine protected area networks in the Coral Triangle: good practices for expanding the Coral Triangle marine protected area system. Coast Manag 42(2):183–205

Wright DH, Reeves JH (1992) On the meaning and measurement of nestedness of species assemblages. Oecologia 92:416–428

Impact of Anthropogenic Activities on the Biodiversity Conservation and Land Degradation of Rema-Kalenga Wildlife Sanctuary, Sylhet, Bangladesh

7

Mahfuza Parveen, Fariha Jesmin, Sagar Mozumder, S. M. Mahmudur Rahman, and A. B. M. Kamal Pasha

Abstract

The Rema Kalenga is the second-largest natural forest in Bangladesh, after the Sundarbans. A safe haven for wildlife, this dry woodland has been designated a "Reserve Forest" by the government of Bangladesh. The study is focused on the effects of anthropogenic activities on the biodiversity as well as the social impacts in the Rema-Kalenga Forest area. The impact of Covid-19 on the biodiversity and local communities also reflected in the study. The study is conducted using remote-sensing tools and survey methods including questionnaire, focus group discussion and field survey. The decrease of forest area and increase of commercial agricultural land is identified from 1994 to 2022 by Landsat image. Anthropogenic activities like illegal tree cutting, vegetation, tourism have significant impact on the biodiversity of the forest area. However, the majority of respondents claim that they preserve the forest against illegal activities and, if necessary, will plant more trees. The Covid-19 reduced human footprint around the forest area, which help to improve the biodiversity of the area. On the other hand, local communities lost their income source due to the absence of tourists, which further forced them to involve in illegal activities. However, the Covid-19 teach the communities a valuable lesson about the dependency of the forest and the nature.

M. Parveen (✉)
Department of Environmental Science and Disaster Management, Daffodil International University, Dhaka, Bangladesh

Center for Environment and Sustainability, Daffodil International University, Dhaka, Bangladesh
e-mail: mahfuza.esdm@diu.edu.bd

F. Jesmin · S. Mozumder · S. M. M. Rahman · A. B. M. Kamal Pasha
Department of Environmental Science and Disaster Management, Daffodil International University, Dhaka, Bangladesh

S. Singh, M. K. Gautam (eds.), *Biodiversity Conservation and Land Degradation Neutrality*, https://doi.org/10.1007/978-981-96-7210-3_7

Keywords
Biodiversity · Wildlife sanctuary · Anthropogenic activities · Covid-19 · Rema-Kalenga

7.1 Introduction

According to Union for Conservation of Nature (IUCN), a protected area (PA) is a clearly defined geographical space, recognized, dedicated and managed, through legal or other effective means, to achieve the long-term conservation of nature with associated ecosystem services and cultural values. The establishment of a protected area (PA) is one of the most current approaches of protecting natural resources, biodiversity and social heritage (Morshed 2013; Msofe et al. 2019). For the conservation of protected areas, especially forest areas, community involvement may increase the productivity of the forest as well as enhance the community development. In Bangladesh, community involvement through community forest management (CFM) is an initiative which was first adopted in 1979 by the Forest Department (FD) on government forest land (Alam 2009; Jashimuddin 2011). Within the field of the management of forest, community forestry (CF) received significant attention in addition to social forestry.

The major goal of community involvement in PA management is to reduce forest degradation and the community development (Chowdhury et al. 2014). In CFM, there is usually a set of rights and entitlements such as access of local communities to the forests and the decision-making process in return for certain delineated roles and responsibilities (Brendler and Carey 1998). These impacts are visible in a variety of aspects, e.g. preservation of biodiversity, reduction in dependency on forest resources, socio-economic upliftment of local communities, women's empowerment, and self-reliance, etc.

In Bangladesh, a total of 19 protected areas (PAs) have been established representing all the four forest types of the country (Chowdhury and Koike 2010). According to Bangladesh Forest Department, the total area of forestland of Bangladesh is 2.52 million ha of which the Forest Department (hereafter FD) manages 1.52 million ha (Chowdhury and Koike 2010). The Rema Kalenga is the largest natural forest in Bangladesh, second only to the Sundarbans. This dry woodland is secure for wild animals because it has been recognized as a "Reserve Forest" by the government. Established in 1982 and subsequently expanded in 1996, this Wildlife Sanctuary is a protected area that supports mixed tropical evergreen and semi-evergreen forests with significant biological significance. The collaborative management in this wild life sanctuary has been underway since 2004 under the nationwide co-management initiative called the Nishorgo Support Project (NSP) by the Forest Department of Bangladesh (Chowdhury et al. 2014; Sohel et al. 2009).

Located in the Habiganj district of the Sylhet division's Chunarughat upazila is the Rema Kalenga Wildlife Sanctuary. The low-lying valleys and a number of hills with varying altitudes are included in the sanctuary. In accordance with Article

23(1) of the Bangladesh Wildlife (Preservation) Order of 1973 (Presidents Order No. 23 of 1973), as modified by the Bangladesh Wildlife (Preservation) (Amendment) Act, 1974, the Rema-Kalenga Wildlife Sanctuary was created. The Order prohibits agriculture, individuals living in or visiting the sanctuary, environmental damage, hunting, trapping, and any kind of wildlife slaughter (inside and within a mile radius of the sanctuary's outer boundary 2). Additionally, the injunction prohibits the entry of any exotic animals into the sanctuary or the release of any domesticated animals. Only in extraordinary circumstances, such as when they are deemed necessary for development, beautification, or other scientific reasons, may the government permit such operations (Bangladesh Wildlife (Preservation) (Amendment) Act, 1974).

The Rema Kalenga Wildlife Sanctuary enriched with diversified biodiversity. Roughly 76% of the forest remains in its natural condition. There are only plantations covering 9% of the forest. A wide variety of vegetation, animals, and birds can be found there. There are 37 mammal species, 18 reptiles, 7 amphibians, and 167 bird species in the PA. About 606 plant species were identified in the sanctuary, where 82 have been identified as medicinal plants (Rana et al. 2010).

The soil in the sanctuary has low pH, clay to sandy loam, and quite fertile. Some soils have a reddish red sandy clay texture with particles of magnified iron ore mixed in. Upper Tertiary rocks, such as siltstones, mudstones, and sandstone, make up the majority of the hills in Rema-Kalenga. These rocks have locally transformed into shales, slates, and limestones (Bhuiyan et al. 2021). Four immigrant populations have lived near and within the refuge for a long time: Tripura, Shantal, Telugu, and Urang. These communities are actively contributing to the protection of the Sanctuary and diversifying, despite their historical reliance on the forest.

Co-management of the Rema-Kalenga Wildlife Sanctuary is being examined to determine how equitable capital assets affected vegetation cover and were incorporated into beneficiary selection. The most vulnerable members of the co-management team were underrepresented because individuals do not regard the forest highly in terms of gathering resources or its monetary value. The local population's way of life and dependence on the forest is indicating overall situations of the area, including patterns of local livelihood, forest conservation, and reliance on the forest. Furthermore, there is still a lot of work to be done because the technique is still in its early phases. More focus on local needs is necessary for a more effective and sustainable co-management system, even though locals are becoming more and more involved in decision-making.

Tree poaching was one of the several anthropogenic activities that were found. Large groups of marauders show up to chop down trees and steal them for cash. People have been known to damage forest trees by clawing and stabbed them with sharp instruments. The snake is indigenous to the woods. Residents have been known to kill snakes that roam the neighbourhood in order to ensure their own and their families' safety.

There are several research projects available at Rema-Kalenga Wildlife Sanctuary. However, there are not many research on human activity at the Rema-Kalenga wildlife refuge. Furthermore, no research has yet been done on how COVID-19 affects the community and wildlife. The study's goals were to determine how human activity affects the forest's biodiversity and how local communities affect the use of forest space and to provide some coverage of how the COVID-19 pandemic has affected both human behaviour and animals.

7.2 Methodology

7.2.1 Study Area

The Rema-Kalenga Wildlife Sanctuary is a protected forest and wildlife sanctuary in Bangladesh. This evergreen, dry forest sanctuary is situated 130 kilometres north of Dhaka (capital city of Bangladesh) in the Chunarughat, Habiganj district, Sylhet, Bangladesh, close to the Indian state of Tripura's border (Fig. 7.1). The Sanctuary was established in 1982 and enlarged in 1996. The wildlife sanctuary has 1795.54 hectares in size as of 2009. The four ranges and a portion of the forest in Habiganj District are Kalenga, Rema, Chanbari, and Rashidpur.

7.2.2 Data Collection

The study was conducted following two methodologies: (i) Remote Sensing based analysis and (ii) survey-based analysis.

(i) *Remote Sensing Based Analysis*

This study used supervised classification to categories the land cover using data from Landsat 5 and Landsat 8 satellite. In supervised classification, the user defines the known land cover and creates the cell signature value for each land cover category. Based on the earlier studies in Dhaka, the principal land cover has been decided to be composed of urbanized areas, water bodies, vegetation, bare soil (Bayes et al. 2013). Table 7.1 lists the specifics of five significant land covers. For Landsat 5 and Landsat 8, the colour combinations RGB = 432 (band 4, 3, and 2) and RGB = 543 (band 5, 4, and 3) have been selected, respectively (Dewan et al. 2012). An immense number of signatures were gathered from composite band images utilizing Google Earth imagery and supplementary data in order to categorize the photos into five classes. After combining signatures from the same class, the Maximum-Likelihood Supervised Classification (MLSC) method was used with a default equal prior probability to all land cover classes (Dewan and Yamaguchi 2009). The remote sensing analysis is influenced by M. I. Hossain and Numata (2021), who analysed the data from 1994 to 2018. The present study added the data of the year 2022 and compared among the year from 1994, 2005, 2013, 2018 and 2022.

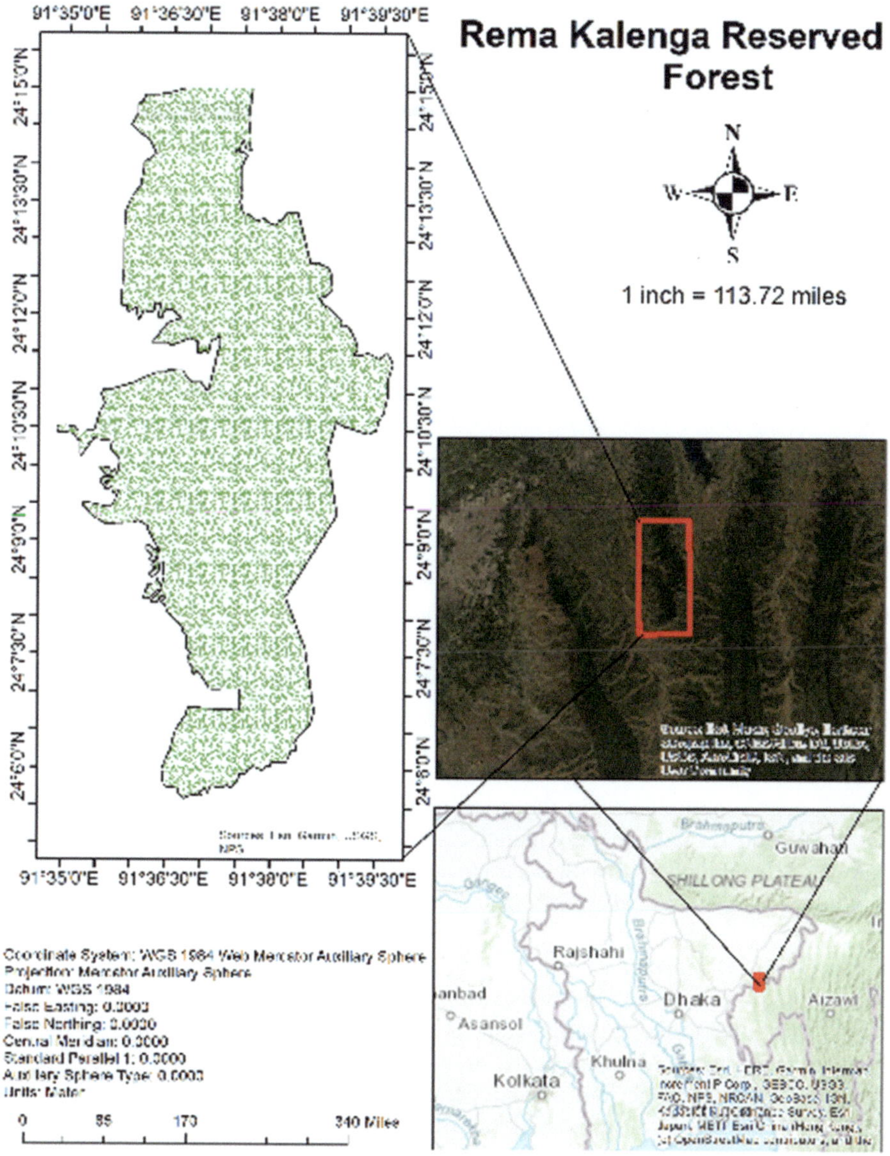

Fig. 7.1 Map of Rema-Kalenga Wildlife Sanctuary, Sylhet, Bangladesh

(ii) *Survey Based Analysis*

The collection of primary data was the main emphasis of the study. Primary data was gathered through focus groups, questionnaire surveys, field surveys, and interviews. Before starting the field study in Chunarughat, Sylhet, approval from the Forest Department was acquired. Forest officials, Co-management Committee

Table 7.1 Description of land cover classification

Land cover class	Description
Built-up area	Every sort of infrastructure, including settlements, roads, and mixed-use and other urban regions as well as houses, and factory areas
Water bodies area	Canals, ponds, lakes, rivers, and other bodies of permanently open water
Vegetation area	Trees, mixed and deciduous forests, scrub, and others
Bare land area	Development sites, under-developed land, archaeological site, free space, and terrain with exposed soil, earth, and sand.
Agricultural land area	Low-lying places, swampy land, and agricultural lands

members, and representatives of the local community met in the Kalenga Range Forest Range Office prior to the data collection. The survey's location included the local market in the Kalenga range as well as the forest's environs. The survey study was conducted on the first quarter of 2022.

Interviews of the Local Community
Interviews with about 100 respondents took place during the data collection phase. Individuals range in age from 18 to 75. There were both male and female participants. Most of them were cultivators, growing lemons and rice. The others are people with jobs, the elderly, and merchants. Women took care of livestock, including goats and cows, as housewives.

Questionnaire of the Local Community
Ninety people participated in the questionnaire process. They were between the ages of 18 and 75. Women were tentative and nervous when filling out the questionnaire, and they failed to provide accurate answers to all of the questions. There were more men than women who participated in this survey.

Focus Group Discussions
Semi-structured in-person interviews were carried out. The study's objectives were briefly explained to the respondents prior to the start of the questionnaire surveys. During the FGD, people congregated in the markets to participate in the conversation. They were very helpful and cooperative along the process. Residents of the Kalenga Range were consulted both briefly and in-depth during this process regarding anthropogenic activities, the social and economic values of the forest, and forest conservation. Women did not attend the market because they were shy, so all of the participants were males.

Direct Observation During the Survey
The study mainly relied on field research and direct observation. From the local market to the forest entrance and then across the Kalenga Forest Range Office, the region was examined during the field survey. The survey covered pathways, local homes, and farmed areas. There were signs for the monitoring area for easy navigation.

7.3 Results

7.3.1 Study Findings

The result section provides an explanation of the study findings. The difficultics and solutions to negative issues such as unlawful tree cutting, animal hunting, management team responsibilities, and tree cutting are discussed, along with current anthropological operations on the Kalenga Range.

(i) *Remote Sensing Analysis*

Forestland was the most common land cover in Rema Kalenga Reserve Forest. According to M. I. Hossain and Numata (2021), Rema Kalenga Reserved Forest had the most forestland in 2005, accounting for 94.42 percent (5885.13 ha) of the total LULC (Land Use/Land Cover) categories given, up from 89.65 percent (5587.90 ha) in 1994. The forest cover then changed dramatically, dropping to 86.68 percent (5402.64 ha) in 2013. In 2022 the total amount of forest was 5062.2 ha which was 81.62% of the total LULC, and it was 5.05% less than 2013 forest cover.Non-commercial agricultural land declined from 8.16 percent (508.81 ha) in 1994 to roughly 3.79 percent (235.98 ha) in 2005, then rose to 9.94 percent (619.27 ha and 14.68 percent (915.21 ha) in 2013 and 2018, respectively (Hossain and Numata 2021). The amount of non-commercial agricultural land in 2022 was (917 ha) which was 14.8% of total LULC, and it shows 4.88% of enhancement of non-commercial agricultural land in comparison to 2013 land cover. Non-commercial agricultural land declined to 3.79 percent (235.98 ha) in 2005 from 8.16 percent (508.81 ha) in 1994, then grew to 9.94 percent (619.27 ha) and 14.8 percent (917 ha) in 2013 and 2022, respectively. From 1994 to 2022, the amount of forest lost is 12.8% which was (823 ha), and the amount of non-commercial agricultural land increased 11.01% which was 682 ha of land. Over the previous 28 years, urbanization has increased. In particular, between 2005 and 2022, cover in this class rose significantly, rising from 1.18 percent (73.40 ha) to 3.58 percent (222.03 ha) of total land cover.

(ii) *Survey Analysis*

For questionnaire survey, total 12 questions were asked by 90 people ranging the age from 18 to 75 years. According to them, around 90% participants know that the area is a protected forest areas in Bangladesh. About 55% of the participants thought that the anthropogenic activities of local people do not impact on the forest areas and wildlife, whereas 10% people thought the opposite. Additionally, 20% of participants has given the opinion that the ecosystem of the protected area is developed by the locals and 10% of them thought that Government should take more initiatives to protect the forests. About 84% participants are visited frequently to the forest because of their dependency on the forests for survival, while 12% participants are visited to the forest for recreation purpose, and rest of the people are not interested to visit the forest.

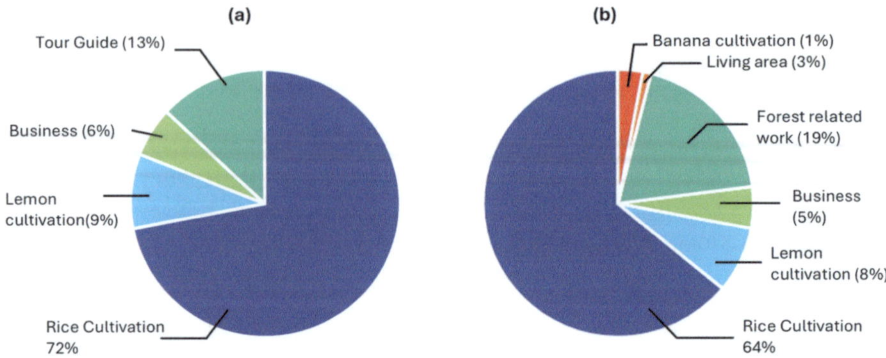

Fig. 7.4 Types of economic activities (**a**) and the uses of lands in Kalenga Range area (**b**)

The participants were also asked about the questions on unauthorized tree cutting inside the forest area. Most of the participants (69%) has given their opinion that tree cutting has been decreased from the previous because of the government's initiatives, local people's awareness and poachers are getting jobs by government and the society.

There are mainly four types of economic activities occurred in the forest area including rice cultivation (72%), lemon cultivation (9%), business (9%) and tour guide (13%) (Fig. 7.4a). Apart from the forest area, the lands of forest premises are used in different purposes. They are rice cultivation (64%), lemon cultivation (8%), business (5%), forest related work (19%), banana cultivation (1%) and living area (3%) (Fig. 7.4b).

The tourists of Rema Kalenga forests have an overall impact surrounding the forest area and the dependents of the forest. About 70% of the participants thought that tourism is profitable for the local and they are not disturbing the wildlife, 12% participants thought that tourists are generating wastes and harmful for the wildlife, and 18% have no comments.

According to G. M. Hossain and Hossain (2014), a total of 304 plant species under 240 genera and 81 families were identified. The major flora and fauna species were identified during the study and are listed at Table 7.2.

7.4 Discussion

Around the world, people's involvement in protected areas has been founded on a vision of human welfare, development, and emancipation that encompasses all available natural resources. Protected areas support local residents in a number of ways, including direct income from environmental preservation, diversification of livelihoods, security of access to resources, and the preservation of ecosystem functions like watershed protection. The villagers surrounding Rema-Kalenga Wildlife Sanctuary were underprivileged and dependent on the sanctuary for their living. In

Table 7.2 The major flora and fauna species were identified during the study at Kalenga Range area

Flora	Fauna
Akashmoni (*Acacia auriculiformis*)	Owl
Eucalyptus (*Eucalyptus camaldulensis*)	Hyena
Hortoki (*Terminalia chebula)*	Monkey
Mahogany (*Chukrasia tabularis*)	Deer
Banana (*Musa* sp.)	Squirrel
Chamkathal (*Artocarpus chama*)	Kite bird
Teak (*Tectona grandis*)	Masranga
Gandharai (*Gardenia jasminoides*)	Tuntuni
*Amloki (*Emblica officinale)	Bhimraj
Basak (*Adhatoda vasica*)	hanuman
Neem (*Azadirachta indica*)	Tia
Lojjaboti (Mimosa)	Moyna
Black fig (*Ficus carica*)	mongoose
Kodom (*Anthocephalus cadamba*)	Snail
Medicinal plants	Snake

the past, they earned a living by illegally felling trees or extracting unsustainable resources, which accelerated the rate of deforestation.

In order to reduce deforestation and integrate public participation at all levels, conservation measures and various management plans have been implemented in Bangladesh's protected areas as policies shift away from net deforestation. However, little research has been done on how land-related policies affect deforestation in protected areas. When a plantation program was put in place between 1994 and 2005, the overall area of forests rose. In the years 2005–2013 and 2013–2018, numerous forest areas were replaced with noncommercial agricultural land areas (Hossain and Numata 2021). In the present study, 12.8% forest decrease and 11.01% commercial agricultural land increase were identified from 1994 to 2022 (Fig. 7.2). The major reasons behind the decrease on forest resource area are harvesting unsustainable resources, unlawfully cutting down trees, agricultural land expansion (rice, fruit), bamboo and fuelwood collection from the forest area (Hossain and Numata 2021; Rahman and Miah 2017). Anthropogenic activities have a relatively smaller impact on forest biodiversity, according to the majority of survey participants. The ecosystem of the forest is unaffected by them. Instead, they protect the forest from unauthorized use and, if required, reforest it (Fig. 7.3a).

Agricultural productivity was hindered by a lack of resources for clearing land and erecting protective fences. The possible risks of predator attacks and livestock feeding also made households reluctant to invest in agriculture. A landscape approach to conservation in the area will be necessary to guarantee the long-term sustainability of all livelihoods and reduce overall impacts on forest resources, even though agricultural improvements were considered as feasible possibilities for

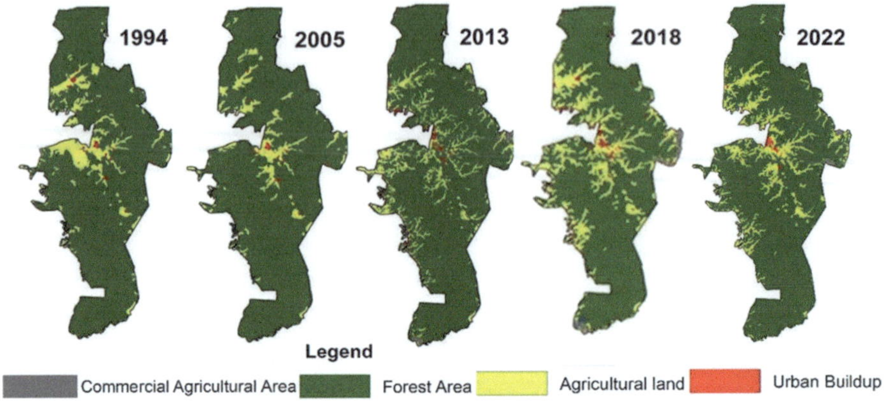

Fig. 7.2 Land use land cover (LULC) changes in 1994, 2005, 2013, 2018 (Hossain and Numata 2021) and 2022

reducing reliance on forests. Forest dependency was assessed based on the quantity of fuelwood and grasses they gathered and the sources from which they obtained them.

Ninety percent (90%) of the estimated 600–648 families that live on the Kalenga Range are poor or extremely impoverished. While gathering natural resources is usually a secondary source of revenue, agriculture is the main source of income for people living inside and outside the sanctuary. The amount of fuel wood collected weekly per family and its local market value were assessed, while the cost of feeding a cow commercial feed (without grass) served as a benchmark for assessing the value of grasses extracted from the forest. Tree-cutting incidents in Rema-Kalenga were common in the past, according to authorities and residents, but have considerably decreased as a result of government initiatives (Fig. 7.3b). Rema-Kalenga is recognized as one of the biggest forests in Bangladesh. The administration should concentrate more on completely eradicating illicit tree cutting, even though it asserts that it has declined in recent years.

The community's noticeable progress in various sociocultural areas seems to be the foundation for the sanctuary's efficient administration. The creation of community-based administrations has made it abundantly evident that local institutional arrangements, such as social norms and conventions, which are intended to foster cooperative solutions, may help resolve problems involving collective action and enhance the efficiency of natural resource use and management. The natural forest covers the majority of the Rema-Kalenga Wildlife Sanctuary. People proclaim that residents do not damage the environment and that the natural forest is vast and deep on its own, with a wide variety of wildlife habitat. After then, plantations are built on the fields. Trees are grown by locals who pay for them themselves. Large tracts of land are used for farming, and other territories are given for forest preservation in order to prevent poaching and forest destruction. The most abundant fields are paddy fields. A lake and streams are also present. Given the demonstrated importance of protected areas for conservation, anthropogenic factors provide challenges

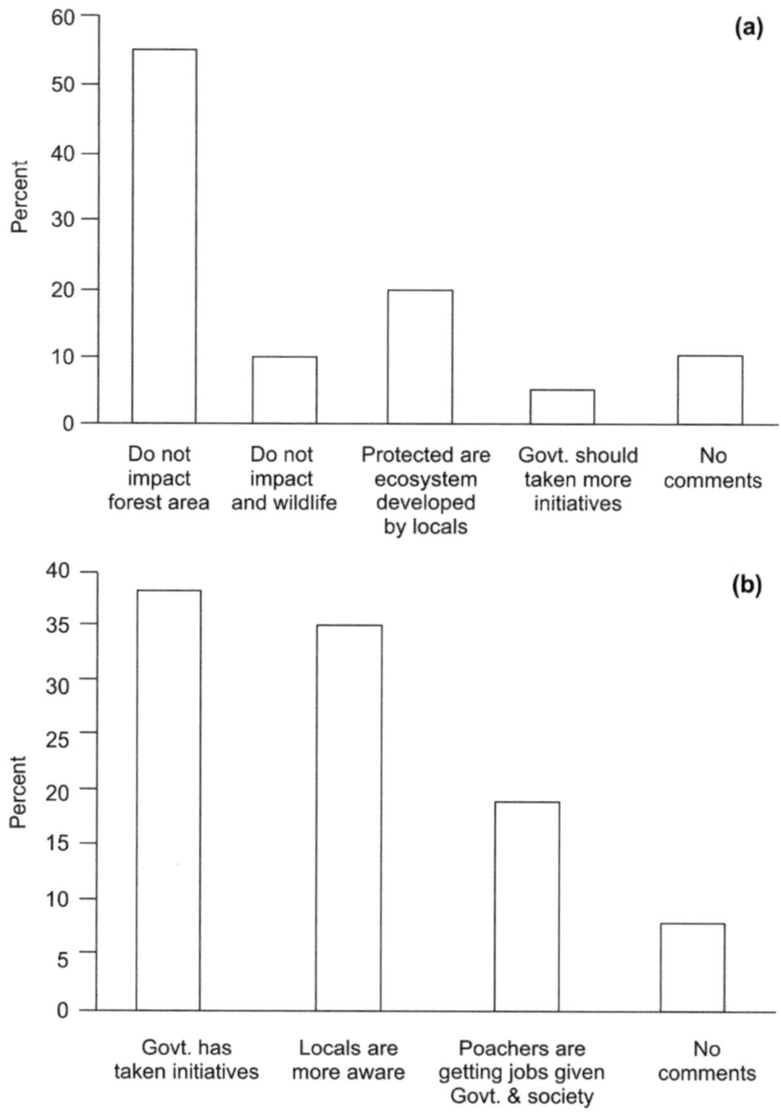

Fig. 7.3 Survey participant's thought on the anthropogenic activities (**a**) and unauthorized tree cutting inside the Kalenga Range area (**b**)

to their implementation, such as the issue of how to balance rural development with environmental preservation.

Bangladeshi woods have always been extremely susceptible to both climate change and human disturbances. In an effort to preserve natural forests, the bulk of state forests have been designated as protected areas (PAs) since liberation (1971), taking the form of national parks and wildlife sanctuaries. On the inside, villagers cultivate vegetables and fruit in their own yards, raise a small herd of cattle, and grow rice in neighbouring irrigated agricultural areas. Some people own their own companies. The great majority of the population works in rice agriculture. A few are employed as tour guides (Fig. 7.4a). Some qualified individuals work as day workers, while others are affiliated with non-governmental groups.

The usage of PAs in tourism essentially entails travel for the purpose of discovering and learning about wild areas. In many industrialized countries, tourism is seen as an eco-friendly means of boosting struggling rural economies and communities. In developing nations such as Bangladesh, where PAs are becoming popular travel destinations for both domestic and foreign wildlife tourists, a similar viewpoint is gaining traction. Rema-Kalenga is a fantastic place to visit. Since tourism is a major source of revenue for the forest's residents, they rely significantly on it. There are many natives that work as tour guides. Locals believe that wildlife will move if there is any disturbance. There has never been a situation when human disturbance has compelled wildlife to relocate. They claim that although certain creatures have gone extinct, human activity is not to blame. They worry a lot about the animals. They are even fed if they visit the living area. They also decided that since monkeys harm agriculture, they should not be allowed there. However, it does not lead to them moving to another place. No noise or disturbance of the animals is made by tourists. Both the managers and the locals aggressively enforce this. According to locals, visitors treat the animals with a great deal of kindness.

Throughout the study, locals were questioned regarding the protected land area. Surveys show that most individuals are aware of the protected land area. The idea of a protected land reserve was new to some of the participants. When asked if they knew it was a protected land area when they moved here, the majority of respondents said "yes." Three categories of land ownership were identified by the informants in the research region: homestead areas, privately owned agricultural lands, and agricultural lands owned by the Forest Department. As part of its efforts to reduce reliance on protected areas, the Forest Department gave local communities access to vacant plain areas within the sanctuary to increase agricultural output. There is no significant manmade activity, according to the inhabitants. The majority of respondents to the study agreed that there should be no endangerment of the forest. The idea that some activities are progressively destroying the forest was not widely accepted.

The Kalenga Range Forest guard claims that the number of people residing in the forest is increasing daily. In light of this, a survey was conducted to determine whether or not residents were allowed to reside in the protected region. The majority of the population believes that many people should not be allowed to live in the woodland area. On the other side, the administration believes that people should be allowed to live there since it allows communities to cooperate in maintaining the forest, so shielding it from illicit activity. But when the family expanded quickly, they started to worry about it. The number of families has increased compared to a few years ago. However, many believe that for the benefit of the forest, fewer people should live on the protected land area.

The effect of lockdown on the behavioural changes of fauna was also observed, as the study was conducted after COVID 19 impact. It was clear that there were more people prior to COVID-19 than there were following it. Wildlife creatures used to spend the most of their time in the forest as a result. But in the past, travellers could see them. Following COVID-19, they started to appear in the surrounding areas as a result of fewer people being there. They become much calmer and less hostile if food is given. However, residents claim that there have been slight alterations in animal behaviour. However, the locals' mindset has somewhat shifted, because during lockdown the tourism sector shut down and the income generating sources shifted to the agricultural sector. Also, the illegal activities increased around the forest area such as poaching.

The study was carried out following the COVID-19 pandemic. According to the survey, the Rema-Kalenga Wildlife Sanctuary was unaffected by COVID-19. Interestingly, no COVID-19 cases were discovered. The population was significantly healthier than in the city. Since tourists had not been to the area since the COVID-19 pandemic, people were not donning masks or taking preventative measures. COVID-19 has solely had an impact on the economy. The region's economic growth was greatly aided by tourists. The pandemic had prevented tourists from coming, and as a result, many people's financial circumstances had gotten worse. Many persons who used to work for NGOs are no longer employed there. They are making their living from agricultural pursuits.

The study is an effort to figure out the overall scenario of the biodiversity and local communities of Rema Kalenga Forest range area, despite some limitations, for example, low access of roads, poor transportation systems, dense forestry, and understanding of local language.

7.5 Conclusion

According to the study, co-management has enhanced forest management, biodiversity protection, and livelihood possibilities for the local population. The co-management approach has led to another important development in forest management: the local community is increasingly involved in decisions about forest management and has a stronger voice. But sometimes the most vocal individuals come from the higher echelons of the community. The results show that co-management is often seen favourably by the local population, especially when it comes to enhancing forest biodiversity, managing forests better, and protecting them from unlawful activities.

From the perspective of the community, tourism is meant to provide benefits that will boost support for the preservation of natural resources. It can also be marketed as a tool for sustainable development and the conservation of wildlife and forests. To manage the forest area, restriction can be strict on the availability of agricultural land and plantation of invasive species in forest areas.

References

Alam M (2009) Evolution of forest policies in Bangladesh: a critical analysis. Int J Soc Forest 2(2):149–166

Bayes A, Md K, Zhu X, Rahman S, Choi K (2013) Simulating land cover changes and their impacts on land surface temperature in Dhaka, Bangladesh

Brendler T, Carey H (1998) Community forestry, defined. J For 96(3):21–23

Bhuiyan MS, Islam S, Haque MMU, Aktar S, Ahmed R (2021) Evaluating capital assets in governing protected area co-management in the Rema-Kalenga Wildlife Sanctuary, Bangladesh. Int For Rev 23(1):16–28

Chowdhury MSH, Koike M (2010) An overview on the protected area system for forest conservation in Bangladesh. J For Res 21:111–118

Chowdhury MSH, Koike M, Izumiyama S (2014) Impact of co-management on rural development: evidence from community survey in and around Rema-Kalenga Wildlife Sanctuary. In: Forest conservation in protected areas of Bangladesh: policy and community development perspectives, pp 111–141

Dewan AM, Yamaguchi Y (2009) Land use and land cover change in Greater Dhaka, Bangladesh: using remote sensing to promote sustainable urbanization. Appl Geogr 29(3):390–401

Dewan AM, Yamaguchi Y, Rahman MZ (2012) Dynamics of land use/cover changes and the analysis of landscape fragmentation in Dhaka Metropolitan, Bangladesh. GeoJournal 77(3):315–330

Hossain GM, Hossain AE (2014) Effect of exotic tree plantation on floristic composition and phytodiversity of Rema-Kalenga wildlife sanctuary, Bangladesh. Jahangirnagar Univ J Biol Sci 3(2):33–47

Hossain MI, Numata S (2021) Effects of land-related policies on deforestation in a protected area: the case study of Rema-Kalenga Wildlife Sanctuary, Bangladesh. Conservation 1(3):168–181

Jashimuddin M (2011) Drivers of land use change and policy analysis in Bangladesh: theory and policy recommendations. No Title

Morshed HM (2013) Benefits and weaknesses of collaborative management: a case study in the Rema-Kalenga Wildlife Sanctuary. In: Fox J, Mustafa MG, Bushley BR, Durand L (eds) Connecting communities and conservation: co-management initiatives implemented by IPAC in wetlands and forests of Bangladesh, pp 102–114

Msofe NK, Sheng L, Lyimo J (2019) Land use change trends and their driving forces in the Kilombero Valley Floodplain, Southeastern Tanzania. Sustain For 11(2):505

Rahman MH, Miah MD (2017) Are protected forests of Bangladesh prepared for the implementation of REDD+? A forest governance analysis from Rema-Kalenga Wildlife Sanctuary. Environments 4(2):43

Rana MP, Sohel MSI, Mukul SA, Chowdhury MSH, Akhter S, Chowdhury Q, Koike M (2010) Implications of ecotourism development in protected areas: a study from Rema-Kalenga wildlife sanctuary, Bangladesh. IForest-Biogeosci Forest 3(1):23

Sohel MSI, Rana MP, Akhter S, Karim SMF, Khan MMR (2009) Evaluation of co-management impact in protected area: field experience from Rema-Kalenga Wildlife Sanctuary, Bangladesh. J Forest Environ Sci 25(1):43–48

Forest Fires and Land Degradation

8

Juhi Agrawal, Anil Kumar, Devesh Dubey, and Kangujam Premkumar

Abstract

Soil is a vital non-renewable resource crucial for ecosystem services like carbon sequestration, nutrient cycling, and plant productivity. However, it is highly vulnerable to degradation, especially in forest ecosystems, due to factors such as deforestation, erosion, and fires. Both natural and human-induced forest fires are increasingly recognized for their significant impacts on soil's chemical, physical, and biological properties. Wildfires, often caused by natural events or human activities, can lead to significant losses in nutrients, reductions in soil organic carbon (SOC), changes in pH, and increased erosion, primarily due to the combustion of organic material. In contrast, prescribed fires, when carefully managed, can help improve soil health by releasing nutrients and enhancing soil structure, though they can still pose risks if fire intensity is too high. The severity, duration, and fuel load of a fire are crucial in determining whether its impact on soil is beneficial or harmful. With climate change driving an increase in wildfire frequency and intensity, understanding the relationship between fire and soil is essential for promoting sustainable forest management and soil conservation practices. This chapter explores the complex interactions between fire intensity, soil characteristics, and ecosystem processes, highlighting the need for adaptive management strategies to mitigate long-term impacts on soil health.

J. Agrawal · A. Kumar
University of Petroleum and Energy Studies, Dehradun, Uttarakhand, India

D. Dubey
Arahas Technologies, Gurugram, Haryana, India

K. Premkumar (✉)
Forest Ecology and Climate Change Division, Forest Research Institute, Dehradun, India

Department of Forestry, North-Eastern Hill University, Tura, Meghalaya, India

S. Singh, M. K. Gautam (eds.), *Biodiversity Conservation and Land Degradation Neutrality*, https://doi.org/10.1007/978-981-96-7210-3_8

Keywords

Soil degradation · Wildfire · Prescribed fire · Nutrient cycling · Soil health

8.1 Introduction

Soil is one of the most invaluable natural resources within both natural and human-environment systems (Aaltonen et al. 2019). It plays a crucial role in functions such as carbon sequestration, nutrient cycling, plant production support and serving as a mineral reservoir (Addo-Fordjour et al. 2020). Despite its significance, soil is considered a non-renewable resource on a human timescale due to its rapid degradation compared to its slow formation process (Adkins et al. 2019). The degradation of soil, enveloping biological, physical, along with chemical characteristics, diminishes its capacity to perform optimally, with the reaction being either temporary or permanent. In forest ecosystems, several key drivers contribute to soil degradation, including soil contamination, erosion, deforestation, and fires (Agbeshie and Abugre 2021).

Fires are especially impactful and are widely recognized as destructive forces in boreal, tropical, and temperate forest ecosystems (MacCarthy et al. 2023). They are regarded as global phenomena, affecting vast land areas (Alexakis et al. 2021), with evidence of fires dating back to the Carboniferous period, approximately 360 million years ago (Verma and Jayakumar 2012). Forest fires exert significant influence on forest ecosystems, altering ecological biodiversity, species composition, regeneration processes, soil hydrophobicity, and nutrient cycling dynamics (Agbeshie et al. 2022; Li et al. 2023; Singh 2022). While some forest fires are an impact of human action, naturally appearing fires constitute only a minor portion of worldwide fires (Knorr et al. 2016).

Forest fires detrimentally influence the physio-chemical along with biological nature of soils, leading to a reduction in the nutrient pool through different procedure such as erosion, ash transmission, oxidation, along with volatilization (Pellegrini et al. 2018). Studies have shown that forest fires can result in a significant short-term reduction in microbial biomass carbon (Akburak et al. 2018), as well as substantial depletion of nutrient pools, including nitrogen (N), magnesium (Mg), and phosphorus (P), through processes of oxidation and volatilization (Raison et al. 1986). Nevertheless, wildfire is a significant reason of nutrient losses and soil degradation, primarily through erosion and volatilization processes (Lal 2015). Wildfires with temperatures exceeding 300 °C have been recognized to destroy soil water repellence, significantly impacting soil erosion and water cycling feature (Mataix-Solera et al. 2011). The impact of fires on soil properties is also significantly influenced by the intensity and density of the fires (Johnston and Barati 2013; Maynard et al. 2014; Pellegrini et al. 2018). Studies suggest that slow-burning, high-intensity fires inflict more extensive damage on soil properties compared to fast-moving fires (Perez-Izquierdo et al. 2021). Additionally, soil surfaces with heavy, dry fuel loads can reach temperatures of 850 °C or higher during fires, resulting in severe effects on soil characteristics (Knicker 2007). Certain nutrients are more vulnerable to fire damage than others. For example, the levels of magnesium (Mg), calcium (Ca), and potassium (K) may increase or remain unchanged after a fire, whereas nitrogen (N) and sulfur (S) typically decrease (DeBano 1991). Temperature plays a crucial role in the volatilization of soil nutrients. For instance, nitrogen begins to volatilize at 200 °C, while calcium requires temperatures as high as 1484 °C to vaporize (Johnston and Barati 2013).

Ongoing severe droughts, higher temperatures, and global warming notably boost the danger of wildfires (Zhang and Biswas 2017). Numerous predictions suggest a potential boost in fire strength, frequency, along with duration in wooded areas, particularly in tropics, due to elevated temperatures (Cavard et al. 2019; Rodriguez et al. 2018). Fire has emerged as an increasingly significant driver of global tree cover loss, rising from approximately 20% in 2001 to nearly 33% today, surpassing other factors such as mining and forestry (MacCarthy et al. 2023). Record-setting forest fires are becoming increasingly prevalent, with 2020, 2021, and 2023 representing the fourth, third, and most severe years for global forest fires, respectively. As a result, the increased likelihood of fires will not only alter forest vegetation but also affect the biological, chemical, and physical properties of the soil (Romeo et al. 2020). While fires exert multifaceted impacts on forest soils, these effects have not been as systematically studied as their influence on vegetation (Agbeshie et al. 2022; Certini 2005). Therefore, understanding changes in soil characteristics following fires is essential for developing forest management operations and sustainable and adaptive soil (Addo-Fordjour et al. 2020).

8.2 The Concept of Wildfires and Prescribed Fires: Definitions and Implications

8.2.1 Wildfire

The term "wildfire" typically refers to fires occurring outside urban areas. According to the Australasian Fire Authorities Council (AFAC), a forest fire is any unintentional vegetation fire, including wildfires and bushfires. In the United States, a forest fire is defined as "any unplanned fire that occurs in wildland areas." In Australia, the term "bushfires" is also used to describe any vegetation fire. In Europe, wildfires or bushfires are commonly referred to as forest fires (European Commission, 2010). Wildfires are recognized and categorized as natural hazards by various global data sets (e.g., EM-DAT (CRED), NatCatSERVICE (Munich RE)), international institutions (such as National Aeronautics and Space Administration (NASA) and the United Nations Office for Disaster Risk Reduction (UNDRR)), political entities (e.g., the European Union), governments (Middlemann and Middelmann 2007), and scientists (McCaffrey et al. 2015; Moritz et al. 2014).

Over the past decade, significant forest fires have caused destructive changes in various regions (Adam 2020). For example, in Canada, between 1970 and 2017, approximately 5.6 million acres of flora were lost each year due to around 8000 forest fire events (Wang et al. 2020). Predictions of ongoing climate change, coupled with more frequent and prolonged droughts, are expected to exacerbate the occurrence of forest fires in many agroecological zones (Caon et al. 2014).

Forest fires, wildfires, or bushfires, regardless of their name, significantly disrupt forest ecosystems. The impact on soil characteristics is influenced by factors such as fire density, duration, and intensity (Bento-Goncalves et al. 2012). Fire intensity, which measures the heat produced per unit area over time, can be categorized as high (over 350 °C), low (below 100 °C), and moderate (up to 250 °C) (Caon et al. 2014). The severity of woodland fires largely depends on factors like moisture content and soil type, as well as the fire's intensity and the fuel load in the ecosystem. Numerous studies have documented the effects of forest fires on soil characteristics (Dooley and Treseder 2012; Hosseini et al. 2017; Ibáñez et al. 2022; Varela et al. 2015).

In prairie ecosystems, an increase in soil nutrients has been observed following low-intensity forest fires, a finding supported by various studies (Fernández-García et al. 2019a; Hosseini et al. 2017). Conversely, medium- to high-intensity forest fires are generally associated with negative impacts on biological and physical soil properties. For instance, forest fires have been reported to adversely affect soil aggregates, water repellence, and alter bulk density and water content (Hiers et al. 2020; Hinojosa et al. 2021; Ibáñez et al. 2022), while also reducing enzymatic activities and microbial biomass (Inbar et al. 2014; James et al. 2018; Jhariya and Singh 2021). Therefore, evaluating the impact of forest fires is crucial, as they cause significant changes in soil biological, chemical, and physical characteristics in both the short and long term.

8.2.2 Prescribed Fires

Prescribed fires are controlled and low-strength fires that are employed to accomplish management objectives (Johnson et al. 2014). These fires serve as a scientific appliance to conduct fuel loads, especially organic material in forest regions, with the aim of preventing or mitigating potential wildfires. Apart from reducing the aggregation of huge fuel loads, prescribed burning is utilized for purposes such as species regeneration (Kemp et al. 2016), training exercises for firefighter (Shaltout and Ismail 2020), and clearing land in primitive subsistence agriculture (Alcañiz et al. 2016). Furthermore, prescribed fires influence ecological processes and soil properties, impact animal and plant biodiversity across various vegetative ecosystems (Fellows et al. 2018), and aid in controlling forest diseases and insect.

Because of their low harshness, prescribed fires typically have positive results on soil characteristics (Hahn et al. 2021; Heydari et al. 2017).

Several studies have recorded the positive impacts of prescribed burning (Alcañiz et al. 2016; Francos et al. 2019; Goberna et al. 2012). In various parts of the world, ecologists (Akburak et al. 2018; Badía et al. 2014; Switzer et al. 2012) have observed increased levels of base cations in soils following prescribed burning, associated to the aggregation of ash on the soil exterior. Research papers (Francos et al. 2019;

Muqaddas et al. 2015) reported higher soil pH in burnt soils after recommended fires. Moreover, electrical conductivity has been found to increment in burnt soil following low-strength prescribed fire because of the discharge of soluble inorganic ions along with the construction of black carbon (Alcañiz et al. 2016). However, despite the benefits associated with prescribed burning, its sources could potentially harm the soil ecosphere if temperature exceeds the permissible threshold for low-strength fire (Francos and Ubeda 2021).

8.3 Fire Effect on Soil Characteristics

Wildfires have a profound impact on the chemical, biological, and physical properties of soils (Panico et al. 2020; Ribeiro-Kumara et al. 2020). The heat transmission during fires, originating from burning biomass within the soil also on the surface, plays a direct role in these transformations (Sazawa et al. 2018). The results of fire on soil characteristics rely upon on aspects such as timespan, frequency, and intensity, collectively known as fire harshness (Alcañiz et al. 2018; Fernández-García et al. 2021; Lucas-Borja et al. 2020). However, alterations in soil characteristics are influenced by soil temperature at the time of burning. The subsequent sections elaborate on the outcome of fire on soil across different ecosystems, with summarized outputs provided in Table 8.1.

8.3.1 Chemical Characteristics

Many research works have documented the influence of fire on soil chemical characteristics. The introduction of ash resulting from the complete or half combustion of organic matter and biomass, also with its integration into the soil, brings about significant differences in soil chemistry (Jhariya and Singh 2021; Li et al. 2021; Maksimova and Abakumov 2015). Soil temperatures exceeding 200 °C during a wildfire cause the incineration of organic matter, ash formation, resulting in the making of char compounds, and subsequent pH elevation.

8.3.2 Soil Organic Carbon (SOC)

SOC receives considerable attention in post-fire soil studies because of its crucial role in soil type assessment (Aaltonen et al. 2019; Hobley et al. 2017). Nevertheless, alterations in SOC levels are changeable along with contingent upon factors such as fire type along with intensity, moisture content, available biomass, and fire duration (Reyes et al. 2015). These variations can lead to various outcomes, including substantial soil organic element loss (Miesel et al. 2015; Zhang and Biswas 2017). SOC degradation typically begins around temperatures of 200–250 °C, with whole combustion occurring at temperatures ranging between 460 and 500 °C (Badía et al. 2014).

Table 8.1 Forest fire impact on soil chemical properties

References	Vegetation type	Location	Characteristics of fire	Characteristic of soil	Soil property	Impact	Causes for impact
Inbar et al. (2014)	*Pinus halepensis*	Northern Israel	Low-moderate severity, WF	Sandy clay loam, Lithic Xerorthenth	OM	Increased	Mixing partially burnt biomass into fire-affected soil increases organic matter content
					CEC	Increased	Because of the increase in OM composition
					pH	Insignificant	–
					EC	Insignificant	–
Muqaddas et al. (2015)	Wet sclerophyll forest	Queensland, Australia	Low intensity, 2 years burning regime, heat release rate of <2500 kWm^{-1}, PF	Sandy, red to yellow Kandosols	Total N	Decreased	As a result of nitrogen volatilization
					Total C	Decreased	Because of CO2 emission following combustion of biomass
					pH	Increased	Because of increased base cations
					Recalcitrant C, N	Decreased	Decrease in overall C
Francos et al. (2019)	*Quercus ilex* and *Pinus halepensis*	Northeast Spain	Temperature of 65 °C on soil surface, maximum fire temperature of 435 °C, PF	Xerorthents	SOM	–	–
					pH	Increased	Heating leads to the denaturation of organic acids

					EC	Increased	Creation of black C after flames
					Extractable K	Increased	Increased base cations
					Available P	–	–
Liu et al. (2018)	Grassland vegetation	Ningxia Hui Autonomous Region, China	Surface head fire, low intensity fire, WF	Calci-Orthic Aridisol	Soil OC	Increased	Elevated levels of pyrogenic carbon resulting from incomplete combustion
					Total N	Increased	Mixing of incomplete residual burnt material with soil
					NO –		
3	Decreased	Rapid revegetation with increased organic N uptake (NO3 –)					
					Total P	–	–
					Available K	Insignificant	–
Alcañiz et al. (2016)	*Pinus halepensis* forest	Montgrí Massif, Catalonia, Spain	flame height <2.5 m, 324 °C, PF	Lithic Xerorthent	pH (IAB)	Insignificant	–
					EC (IAB)	Increased	Release of soluble inorganic ions following burning

(continued)

Table 8.1 (continued)

References	Vegetation type	Location	Characteristics of fire	Characteristic of soil	Soil property	Impact	Causes for impact
					Total C (IAB)	Increased	Formation of black C because of low fire and addition of ash content to the soil
					Total N (IAB)	Increased	–
					Available P (IAB)	Increased	Addition of ash into the soil and vegetation burning
					Extractable cations		
(IAB)	Increased	Low fire severity					
Badía et al. (2014)	Aleppo pine (*Pinus halepensis*) forests	Montes de Zuera, Northeast Spain	high burn severity, affected 2700 ha, WF	Rendzic Phaeozem	SOC	Decreased	Soil losses resulting from severe combustion
					pH	–	–
					EC	Increased	Addition of basic cations
					NO –		
3	Increased	Organic N transformation					
					NH +		
4	Increased	Mineralization of organic N to mineral N					

					CEC	Increased	Due to the increased OM content and inorganic ions
					Available P	Increased	Mineralization of organic P and dissolution of P from ash beds
Meira-Castro et al. (2015)	*Pinus pinaster*						
vegetation	Northern Portugal	Fire spread					
10–15 mh^{-1}, PF	Umbric Leptosol and Umbric Cambisol	pH	–	–			
					SOM	–	–
Goberna et al. (2012)	Shrubland						
(Rosmarinus officinalis) vegetation	Valencia, Spain	Fire temp. of 611 °C, soil					
surface temp. of 338 °C, PF	Humic Leptosols	pH	–	–			
					Total OC	–	–
					NO3–	Increased	Increased nitrification
					NH4+	Increased	Mineralization of organic N

(continued)

Table 8.1 (continued)

References	Vegetation type	Location	Characteristics of fire	Characteristic of soil	Soil property	Impact	Causes for impact
					Available P	Increased	Mineralization of organic P
Bennett et al. (2014)	Eucalyptus forest	Victoria, Australia	High intensity, 259 kWm^{-1}, PF	Kandosols and Dermosols	Carbon stocks	Decreased	Combustion of organic matter
Akburak et al. (2018)	Quercus frainetto						
forest	Istanbul, Turkey	Low intensity, burning					
–			+				

Where *NO3* Nitrate, *SOC* Soil organic carbon, *OM* Organic matter, *NH4* Ammonium, *CEC* Cation exchange capacity, *SOM* Soil organic matter, *C* Carbon, *CO2* Carbon dioxide, *P* Phosphorus, *GHG* Greenhouse gas, *EC* Electrical conductivity, *WF* Wildfires, *PF* Prescribed fires, *MBN* Soil microbial biomass nitrogen, *OC* Organic carbon, *MBC* Microbial biomass carbon, *N* Nitrogen

The influence of fires on soils displays significant variability, indicating that low-intensity fires may conclude in either minimal or increased differences in soil organic carbon (SOC), while high-intensity fires typically lead to declined SOC levels (Caon et al. 2014). For instance, in their investigation of wildfire impacts on Aleppo pine forests in Spain, Badía et al. (2014) observed a 27.9% decrement in SOC in the 1 cm soil layer following a highly devasted fire. Ecologist (Rodriguez-Cardona et al. 2020) documented a 21.0% decrease in SOC in a Lithic Haplocalcid soil after a medium- to high-strength forest fire happened in Spain (*P. halepensis* forest). The reduction in SOC following high-strength fire may be associated to aspects such as the elevated rates of carbon mineralization, volatilization, carbon solubilization, and the combustion of soil organic matter (SOM) because of gained pH resulting from nutrient-dense ash (Akburak et al. 2018; Alcañiz et al. 2016; Fernández-García et al. 2019a; Gao et al. 2018; Hu et al. 2020; Liu et al. 2018; Sánchez Meador et al. 2017; Santín et al. 2018).

8.3.3 Soil Reaction (pH)

At the time of forest fires, damages happen in organic and biomass layers, leading to the inclusion of nutrition-dense ash over the soil exterior. This ash deposition gains soil pH as well as significantly impacts soil biological and chemical characteristics. Soils with larger burn severity, typically characterized by temperatures exceeding 450 °C, exhibit elevated pH levels (Knicker 2007). Numerous investigators have recorded pH increases after fires (Akburak et al. 2018; Dzwonko et al. 2015; Hahn et al. 2021; Hinojosa et al. 2021). In a research paper (Alcañiz et al. 2016), it was reported that a significant pH raised following prescribed forest burning in Spain (*P. halepensis*). Similarly, it was observed a pH increase following a fire in a forest in Turkey (*Q. frainetto*) (Akburak et al. 2018). These notable pH increments, specifically after high-intensity fires, may result from the inclusion of nutrition- dense ash, which includes base positive ions following the denaturation of organic groups along with combustion of organic material from organic matter (Alcañiz et al. 2016). However, the extent of pH increase may rely on aspects such as nutrition contents, the amount of deposited ash, and initial pH levels (Neary et al. 2005).

Contrarily, some reports indicate no change in pH following fires (Badía et al. 2014; Downing et al. 2017; Fernández-Fernández et al. 2015; Fultz et al. 2016). This could be attributed to post-fire environmental circumstances, such as erosion (Fernández-García et al. 2019b), and low to moderate fire intensities (Valkó et al. 2016). Soil pH plays a critical role in nutrient availability (Alcañiz et al. 2018; Roques et al. 2013).

8.3.4 Nitrogen (N)

N is recognized as one of the most confining nutrition in both forest ecosphere along with cropping (Neary et al. 2005). It happens in the soil matrix in two forms: organic N (Agbeshie et al. 2020) and available or mineral N (e.g., NH_4+ and NO_3-). Forest fires result in the deduction of nitrogen primarily through volatilization, occurring when soil temperatures go above 200 °C (Zhang and Biswas 2017). Additionally, nitrogen loss after a fire may occur due to leaching and erosion (Caon et al. 2014; Cheng et al. 2021; Huddell et al. 2020). Although, the impacts of woodland fires on soil nitrogen yield conflicting effects that are still to be fully accepted.

For instance, field studies in Australia (Qiu et al. 2021) demonstrated that low-strength prescribed fires led to a decrement in whole nitrogen in the humus of a wet sclerophyll forest. In research papers (Dzwonko et al. 2015; Muqaddas et al. 2015), it has been observed significant decreases in whole nitrogen in the topsoil of *P. halepensis* and *Q. ilex* forests in Spain following wildfires. According to a study (Hahn et al. 2021), a prescribed fire in USA (Fishburn forest) resulted in a 33 percent decrease in whole nitrogen.

A standardized analysis (Sánchez Meador et al. 2017) highlighted the influence of wildfires on soil nitrogen, showing a decrease in complete nitrogen in the A horizon (Fraterrigo and Rembelski 2021; Muqaddas et al. 2015).

8.3.5 Physical Attributes

Soil physical characteristics play a crucial role in influencing plant development, much like soil chemical aspect. Numerous investigators have investigated the results of wildfires on soil physical characteristics, focusing on parameters such as porosity, water content and repellence, aggregate stability, bulk density, and texture (Albalasmeh et al. 2013; Busse et al. 2021; Dove et al. 2020; Thomaz and Fachin 2014).

8.3.6 Soil Water Repellence (SWR)

SWR is a notable soil physical property that can be influenced by fire (Bodí et al. 2012; Robichaud et al. 2016). SWR reduces water infiltration and increases runoff, which in turn amplifies soil erosion (Weninger et al. 2019). Forest fires either can create, increase, or decrease soil water repellence (Pereira et al. 2013; Scharenbroch et al. 2012). Ecologists (Granged et al. 2011; Weninger et al. 2019) reported an increase in soil water repellency following a wildfire in a Mediterranean karst forest in Croatia. However, in a study (Varela et al. 2015), it was observed a decrease in soil water repellency after a medium-intensity wildfire in a forest in northwest Spain.

The increase in soil water repellence is associated to the formation of a hydrophobic layer on mineral elements at the time of organic matter combustion

(Mataix-Solera et al. 2011). During the fire, hydrophobic compounds move downward through the soil and condense on cooler particles, forming a hydrophobic coating. However, Zavala and colleagues (Zavala et al. 2010) observed that soil water repellency typically increases as soil temperature rises to 200 °C but diminishes at temperatures above 300 °C (Inbar et al. 2014; Mataix-Solera et al. 2011). Factors influencing these include the interaction of organic matter and moisture content, where high moisture and low organic matter levels enhance soil water repellency after fires (González-Pelayo et al. 2015; Mao et al. 2019); the depth of the soil profile, with denser profiles exhibiting prolonged repellency (Rodríguez-Alleres et al. 2012); and surface temperatures, as high surface heat during burning can completely eliminate soil hydrophobicity (Badía et al. 2014; Plaza-Álvarez et al. 2018).

Erosion along with infiltration procedures is jointly linked to soil water repellence, with increased repellence leading to increased overland flow and decreased water infiltrability. For instance, ecologists (Hosseini et al. 2017) realized sediment loss and gained runoff identical with expanded water repellence in Portugal following a forest fire. Inbar and colleagues (Inbar et al. 2014) also show that expanded soil damage on burned plots below *P. halepensis* after forest fire, attributing it to a collapse in soil accumulates along with expanded water repellence, resulting in increased runoff.

8.3.7 Soil Texture

Soil texture, representing the distribution of particle sizes in the soil, consists of the corresponding proportions of inorganic components smaller than 2 mm in mineral soil. Forest fires generally have essential consequence on soil character due to the high temperature thresholds of silt, clay, along with sand particles (Alcañiz et al. 2016). Clay elements are extra susceptible to fire effects because their temperature threshold (400–800 °C) is lower than that of silt and sand (1414 °C) (Neary et al. 2005). Consequently, clay elements are more influenced in terms of texture. For instance, in a pine woodland on sandy clay loam, ecologists (Inbar et al. 2014) observed a slight decrement in clay elements at a depth of 2–5 cm after a wildfire, accompanied by an increase in silt particles. Similarly, it was documented that there was a decrease in clay content in the A horizon under oak forest after a high-severity forest fire, noting a subsequent increase in sand elements (Heydari et al. 2017).

Despite these shifts in particle composition, most investigators do not report a significant difference in soil character. However, the increment in silt and sand elements following a forest fire can be associated to the breakdown and demolition of the clay structure formation (Alcañiz et al. 2018). Consequently, the breakdown of clay elements promotes the accumulation of smaller elements, leading to the making of bigger or coarser silt along with sand particles.

8.4 Conclusion

The impact of wildfires on soil characteristics largely depends on the severity of the burn, which includes both fire duration and intensity. Both prescribed fires and wildfires influence chemical, physical, and biological characteristics of soil. Fire leads to the partial or complete combustion of organic matter, accompanied by the deposition of ash on the soils surface. Consequently, the blending of partly burned organic materials and ash into the soil originally modifies soil chemistry. Although the effects of fire on soils chemical characteristics vary widely, low-intensity fires often cause minimal effect or even an increase in available nutrients (such as K^+, Ca^{2+}, Mg^{2+}, PO_4^{3-}, NH_4^+) and pH because of its ash deposition. Conversely, higher-strength fires tend to decrease of soil organic matter along with whole nitrogen because of its combustion along with volatilization damages.

It is essential to carefully manage burned areas to enhance nutrient input from ash deposits and mineralization, while also mitigating high-intensity wildfires and minimizing volatilization losses. This analysis highlights that forest fires typically have a negative impact on soil's biological and physical properties. Soils that become water-repellent after a fire reduce water infiltration, leading to excessive runoff and a higher risk of erosion. Soil microbiotas are highly sensitive to ecological changes and are completely destroyed at soil temperatures above 120 °C. Given the present trends of extreme droughts, global warming, and higher temperatures, the possibility of wildfires is expected to increase significantly. Therefore, the usage of low- strength fires, such as prescribed burning, as an executive device to decrease fuel load in woodland zones is largely prescribed to mitigate the occurrence of woodland fires. Comprehensive soil and forest management strategies are essential for restoring fire-disturbed soil ecosystems. Furthermore, further investigation in tropical climates is imperative to better comprehend the dynamics of woodland fires on soil characteristics across various ecosphere.

References

Aaltonen H, Koster K, Koster E, Berninger F, Zhou X, Karhu K, Biasi C, Bruckman V, Palviainen M, Pumpanen J (2019) Forest fires in Canadian permafrost region: the combined effects of fire and permafrost dynamics on soil organic matter quality. Biogeochemistry 143:257–274

Adam D (2020) Largest wildfires of the decade. https://blog.batchgeo.com/largest-wildfires/

Addo-Fordjour P, Kadan F, Rahmad ZB, Fosu D, Ofosu-Bamfo B (2020) Wildfires cause shifts in liana community structure and liana-soil relationships in a moist semi-deciduous forest in Ghana. Folia Geobot 55(4):273–287

Adkins J, Sanderman J, Miesel J (2019) Soil carbon pools and fluxes vary across a burn severity gradient three years after wildfire in Sierra Nevada mixed-conifer forest. Geoderma 333:10–22

Agbeshie AA, Abugre S (2021) Soil properties and tree growth performance along a slope of a reclaimed land in the rain forest agroecological zone of Ghana. Sci Afr 13:e00951

Agbeshie AA, Logah V, Opoku A, Tuffour HO, Abubakari A, Quansah C (2020) Mineral nitrogen dynamics in compacted soil under organic amendment. Sci Afr 9:e00488

Agbeshie AA, Abugre S, Atta-Darkwa T, Awuah R (2022) A review of the effects of forest fire on soil properties. J For Res 33(5):1419–1441

Akburak S, Son Y, Makineci E, Cakir M (2018) Impacts of low-intensity prescribed fire on microbial and chemical soil properties in a Quercus frainetto forest. J For Res 29:687–696

Albalasmeh AA, Berli M, Shafer DS, Ghezzehei TA (2013) Degradation of moist soil aggregates by rapid temperature rise under low intensity fire. Plant Soil 362:335–344

Alcañiz M, Outeiro L, Francos M, Farguell J, Úbeda X (2016) Long-term dynamics of soil chemical properties after a prescribed fire in a Mediterranean forest (Montgrí Massif, Catalonia, Spain). Sci Total Environ 572:1329–1335

Alcañiz M, Outeiro L, Francos M, Úbeda X (2018) Effects of prescribed fires on soil properties: a review. Sci Total Environ 613-614:944–957. https://doi.org/10.1016/j.scitotenv.2017.09.144

Alexakis D, Kokmotos I, Gamvroula D, Varelidis G (2021) Wildfire effects on soil quality: application on a suburban area of West Attica (Greece). Geosci J 25:243–253

Badía D, Martí C, Aguirre AJ, Aznar JM, González-Pérez J, De la Rosa J, León J, Ibarra P, Echeverría T (2014) Wildfire effects on nutrients and organic carbon of a Rendzic Phaeozem in NE Spain: changes at cm-scale topsoil. Catena 113:267–275

Bennett LT, Aponte C, Baker TG, Tolhurst KG (2014) Evaluating long-term effects of prescribed fire regimes on carbon stocks in a temperate eucalypt forest. Forest Ecol Manag 328:219–228

Bento-Goncalves A, Vieira A, Ubeda X, Martin D (2012) Fire and soils: key concepts and recent advances. Geoderma 191:3–13

Bodí MB, Doerr SH, Cerdà A, Mataix-Solera J (2012) Hydrological effects of a layer of vegetation ash on underlying wettable and water repellent soil. Geoderma 191:14–23

Busse M, Zhang J, Fiddler G, Young D (2021) Compaction and organic matter retention in mixed-conifer forests of California: 20-year effects on soil physical and chemical health. Forest Ecol Manag 482:118851

Caon L, Vallejo VR, Ritsema CJ, Geissen V (2014) Effects of wildfire on soil nutrients in Mediterranean ecosystems. Earth Sci Rev 139:47–58

Cavard X, Bergeron Y, Paré D, Nilsson M-C, Wardle DA (2019) Disentangling effects of time since fire, overstory composition and organic layer thickness on nutrient availability in Canadian boreal forest. Ecosystems 22:33–48

Certini G (2005) Effects of fire on properties of forest soils: a review. Oecologia 143:1–10

Cheng Y, Li P, Xu G, Wang X, Li Z, Cheng S, Huang M (2021) Effects of dynamic factors of erosion on soil nitrogen and phosphorus loss under freeze-thaw conditions. Geoderma 390:114972

DeBano LF (1991) The effect of fire on soil properties. Paper presented at the Proceedings management and productivity of western-Montane. Forest Soils

Dooley SR, Treseder KK (2012) The effect of fire on microbial biomass: a meta-analysis of field studies. Biogeochemistry 109:49–61

Dove NC, Safford HD, Bohlman GN, Estes BL, Hart SC (2020) High-severity wildfire leads to multi-decadal impacts on soil biogeochemistry in mixed-conifer forests. Ecol Appl 30(4):e02072

Downing TA, Imo M, Kimanzi J, Otinga AN (2017) Effects of wildland fire on the tropical alpine moorlands of Mount Kenya. Catena 149:300–308

Dzwonko Z, Loster S, Gawroński S (2015) Impact of fire severity on soil properties and the development of tree and shrub species in a Scots pine moist forest site in southern Poland. Forest Ecol Manag 342:56–63

Fellows AW, Flerchinger GN, Lohse KA, Seyfried MS (2018) Rapid recovery of gross production and respiration in a mesic mountain big sagebrush ecosystem following prescribed fire. Ecosystems 21(7):1283–1294

Fernández-Fernández M, Gómez-Rey MX, González-Prieto SJ (2015) Effects of fire and three fire-fighting chemicals on main soil properties, plant nutrient content and vegetation growth and cover after 10 years. Sci Total Environ 515:92–100

Fernández-García V, Marcos E, Fernández-Guisuraga JM, Taboada A, Suárez-Seoane S, Calvo L (2019a) Impact of burn severity on soil properties in a Pinus pinaster ecosystem immediately after fire. Int J Wildland Fire 28(5):354–364

Fernández-García V, Miesel J, Baeza MJ, Marcos E, Calvo L (2019b) Wildfire effects on soil properties in fire-prone pine ecosystems: indicators of burn severity legacy over the medium term after fire. Appl Soil Ecol 135:147–156

Fernández-García V, Marcos E, Huerta S, Calvo L (2021) Soil-vegetation relationships in Mediterranean forests after fire. Forest Ecosyst 8:1–13

Francos M, Ubeda X (2021) Prescribed fire management. Curr Opin Environ Sci Health 21:100250. https://doi.org/10.1016/j.coesh.2021.100250

Francos M, Stefanuto E, Úbeda X, Pereira P (2019) Long-term impact of prescribed fire on soil chemical properties in a wildland-urban interface. Northeastern Iberian Peninsula. Sci Total Environ 689:305–311

Fraterrigo JM, Rembelski MK (2021) Frequent fire reduces the magnitude of positive interactions between an invasive grass and soil microbes in temperate forests. Ecosystems 24(7):1738–1755

Fultz LM, Moore-Kucera J, Dathe J, Davinic M, Perry G, Wester D, Schwilk DW, Rideout-Hanzak S (2016) Forest wildfire and grassland prescribed fire effects on soil biogeochemical processes and microbial communities: two case studies in the semi-arid Southwest. Appl Soil Ecol 99:118–128

Gao C, He J, Cong J, Zhang S, Wang G (2018) Impact of forest fires generated black carbon deposition fluxes in Great Hinggan Mountains (China). Land Degrad Dev 29(7):2073–2081

Goberna M, García C, Insam H, Hernández M, Verdú M (2012) Burning fire-prone Mediterranean shrublands: immediate changes in soil microbial community structure and ecosystem functions. Microb Ecol 64:242–255

González-Pelayo O, Gimeno-García E, Ferreira C, Ferreira A, Keizer JJ, Andreu V, Rubio JL (2015) Water repellency of air-dried and sieved samples from limestone soils in central Portugal collected before and after prescribed fire. Plant Soil 394:199–214

Granged AJP, Zavala LM, Jordán A, Bárcenas-Moreno G (2011) Post-fire evolution of soil properties and vegetation cover in a Mediterranean heathland after experimental burning: a 3-year study. Geoderma 164(1–2):85–94

Hahn GE, Coates TA, Aust WM (2021) Soil chemistry following single-entry, dormant season prescribed fires in the Ridge and Valley Province of Virginia, USA. Commun Soil Sci Plant Anal 52(17):2065–2073

Heydari M, Rostamy A, Najafi F, Dey DC (2017) Effect of fire severity on physical and biochemical soil properties in Zagros oak (Quercus brantii Lindl.) forests in Iran. J For Res 28:95–104

Hiers JK, O'Brien JJ, Varner JM, Butler BW, Dickinson M, Furman J, Gallagher M, Godwin D, Goodrick SL, Hood SM (2020) Prescribed fire science: the case for a refined research agenda. Fire Ecol 16:1–15

Hinojosa MB, Albert-Belda E, Gómez-Muñoz B, Moreno JM (2021) High fire frequency reduces soil fertility underneath woody plant canopies of Mediterranean ecosystems. Sci Total Environ 752:141877

Hobley EU, Brereton AJLG, Wilson B (2017) Forest burning affects quality and quantity of soil organic matter. Sci Total Environ 575:41–49

Hosseini M, Geissen V, González-Pelayo O, Serpa D, Machado AI, Ritsema C, Keizer JJ (2017) Effects of fire occurrence and recurrence on nitrogen and phosphorus losses by overland flow in maritime pine plantations in north-central Portugal. Geoderma 289:97–106

Hu M, Song J, Li S, Li Z, Hao Y, Di M, Wan S (2020) Understanding the effects of fire and nitrogen addition on soil respiration of a field study by combining observations with a meta- analysis. Agric Forest Meteorol 292:108106

Huddell AM, Galford GL, Tully KL, Crowley C, Palm CA, Neill C, Hickman JE, Menge DNL (2020) Meta-analysis on the potential for increasing nitrogen losses from intensifying tropical agriculture. Glob Chang Biol 26(3):1668–1680

Ibáñez TS, Wardle DA, Gundale MJ, Nilsson M-C (2022) Effects of soil abiotic and biotic factors on tree seedling regeneration following a boreal forest wildfire. Ecosystems 25(2):471–487

Inbar A, Lado M, Sternberg M, Tenau H, Ben-Hur M (2014) Forest fire effects on soil chemical and physicochemical properties, infiltration, runoff, and erosion in a semiarid Mediterranean region. Geoderma 221:131–138

James JA, Kern CC, Miesel JR (2018) Legacy effects of prescribed fire season and frequency on soil properties in a Pinus resinosa forest in northern Minnesota. Forest Ecol Manag 415:47–57
Jhariya M, Singh L (2021) Effect of fire severity on soil properties in a seasonally dry forest ecosystem of Central India. Int J Environ Sci Technol 18(12):3967–3978
Johnson DW, Walker RF, Glass DW, Stein CM, Murphy JB, Blank RR, Miller WW (2014) Effects of thinning, residue mastication, and prescribed fire on soil and nutrient budgets in a Sierra Nevada mixed-conifer forest. For Sci 60(1):170–179
Johnston MD, Barati M (2013) Calcium and titanium as impurity getter metals in purification of silicon. Separat Purif Technol 107:129–134
Kemp KB, Higuera PE, Morgan P (2016) Fire legacies impact conifer regeneration across environmental gradients in the US northern Rockies. Landsc Ecol 31:619–636
Knicker H (2007) How does fire affect the nature and stability of soil organic nitrogen and carbon? A review. Biogeochemistry 85(1):91–118
Knorr W, Arneth A, Jiang L (2016) Demographic controls of future global fire risk. Nat Clim Chang 6(8):781–785
Lal R (2015) Restoring soil quality to mitigate soil degradation. Sustain For 7(5):5875–5895
Li X-Y, Jin H-J, Wang H-W, Marchenko SS, Shan W, Luo D-L, He R-X, Spektor V, Huang Y-D, Li X-Y (2021) Influences of forest fires on the permafrost environment: a review. Adv Clim Chang Res 12(1):48–65
Li T, Cui L, Liu L, Chen Y, Liu H, Song X, Xu Z (2023) Advances in the study of global forest wildfires. J Soils Sedimen 23(7):2654–2668
Liu J, Qiu L, Wang X, Wei X, Gao H, Zhang Y, Cheng JJP, soil. (2018) Effects of wildfire and topography on soil nutrients in a semiarid restored grassland. Plant Soil 428:123–136
Lucas-Borja ME, Ortega R, Miralles I, Plaza-Álvarez PA, González-Romero J, Peña-Molina E, Moya D, Zema DA, Wagenbrenner JW, De las Heras J (2020) Effects of wildfire and logging on soil functionality in the short-term in Pinus halepensis M. forests. Eur J For Res 139:935–945
MacCarthy J, Richter J, Tyukavina S, Weisse M, Harris N (2023) The latest data confirms: Forest fires are getting worse. World Resources Institute. Available online: https://www.wri.org/insights/global-trends-forest-fires
Maksimova E, Abakumov E (2015) Wildfire effects on ash composition and biological properties of soils in forest–steppe ecosystems of Russia. Environ Earth Sci 74:4395–4405
Mao J, Nierop KG, Dekker SC, Dekker LW, Chen B (2019) Understanding the mechanisms of soil water repellency from nanoscale to ecosystem scale: a review. J Soils Sedimen 19:171–185
Mataix-Solera J, Cerdà A, Arcenegui V, Jordán A, Zavala LM (2011) Fire effects on soil aggregation: a review. Earth Sci Rev 109(1–2):44–60
Maynard D, Paré D, Thiffault E, Lafleur B, Hogg K, Kishchuk B (2014) How do natural disturbances and human activities affect soils and tree nutrition and growth in the Canadian boreal forest? Environ Rev 22(2):161–178
McCaffrey S, Toman E, Stidham M, Shindler B (2015) Social science findings in the United States. In: Wildfire hazards, risks and disasters (pp 15–34). Elsevier: Amsterdam, Netherlands
Meira-Castro A, Shakesby R, Espinha Marques J, Doerr S, Meixedo J, Teixeira J, Chaminé HI (2015) Effects of prescribed fire on surface soil in a Pinus pinaster plantation, northern Portugal. Environ Earth Sci 73:3011–3018
Middlemann MH, Middelmann M (2007) Natural hazards in Australia: identifying risk analysis requirements: geoscience Australia
Miesel JR, Hockaday WC, Kolka RK, Townsend PA (2015) Soil organic matter composition and quality across fire severity gradients in coniferous and deciduous forests of the southern boreal region. J Geophys Res Biogeo 120(6):1124–1141
Moritz MA, Batllori E, Bradstock RA, Gill AM, Handmer J, Hessburg PF, Leonard J, McCaffrey S, Odion DC, Schoennagel T (2014) Learning to coexist with wildfire. Nature 515(7525):58–66
Muqaddas B, Zhou X, Lewis T, Wild C, Chen C (2015) Long-term frequent prescribed fire decreases surface soil carbon and nitrogen pools in a wet sclerophyll forest of Southeast Queensland, Australia. Sci Total Environ 536:39–47

Neary DG, Ryan KC, DeBano LF (2005) Wildland fire in ecosystems: effects of fire on soils and water Gen. Tech. Rep. RMRS-GTR-42-vol 4(250):42

Panico SC, Ceccherini MT, Memoli V, Maisto G, Pietramellara G, Barile R, De Marco A (2020) Effects of different vegetation types on burnt soil properties and microbial communities. Int J Wildland Fire 29(7):628–636

Pellegrini AF, Ahlström A, Hobbie SE, Reich PB, Nieradzik LP, Staver AC, Scharenbroch BC, Jumpponen A, Anderegg WR, Randerson JT (2018) Fire frequency drives decadal changes in soil carbon and nitrogen and ecosystem productivity. Nature 553(7687):194–198

Pereira P, Úbeda X, Mataix-Solera J, Martin D, Oliva M, Novara A (2013) Short-term spatio-temporal spring grassland fire effects on soil colour, organic matter and water repellency in Lithuania. Solid Earth Discuss 5(2):2119

Perez-Izquierdo L, Clemmensen KE, Strengbom J, Granath G, Wardle DA, Nilsson MC, Lindahl BD (2021) Crown-fire severity is more important than ground-fire severity in determining soil fungal community development in the boreal forest. J Ecol 109(1):504–518

Plaza-Álvarez PA, Lucas-Borja ME, Sagra J, Moya D, Alfaro-Sánchez R, González-Romero J, De las Heras J (2018) Changes in soil water repellency after prescribed burnings in three different Mediterranean forest ecosystems. Sci Total Environ 644:247–255

Qiu L, Zhu H, Liu J, Yao Y, Wang X, Rong G, Zhao X, Shao M, Wei XJA, Ecosystems, and Environment. (2021) Soil erosion significantly reduces organic carbon and nitrogen mineralization in a simulated experiment. Agric Ecosyst Environ 307:107232

Raison R, Woods P, Jakobsen B, Bary G (1986) Soil temperatures during and following low- intensity prescribed burning in a Eucalyptus pauciflora forest. Soil Res 24(1):33–47

Reyes O, García-Duro J, Salgado J (2015) Fire affects soil organic matter and the emergence of Pinus radiata seedlings. Ann For Sci 72:267–275

Ribeiro-Kumara C, Pumpanen J, Heinonsalo J, Metslaid M, Orumaa A, Jõgiste K, Berninger F, Köster K (2020) Long-term effects of forest fires on soil greenhouse gas emissions and extracellular enzyme activities in a hemiboreal forest. Sci Total Environ 718:135291

Robichaud PR, Wagenbrenner JW, Pierson FB, Spaeth KE, Ashmun LE, Moffet CA (2016) Infiltration and interrill erosion rates after a wildfire in western Montana, USA. Catena 142:77–88

Rodriguez J, Gonzalez-Perez JA, Turmero A, Hernandez M, Ball AS, Gonzalez-Vila FJ, Arias ME (2018) Physico-chemical and microbial perturbations of Andalusian pine forest soils following a wildfire. Sci Total Environ 634:650–660

Rodríguez-Alleres M, Varela M, Benito E (2012) Natural severity of water repellency in pine forest soils from NW Spain and influence of wildfire severity on its persistence. Geoderma 191:125–131

Rodriguez-Cardona BM, Coble A, Wymore AS, Kolosov R, Podgorski D, Zito P, Spencer R, Prokushkin A, McDowell WH (2020) Wildfires lead to decreased carbon and increased nitrogen concentrations in upland arctic streams. Sci Rep 10(1):8722

Romeo F, Marziliano PA, Turrion MB, Muscolo A (2020) Short-term effects of different fire severities on soil properties and Pinus halepensis regeneration. J For Res 31:1271–1282

Roques S, Kendall S, Smith K, Price PN, Berry P (2013) A review of the non-NPKS nutrient requirements of UK cereals and oilseed rape: HGCA

Sánchez Meador A, Springer JD, Huffman DW, Bowker MA, Crouse JE (2017) Soil functional responses to ecological restoration treatments in frequent-fire forests of the western United States: a systematic review. Restor Ecol 25(4):497–508

Santín C, Otero XL, Doerr SH, Chafer CJ (2018) Impact of a moderate/high-severity prescribed eucalypt forest fire on soil phosphorous stocks and partitioning. Sci Total Environ 621:1103–1114

Sazawa K, Yoshida H, Okusu K, Hata N, Kuramitz H (2018) Effects of forest fire on the properties of soil and humic substances extracted from forest soil in Gunma, Japan. Environ Sci Pollut Res 25:30325–30338

Scharenbroch BC, Nix B, Jacobs KA, Bowles ML (2012) Two decades of low-severity prescribed fire increases soil nutrient availability in a Midwestern, USA oak (Quercus) forest. Geoderma 183:80–91

Shaltout RE, Ismail MA (2020) Simulation of fire dynamics and firefighting system for a full-scale passenger rolling stock. In: Sustainable rail transport (pp 209–228). Springer International Publishing

Singh S (2022) Forest fire emissions: a contribution to global climate change. Front Forests Global Change 5:925480

Switzer JM, Hope GD, Grayston SJ, Prescott CE (2012) Changes in soil chemical and biological properties after thinning and prescribed fire for ecosystem restoration in a Rocky Mountain Douglas-fir forest. Forest Ecol Manag 275:1–13

Thomaz EL, Fachin PA (2014) Effects of heating on soil physical properties by using realistic peak temperature gradients. Geoderma 230:243–249

Valkó O, Deák B, Magura T, Török P, Kelemen A, Tóth K, Horváth R, Nagy DD, Debnár Z, Zsigrai G (2016) Supporting biodiversity by prescribed burning in grasslands—a multi-taxa approach. Sci Total Environ 572:1377–1384

Varela ME, Benito E, Keizer JJ (2015) Influence of wildfire severity on soil physical degradation in two pine forest stands of NW Spain. Catena 133:342–348

Verma S, Jayakumar S (2012) Impact of forest fire on physical, chemical and biological properties of soil: a review. Proc Int Acad Ecol Environ Sci 2(3):168

Wang Y, Zheng J, Liu X, Yan Q, Hu Y (2020) Short-term impact of fire-deposited charcoal on soil microbial community abundance and composition in a subtropical plantation in China. Geoderma 359:113992

Weninger T, Filipović V, Mešić M, Clothier B, Filipović L (2019) Estimating the extent of fire induced soil water repellency in Mediterranean environment. Geoderma 338:187–196

Zavala LM, Granged AJ, Jordán A, Bárcenas-Moreno G (2010) Effect of burning temperature on water repellency and aggregate stability in forest soils under laboratory conditions. Geoderma 158(3–4):366–374

Zhang Y, Biswas A (2017) The effects of forest fire on soil organic matter and nutrients in boreal forests of North America: a review. In: Adaptive soil management: from theory to practices (pp 465–476) Springer, Singapore

Land Management for Biodiversity Conservation and Sustainable Development Goals

9

Sodgerel Purevee, Tserennadmid Bataa, Amarmend Chuluunbaatar, Odbayar Otgonbaatar, Aakash Kumar, and Harish Bahadur Chand

Abstract

The degradation of land due to unsustainable practices is a key factor in biodiversity loss and environmental degradation. In Mongolia, overgrazing, land use changes, and climate impacts are increasing desertification and land degradation, disrupting not only natural ecosystems but also deeply impacting the livelihoods and socio-economic well-being of rural communities. This chapter focuses on the role of sustainable land management (SLM) in overcoming these challenges, emphasizing how Mongolia's rangelands, forests, and agricultural lands are undergoing ecological changes. The environmental sustainability index (ESI) and other land use assessments show the urgent need for integrated land management practices that balance human activities with ecological preservation. Through case studies, such as the Darkhan-Uul province, this chapter reflects the growing pressures on land resources and the important steps needed to rehabilitate ecosystems while supporting rural development. SLM practices can play a key role in achieving the United Nations' Sustainable Development Goals, particularly those related to life on land, water conservation, and climate action.

S. Purevee · T. Bataa · A. Chuluunbaatar · O. Otgonbaatar
School of Agroecology and Business, Mongolian Agricultural University of Life Science, Ulaanbaatar, Mongolia

A. Kumar (✉)
Forest Research Institute (Deemed to be) University, Dehradun, Uttarakhand, India

Himalayan Institute for Learning and Leadership (HILL), UPES, Dehradun, Uttarakhand, India

H. B. Chand
Forest Ecology and Climate Change Division, Forest Research Institute, Dehradun, Uttarakhand, India

S. Singh, M. K. Gautam (eds.), *Biodiversity Conservation and Land Degradation Neutrality*, https://doi.org/10.1007/978-981-96-7210-3_9

Keywords

Sustainable land management · Biodiversity conservation · Land degradation · Environmental sustainability index · Rangeland degradation

9.1 Introduction

Land use degradation is one of the biggest causes of biodiversity loss. Deforestation, urbanization, agricultural expansion, and other anthropogenic activities have greatly altered and fragmented our landscapes, including forests and grasslands (IPCC 2018). Such disruptions to the land bring about changes in the global atmospheric concentration of carbon dioxide and affect local, regional, and global climates by changing the energy balance on Earth's surface. Such changes feedback on biodiversity, leading to further degradation and deterioration in the health of terrestrial ecosystems (IPCC 2018; Dash 2008).

More than 90% of Mongolia's territory is vulnerable to desertification because of a rise in human activities leading to land use change and climate change (Dash 2008). Approximately 1.24 million square kilometers of grassland or natural areas disappear every year in Mongolia, as it is converted into artificial land (Chantsallkham 2009).

Studies on land degradation indicate that its severity and extent are increasing in many parts of Mongolia, particularly due to overgrazing, which has degraded more than 72% of the country's grasslands and rangelands (FAO 2023; Wang et al. 2019). This degradation poses significant challenges to the health of Mongolian grasslands and rangelands, threatening the environmental conditions that support their productivity and biodiversity. Land degradation and land use change result in widespread loss of native vegetation, causing substantial environmental damage and significant social and economic harm. These changes have direct and indirect impacts on a large portion of the Mongolian population. To address these challenges, comprehensive plans must be developed, and environmental protection measures must be implemented, focusing on sustainable land management (SLM) practices to achieve land degradation neutrality (LDN) and promote ecosystem health.

The shift in land utilization in Mongolia has led to positive developments in rural areas and their socio-economic conditions. However, this has also had adverse effects on ecosystem services, largely due to land use planning (LUP) being based on poorly designed scientific plans. Traditional methods of managing natural ecosystems are now facing new challenges brought on by land degradation, requiring a new balance between rural development and natural resource conservation. The significant role of land use change and degradation in contributing to climate change in Mongolia remains a critical gap in global warming mitigation efforts, which primarily focus on reducing emissions of greenhouse gases. Mitigating climate change could be more effectively achieved by regulating land use change and neutralizing land degradation, rather than relying solely on emissions reductions (ALAGAC 2008, 2023). Even small shifts, such as 100 square kilometers of urban expansion or deforestation, can disrupt local rainfall patterns and trigger a chain of climate-related disturbances. Thus, both scientific research and public policy must evolve to address all components of human-induced climate change (ALAGAC 2008, 2023).

According to the Food and Agriculture Organization (FAO) (2023), the productivity and sustainability of a land use system depend on the dynamic interaction between land resources, climate, and human activities. In the context of climate change and variability, it is crucial to select appropriate land uses that align with specific biophysical and socio-economic conditions. Additionally, adopting SLM practices plays a vital role in minimizing land degradation, restoring degraded landscapes, promoting the sustainable utilization of land resources (such as soil, water, and biodiversity), and enhancing resilience against environmental challenges (ALGAC 2023).

SLM is essential for attaining the United Nations' Sustainable Development Goals (SDGs) and several other ongoing initiatives. Responsible land management for the conservation of biodiversity holds the potential to contribute significantly to achieving several SDGs. Life on land (SDG 15), life below water (SDG 14), clean water (SDG 6), and climate action (SDG 13) are just a few examples.

9.1.1 Land Use Change in Mongolia

Mongolia is situated at the hinterland of the Mongolian Plateau, which has a varied, fragile geographical environment in arid and semi-arid areas (Densambuu et al. 2018). Understanding the land cover patterns and the changes that have been happening to it is of great significance for the natural resources, environment, ecology, and sustainable development of the Mongolia Plateau and Northeast Asia (CEIC n.d.).

Agricultural land dominates Mongolia's landscape, covering 134.1 million hectares, which accounts for 85.7% of the total land area. Urban settlements occupy 1.23 million hectares (0.8%), while roads and networks span 0.87 million hectares (0.6%). Forests cover 18.2 million hectares (11.6%), and water bodies encompass 20.2 million hectares (1.3%) (Hao et al. 2023) (Fig. 9.1).

Between 1961 and 2021, the areas of forests, permanent meadows, and pastures showed a declining trend (CEIC n.d.; Tsalis et al. 2022). In contrast, the areas of arable and cropland, urban land, bare land, and paved roads exhibited an increasing trend (CEIC n.d.; Tsalis et al. 2022; Wang et al. 2022) (Figs. 9.2 and 9.3).

Several studies have reported a declining trend in Mongolia's vegetative cover, accompanied by severe desertification (Sleeter et al. 2018; FAO 2023; Wang et al. 2019; CEIC n.d.). Forest land has decreased by 2.5% (Tsalis et al. 2022). Some findings are consistent with those of Tsalis et al. (2022) and Tsogtbaatar (2004) but differ from Metternicht (2017). This variation may result from differences in methodology, as Metternicht (2017) used the NDVI threshold method to classify land types, a process influenced by factors such as time, imagery, and location. Key drivers of land use changes include livestock grazing, natural disasters, socio-economic transitions, and the overexploitation of natural resources.

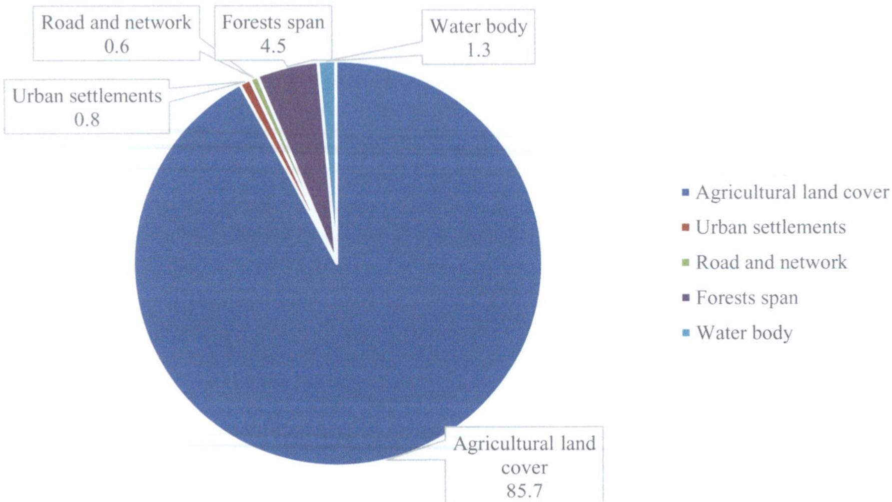

Fig. 9.1 Land use distribution in Mongolia (2023)

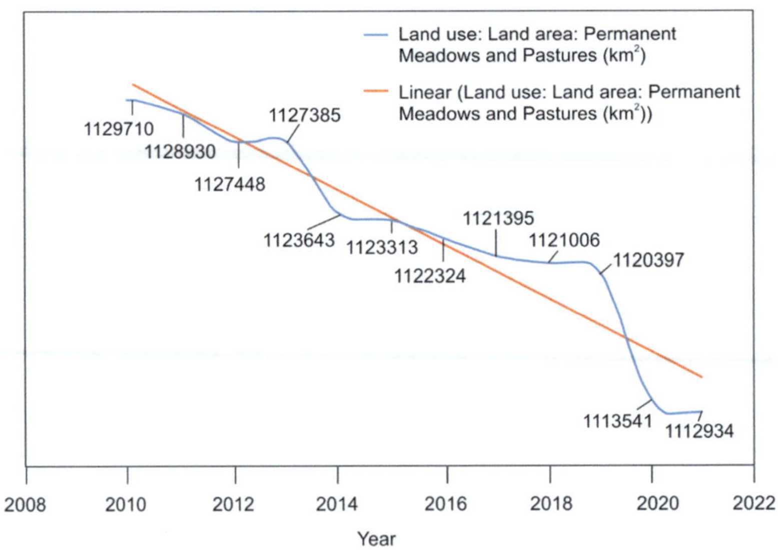

Fig. 9.2 Land use changes of permanent meadows and pasture. (Source: CEIC data Mongolia)

9.1.2 Contribution of LUP to Sustainable Land Utilization and Management

There is a growing global recognition of the need for more sustainable strategies in planning and managing natural landscapes. Developing innovative tools to effectively incorporate sustainable principles into the management and planning of natural systems has become essential. The spatial aspect of sustainability highlights the

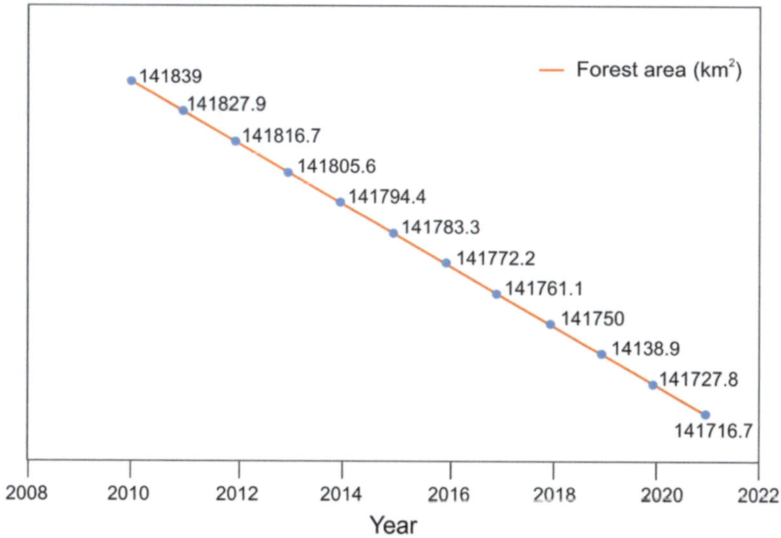

Fig. 9.3 Trend of forest area (2010–2021). (Source: CEIC data Mongolia)

dynamic processes and interconnections between diverse land uses, ecosystems, and habitats across various scales and timeframes (Sanz et al. 2017). As a result, ecological knowledge is crucial in planning for sustainability. The FAO (2023) emphasizes that SLM should incorporate scientifically validated methods for soil and water conservation, as well as natural resource management, within the framework of integrated landscape management (ILM) (ALGAC 2008). ILM adopts a holistic approach to fostering productive and resilient ecosystems by balancing social, economic, physical, and biological needs and values. Through this integration, ILM contributes to SDGs by promoting ecosystem health and long-term sustainability (ALGAC 2023).

Land management involves LUP and policy. Together, these two elements help in effective land resource management to ensure sustainable development and the well-being of communities and ecosystems (Volkov 1998). LUP, spatial planning, regional or territorial planning, and the more recent ecosystem-based or environmental LUP all contribute to promoting sustainable land use and effective resource management (Volkov 1998). The LUP in Mongolia is implemented as the following activities:

- *State land use master planning (16–20 years):* A systematic evaluation of land and water potential, land use alternatives, and socio-economic conditions is essential for identifying and implementing the most suitable land use options at the national level.
- *Municipal LUP (12–16 years):* Interdisciplinary and comprehensive approach aimed at fostering balanced regional development and structuring the physical organization of space in accordance with a long-term, overarching strategic framework.

- *LUP for the current year:* Spatial development of a specified territory, determining priorities for land use, ensuring protection of resources, and establishing foundational principles for effective territorial management.
- *Rangeland management planning:* Rangeland management planning in pastoral areas focuses first on large-scale grazing patterns, which can lead to substantial improvements in feed availability.
- *Urban development master plan:* Determine areas for urban–industrial development and infrastructure.

  The shortcomings of this planning are the following issues:
- The LUP system is weak in its ability to ensure increasing attention to natural conservation and ecological sustainability. Currently, socio-economic interests dominate contemporary LUP, which prioritizes baseline studies, spatial socio-economic considerations, and infrastructure needs over natural conservation efforts. But this perspective has shifted land system science from focusing solely on dramatic land-cover changes to emphasizing subtle transformations in human interactions with the natural environment, this approach encompasses critical aspects such as land management and the enhancement of diverse ecosystem services.
- There is a lack of ILM planning in the land use system stemming from insufficient participation and coordination among multiple key stakeholders from related sectors, such as the Department of Agriculture, Environment and Water Basin Administration, Forest Society, and others.
- The current LUP systems struggle to comprehensively address local issues, including their causes, effects, vulnerabilities, and potential risks, alongside outputs, responsibilities, key performance indicators, required investment, and ecological outcomes.

Ineffective land management in Mongolia stems from flawed decisions by stakeholders at various levels, resulting in significant impacts on the ecosystems of the Mongolian Plateau. LUP serves as a critical tool for balancing competing and often conflicting land uses while fostering sustainable practices. Evidence underscores the importance of LUP, spatial planning, territorial planning, and ecosystem-based or environmental LUP as effective strategies for improving land governance. These approaches enhance economic opportunities through sustainable land resource management and support the development of land use options that align conservation goals with development objectives (Volkov 1998).

9.1.3 Environmental Sustainability Indexing; A Measure of the Balance Between Anthropogenic Pressure and Ecological Relevance; A Representative Case Study in Mongolia

SLM embodies a comprehensive approach aimed at maintaining ecosystem services within long-term productive ecosystems. This is achieved by integrating biophysical, socio-cultural, and economic considerations and values. Scientific research indicates that widespread adoption of SLM practices can play a pivotal role in preventing, reducing, or reversing land degradation while promoting LDN. Additionally, these practices contribute to climate change adaptation and mitigation, safeguard biodiversity, support the attainment of multiple SDGs, and enhance human well-being (NSOM n.d.). Growing evidence highlights the significant benefits of implementing SLM technologies and practices as land-based solutions. These approaches incorporate the principles of both the ecosystem-based approach and landscape approaches (Densambuu et al. 2018; ALGAC 2008; Sanz et al. 2017). By focusing on these methods, it is possible to simultaneously improve food security and livelihoods, protect or enhance biodiversity, facilitate adaptation to climate change, and contribute to climate change mitigation efforts (NSOM (n.d.)).

This study employs the environmental sustainability index (ESI) (Densambuu et al. 2018; Sanz et al. 2017) to assess land use and ecosystem stability, a recently developed index for analyzing environmental value based on land use. The ESI quantifies the environmental value of a specific area by considering its land use characteristics. This index is particularly useful for land use assessments and can be applied to evaluate the effectiveness of corrective measures implemented through territorial, urban, or environmental planning instruments (Densambuu et al. 2018). Researchers such as Densambuu et al. (2018) and Sodgerel (2019) employed comprehensive methodologies and conducted insightful analyses in this regard. Land cover categories are multiplied by individually selected reduction coefficients to account for the ecological importance ("quality") of various land cover types. The ESI ranges from 0 (for heavily anthropogenic areas) to 1 (for ecologically valuable areas). Thus, a higher ESI indicates greater ecological relevance in the area.

The same methodology was applied to analyze land use change in Darkhan-Uul province across different time periods. Using data from 2008, 2023, and 2030, the methodology proved effective in analyzing past trends, assessing current conditions, and predicting future land use shifts. The assessment of the environmental sustainability of Darkhan-Uul province shows a moderate sustainability trend. However, the sustainability index scores decreased from 0.65 in 2008 to 0.63 in 2023. The results suggest a further decline to 0.55 in 2030 (Fig. 9.4). The presence of diverse land use in the region likely affected ecological sustainability, but the primary reason for the decline in the sustainability index is the increasing human-induced transformations of Darkhan-Uul province's ecosystems and landscapes, leading to changes in land cover.

Most developed countries such as Russia, Germany, and Eastern Europe are prioritizing transforming barren land into green areas rather than construction and industrialization, supported by citizen initiatives and public organizations (Densambuu et al. 2018). Some Eastern European countries such as Austria and Poland have prioritized creating special protected areas (SPAs), with 20% of the total area allocated under these SPAs to maintain the ecological–landscape balance (Densambuu et al. 2018). Particularly, preservation and SPAs provide ecological education, making it a more integrative approach that focuses on both social and ecological systems (Densambuu et al. 2018; Sodgerel 2019].

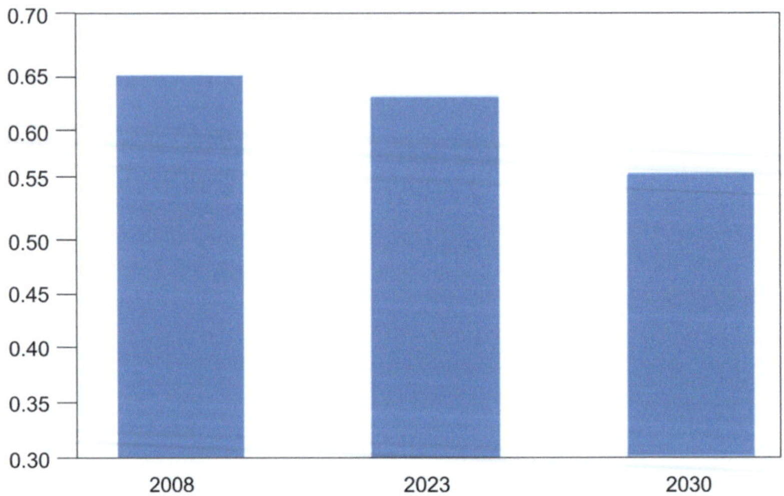

Fig. 9.4 Sustainability assessment of land use change in Darkhan-Uul province, Mongolia

The countries have different land use status classification systems. Densambuu et al. (2018) suggested that maintaining the appropriate proportion of land use in forest and forest-steppe zones is crucial for preserving landscape–ecological balance and maximizing the economic and ecological benefits of natural resources. Their suggestion for proportionate allocations is as follows:

- Agricultural land including rural settlement—22.5%.
- Non-Agricultural land including industry, road transport, and others—2.5%.
- Forest land—18%.
- Natural protected land—57%.

This proportion represents the optimal ratio of land use that should be envisioned when strategizing for sustainable LUP.

Current land use allocations in Darkhan-Uul province are very different from the proportion outlined above. There is 8.8% non-agricultural land, 67% agricultural land, and 21.9% forest land, of which only 0.67% are locally protected land (Nyambayar et al. 2011).

The human-induced land use index for Darkhan-Uul province suggests that human footprints are expected to increase significantly until 2030 in the province (Fig. 9.5). The decline over a period indicates declining ecological sustainability in the province.

9.1.4 The Impact of Land Use Change: Rangeland Degradation Case

Healthy rangelands offer the greatest variety of ecosystem services and functions. The loss of rangeland health through rangeland degradation leads to loss of

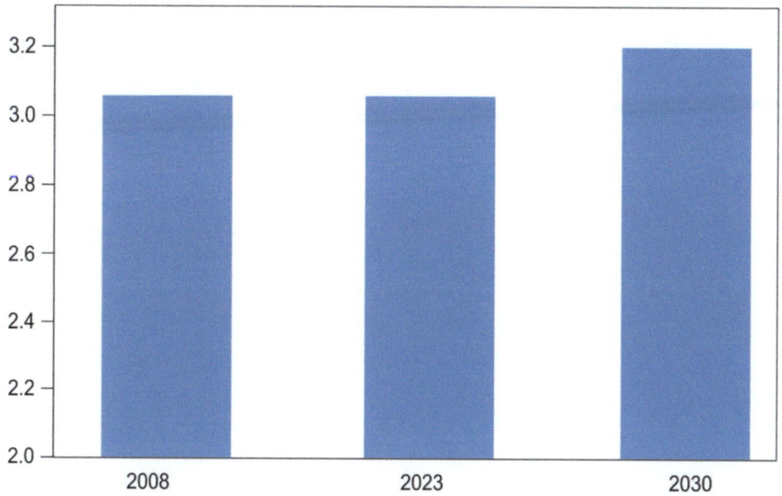

Fig. 9.5 An assessment of human-induced land use in Darkhan-Uul province, Mongolia

ecosystem services and functions. Over 90% of the total area in Mongolia is vulnerable to desertification as a result of climate change, land use, and human activities (FAO 2023). Studies have suggested that land degradation is increasing in severity and extent in many parts of Mongolia. Especially, in the case of grassland-rangeland, more than 72% of the total area is undergoing degradation, mainly from overgrazing (Wang et al. 2019). This is a challenge to the set of environmental conditions that sustain the productivity and biodiversity of Mongolian grassland and rangeland ecosystems.

The number of livestock in Darkhan-Uul province has increased dramatically since 2008 (Nyambayar et al. 2011), and it is estimated to double by 2030 (Oyuntsetseg et al. 2018). Increasing livestock numbers beyond the carrying capacity of rangeland can lead to acute pressure on these fragile ecosystems which will further exacerbate the rangeland degradation. It was seen in Darkhan-Uul province that the degree of degradation differs among seasonal pastures and ecological site groups (Fig. 9.6). An ecological site represents a conceptual categorization of the landscape, defined as a unique type of land distinguished by recurring characteristics such as soil properties, landforms, geological features, and climatic conditions. This differentiation is based on the site's capacity to support specific types and quantities of vegetation and its consistent response to management interventions and natural disturbances (Wang et al. 2019).

The plant community is altered to a negative trend and may be rapidly recovered (three to five growing seasons) with favorable climatic conditions or a change in management (e.g., stocking rate reduction, seasonal deferment, and rotation). Most of the summer rangeland represents a significant loss of important ecosystem services (and are clearly related to anthropogenic drivers) and the temporal aspects of the potential recovery.

Fig. 9.6 (**a** and **b**) Heavy degraded rangeland for grazing summer–autumn season in Orkhon sum, Darkhan-Uul province. (**c** and **d**) Heavy degraded rangeland for grazing summer–autumn season in Khongor sum, Darkhan-Uul province In winter–spring, pastures are influenced by perennial grasses, with Stipa grass being particularly diverse and densely covering the area. However, in summer and autumn, although vegetation cover reaches 78–80% in some areas, there is a decline in mean grass cover and increases in sedges and *Artemisia adamsii* (semi-shrub) cover. Aboveground biomass, however, generally does not vary significantly between seasons because the increase in grazing-tolerant species compensates decline in grazing-sensitive species (Fig. 9.6). But increase in sedges and *Artemisia* sp. is a key rangeland degradation indicator, which was shown in a set of monitoring sites (Wang et al. 2019).

As a first step toward sustainable rangeland management, the Green Gold Project of the Swiss Agency for Development and Cooperation partnered with government ministries and universities to develop a comprehensive approach known as resilience-based rangeland management. This initiative aims to promote management changes and conduct a detailed, field-oriented assessment of rangeland health across Mongolia. The application of internationally accepted procedures for calculating a resilient carrying capacity for Mongolian rangeland, along with a recent national study, shows that the carrying capacity has been greatly exceeded in many areas, with overgrazing identified as a primary factor contributing to rangeland degradation in Mongolia (Wang et al. 2019). High grazing pressure severely disrupts the rangeland ecosystem and causes significant changes in vegetation structure (FAO 2023). During the summer, herders migrate along riverbanks, grazing in one place for extended periods without rotating pastures, which is a major cause of degradation. As a result, these disturbances lead to changes in vegetation communities, affecting species numbers, species richness, vegetation cover, biomass, and the yield of beneficial pasture plants.

A list of 32 extremely rare and rare plants found in the territory of Darkhan-Uul province has been compiled from the 133 species of extremely rare plants with restricted use listed in the Law on Natural Plants of Mongolia (Table 9.1).

Eleven species of extremely rare plants such as *Solidago dahurica, Gentiana macrophylla, Vincetoxicum sibiricum, Lilium martagon, Cypripedium guttatum, Cypripedium macranthum, Cypripedium calceolus, Rhodiola rosea, Hedysarum fruticosum, Rhododendron dahuricum, and Caryopteris mongolica*, along with 21 species of rare plants such as *Valeriana officinalis, Saposhnikovia divaricata, Trifolium repens, Adonis sibirica, Anemone reflexa, Paeonia anomala, Stellaria dichotoma, Sparganium glomeratum, Chelidonium majus, Allium victorialis, Cacalia hastata, Gentiana barbata, Ephedra sinica, Glycyrrhiza uralensis, Polygonum viviparum, Pteridium aquilinum,* and *Malus baccata*, have been distributed in the flora of Darkhan-Uul province. Three species are included in the International Union for Conservation of Nature (IUCN) Red Data Book and five species are included in the Red List of Mongolia. Medicinal and honey-pollinated plant species comprise 37–41% of the flora of Darkhan-Uul province. In recent years, there has been an observed decline in the distribution of medicinal and other useful plant species, leading toward rarity. The decline in the distribution of these plants is due to local citizens and enterprises collecting and using important medicinal and food plants, as well as the overlap of areas with rare and endangered plants with pastures, hayfields, and mining zones.

The following recommendations are made for the protection of local and regional floristic diversity ensuring the preservation of vegetation cover to meet SDGs as in Table 9.2.

9.2 Conclusion and Recommendation

Integrated LUP and SLM are essential for preserving natural ecosystems, maintaining biodiversity, and ensuring the sustainability of ecosystem services. In Mongolia, addressing ecological degradation must be prioritized alongside socio-economic considerations to achieve a balanced approach to land management. Different land use types, such as pastures, forests, hayfields, and water bodies, play a crucial role in local ecosystem health and sustainability. However, urgent management reforms are needed to restore degraded rangelands, enhance resilience to climate and land use changes, and ensure food security and pastoral production. Some of the recommendations are as follows:

- Tailor land use structures to local conditions to ensure environmental sustainability.
- Foster collaboration across sectors to integrate ecological expertise, responsibilities, and performance indicators into land management frameworks.

Table 9.1 The status of very rare and rare plants of use and their distribution in Darkhan-Uul Province

Name of rare and very rare species	Status					Forms of use					
	Very rare	Rare	Red book	Red list	Sub endemic	Medical plants	Food plants	Ornamental	Trading	Pasture and grassland	Hay field
Solidago dahurica Kitag.	+					+					
Gentiana macrophylla Pall.	+					+					
Gentiana barbata Froel.		+				+		+			
Caryopteris mongolica Bge.	+			+	+	+					
Vincetoxicum sibiricum (L.) Decne.	+					+					+
Lilium martagon L.	+					+					
Rhodiola rosea L.	+			+		+					
Cypripedium macranthum Sw.	+		+	+		+					
Cypripedium guttatum Sw.	+					+					
Cypripedium calceolus L.	+		+	+		+					
Hedysarum fruticosum Pall.	+					+				+	
Rhododendron dahuricum L.	+					+					
Valeriana officinalis L.		+				+					
Adonis sibirica Patr.ex Ldb.		+				+					
Anemone reflexa Steph.		+									
Paeona anomala L.		+		+		+					
Stellaria dichotoma L.		+				+				+	+
Trifolium repens L.		+				+				+	+
Chelidonium majus L.		+				+					
Allium anisopodium L.		+				+	+			+	+
Allium odorum L.		+				+	+			+	+
Allium victorialis L.		+				+	+		+		
Cacalia hastata L.		+				+			+		
Ephedra sinica Stapf.		+				+			+		

Glycyrrhiza uralensis Fisch.		+				+			+	+	+
Iris flavissima Pall.		+				+					
Malus baccata (L.) Borkh.		+				+		+			
Polygonum viviparum L.		+				+	+			+	+
Saposhnikovia divaricata Schischk.		+	+			+			+	+	+
Sedum aizoon L.		+				+					
Sparganium glomeratum Laest.		+									
Pteridium aquilium (L.) Kuhn.		+				+					
Total	11	21	3	5	1	30	4	2	5	8	8

Table 9.2 SDGs, action required, and implementation

SLM to SDGs 15	Actions required	Implementation
Protection and land conservation SDG 15.4: Ensure conservation of mountain ecosystems SDG 15.5: Protect biodiversity and natural habitats SDG 15.8: Prevent invasive alien species on land and in water ecosystems	Maintain the long-term adaptive opportunities for vulnerable biodiversity to climate change by increasing SPAs through the better management of protected areas' border and connectivity.	On December18, 2015, according to the resolution of the meeting of citizens' representatives of Khongor soum, Darkhan-Uul province, 10,313 hectares of Khustiin bulag and 14604.09 hectares of Songino and 20,000 hectares of Muul Mountains were taken under local protection.
SDG 15.7: Eliminate poaching and trafficking of protected species SDG 15.6: Promote access to genetic resources and fair sharing of the benefits	To ensure the appropriate use of protected species and take urgent action to end poaching and trafficking of protected species of rare flora on the responsibilities of forest communities in Darkhan-Uul province.	Environmental management plan is being developed based on the plant census. The prices for harvesting plants and medicinal purposes have been revised.
SDG 15.3: End desertification and restore degraded land SDG 15.9: Integrate ecosystem and biodiversity in governmental planning	Resilience-based rangeland management and planning are essential for the protection of local species and biodiversity. Increase the productivity of the animal husbandry sector while ensuring the sustainable development of the sector and reducing the impacts and risks associated with climate change.	Range LUP was developed by the ADB project in 2019, but lack of the implementation is a challenge.
SDG 15.1,3,4,5,8,9	Integrated LUP is focused on natural conservation issues to protect biodiversity. Planting native species for degraded land for biological restoration.	Local government addressed that restriction of the mining area and technical restoration of 187 hectares of damaged areas were carried out.

- Prioritize strengthening conservation efforts, especially for endangered species and fragile ecosystems.
- Implement comprehensive mitigation strategies for mining, urbanization, and other land use activities to reduce land degradation.
- Incorporate ecological priorities into LUP and SLM to align ecological objectives with SDGs.
- Promote long-term environmental resilience and resource sustainability through these integrated approaches.

References

Administration of Land Affairs, Geodesy, and Cartography (ALAGAC) (2008) Report of land use in Mongolia, Ulaanbaatar

Administration of Land Affairs, Geodesy, and Cartography (ALAGAC) (2023) Report of land use in Mongolia, Ulaanbaatar

Chantsallkham J (2009) Sustainable rangeland management in Mongolia: the role of herder community institutions. Land Restoration Training Programme Final Project, Reykjavík

Dash D (2008) Land use principles and methodology, vol 1. UBALAGAC

Densambuu B, Sainnemekh S, Bestelmeyer B, Budbaatar U (2018) National report on the rangeland health of Mongolia. Green Gold-Animal Health Project, SDC, and Mongolian National Federation of Pasture User Groups, Ulaanbaatar

FAO (2023) Sustainable land management. FAO of the UN. Retrieved September 8, 2024, from https://www.fao.org/land-water/land/sustainable-land-management/en/

Global Economic Data, Indicators, Charts & Forecasts (n.d.) CEIC data Mongolia land use: Land area. Retrieved from https://www.ceicdata.com/

Hao J, Lin Q, Wu T, Chen J, Li W, Wu X, Hu G, La Y (2023) Spatial–temporal and driving factors of land use/cover change in Mongolia from 1990 to 2021. Remote Sens 15(7):1813. https://doi.org/10.3390/rs15071813

IPCC (2018) Global Warming of 1.5°C. An IPCC Special Report on the impacts of global warming of 1.5°C above pre-industrial levels and related global greenhouse gas emission pathways, in the context of strengthening the global response to the threat of climate change, sustainable development, and efforts to eradicate poverty. In: Masson-Delmotte V, P Zhai, H-O Pörtner, D Roberts, J Skea, PR Shukla, A Pirani, W Moufouma-Okia, C. Péan R Pidcock S Connors, JBR Matthews, Y Chen, X Zhou, MI Gomis, E Lonnoy T Maycock, M Tignor, and T Waterfield (eds.), In Prss. Intergovernmental Panel on Climate Change, Geneva, Switzerland. Online available at: https://www.ipcc.ch/site/assets/uploads/sites/2/2019/06/SR15_Full_Report_High_Res.pdf

Metternicht G (2017) LUP, 1st edn. UNCCD, pp 24–34. https://www.unccd.int/resources/global-land-outlook/overview

National Statistics Office of Mongolia (NSOM) (n.d.) Statistical tables on land use in Mongolia. https://www.1212.mn/mn/statistic/statcate/573054/table-view/DT_NSO_1001_021V1

Nyambayar D, Oyuntsetseg B, Tungalag R (2011) Mongolian red list and conservation action plans of plants. Admon Printing, Ulaanbaatar

Oyuntsetseg B, Baasanmunkh S, Nyambayar D, Batkhuu NO, Lee CK, Chung GY, Choi HJ (2018) The conservation status of 100 rare plants in Mongolia. Korea National Arboretum, Pocheon

Sanz MJ, de Vente J, Chotte J-L, Bernoux M, Kust G, Ruiz I, Almagro M, Alloza J-A, Vallejo R, Castillo V, Hebel A, Akhtar-Schuster M (2017) Sustainable land management contribution to successful land-based climate change adaptation and mitigation. United Nations Convention to Combat Desertification (UNCCD), Bonn

Sleeter BM, Loveland T, Domke G, Herold N, Wickham J, Wood N (2018) Land cover and land-use change. In: Reidmiller DR, Avery CW, Easterling DR, Kunkel KE, Lewis KLM, Maycock TK, Stewart BC (eds) Impacts, risks, and adaptation in the United States: fourth National Climate Assessment, volume II. U.S. Global Change Research Program, pp 202–231. https://doi.org/10.7930/NCA4.2018.CH5

Sodgerel P (2019) Development of rangeland management involved public communities. Asian Development Bank (ADB), Project of Higher Education Reform, Darkhan HAAIS

Tsalis TA, Malamateniou KE, Koulouriotis D, Nikolaou IE (2022) Land cover change analysis to assess sustainability of development in the Mongolian Plateau over 30 years. Sustainability 14:6129. https://doi.org/10.3390/su14106129

Tsogtbaatar J (2004) Deforestation and reforestation needs in Mongolia. In: Forest ecology and management (Vol. 201, Issue 1, pp. 57–63). Elsevier BV. https://doi.org/10.1016/j.foreco.2004.06.011

Volkov SN (1998) Land management in the context of land reform (pp 209–216)
Wang J, Cheng K, Liu Q, Zhu J, Ochir A, Davaadorj D, Li G, Wei H, Sonomdagva C, Oyunchimeg N, Baterdene A (2019) Land cover patterns in Mongolia and their spatiotemporal changes from 1990 to 2010. Arab J Geosci 12:4893. https://doi.org/10.1007/s12517-019-4893-z
Wang J, Wei H, Cheng K, Ochir A, Shao Y, Yao J, Wu Y, Han X, Davaasuren D, Chonokhuu S (2022) Updatable dataset revealing decade changes in land cover types in Mongolia. Geosci Data J 9:341–354. https://doi.org/10.1002/gdj3.123

Role of Sustainable Land Management Practices in Biodiversity Conservation

10

Moses Fayiah and Abhishek Kumar

Abstract

Sustainable land management (SLM) practices are vital for upholding the balance between biodiversity conservation and effective land resource productivity. SLM practices have the potential to support different forms of life by protecting natural ecosystems from destruction and degradation. Biodiversity is considered a critical element for safeguarding food security and enhancing livelihoods, especially for communities living on the forest edge. This chapter aims to explore the role of SLM practices in supporting biodiversity conservation through habitat and ecosystem restoration. A more effective conceptual approach to SLM can enhance land productivity and support biodiversity conservation. Popularizing SLM concepts and practices will support livelihood activities and stimulate biodiversity. There is a complex nexus between SLM, biodiversity, and sustainable development goals (SDGs). These interconnected linkages are crucial for supporting food security, maintaining ecological balance, and promoting the sustainable development of ecosystems. Out of the 17 SDGs, seven are closely linked to land resource productivity, while SDG 15 is dedicated to supporting life on Earth by addressing land degradation, deforestation, and biodiversity decline through integrated farming practices. This chapter delves into the linkage between SLM and biodiversity and how food security is associated with SLM

M. Fayiah (✉)
Department of Forestry and Wood Science, School of Natural Resources Management, Njala University, Njala, Sierra Leone
e-mail: mfayiah@njala.edu.sl

A. Kumar
Department of Biotechnology, Graphic Era (Deemed to Be University), Dehradun, Uttarakhand, India
e-mail: abhishek259kumar@geu.ac.in

S. Singh, M. K. Gautam (eds.), *Biodiversity Conservation and Land Degradation Neutrality*, https://doi.org/10.1007/978-981-96-7210-3_10

practices. It also talks about how sustainable farming, agriculture, habitat restoration, and wetland conservation contribute to preserving biodiversity.

Keywords

Sustainable · Land · Biodiversity · Ecosystem · Management · Conservation

10.1 Introduction

Sustainable land management (SLM) practices play a crucial and important role in biodiversity conservation by promoting the health and resilience of ecosystems, conserving habitats, and supporting the various array of species that depend on these environments (Izquierdo-Tort et al. 2023). Biodiversity conservation is greatly influenced by SLM practices approaches and policies of any region. SLM actions and policies create opportunities for biodiversity to increase and flourish in any given landscape. The survival of humans and the prosperity of global economies greatly rely on sustainable land resources such as water, soil, minerals, and biodiversity among others. However, land resources are being threatened by massive land degradation due to different land use pressures as a result of population increase and industrialization. According to the United Nations Environment Programme (UNEP 2021a), about 12 million productive lands are being degraded on a yearly basis, while two billion hectares of lands have been degraded globally. This degradation has affected 20–40% of land use such as forests, wetlands, and croplands (UNEP 2021a, b).

An increase in ecosystem services results from an increase in biodiversity conservation and sustainability of natural resources. SLM is a holistic approach to managing land resources in a way that balances the need for environmental health, economic profitability, and social equity. It involves practices that protect soil quality, preserve biodiversity, and maintain ecosystem services while supporting the livelihoods of communities. SLM is crucial because it helps combat land degradation, ensures food security, and mitigates the impacts of climate change. By promoting the efficient use of natural resources, SLM enhances resilience against environmental challenges and supports the long-term sustainability of ecosystems, ultimately contributing to the achievement of sustainable development goals (SDGs). SLM techniques have the ability to lead the way in the preservation of rich biodiversity, particularly in natural areas (Izquierdo-Tort et al. 2023).

10.1.1 Biodiversity Importance in the Twenty-First Century

The rapid advancement of science and technology makes the conservation of biodiversity a critical priority for future generations. For instance, deforestation, population growth, and urbanization serve as a great threat to biodiversity increase and preservation. Regardless of these threats, biodiversity still plays a fundamental role in livelihood and survival, especially in forest edge communities (Grumbine and Xu 2021; Fayiah and Fayiah 2022). Biodiversity is important because it helps in the multi-ecosystem functional maintenance through decomposition, primary production nutrient cycling, and other complex ecological interactions that enhance diverse ecosystem services for both wildlife and human populations. Biodiversity plays a crucial role in agriculture by ensuring a variety of crops and livestock, which can lead to more resilient and productive farming systems. This variety helps in pest and disease management, improves soil health, and enhances crop yields. The presence of a diverse array of wildlife species enhances the productivity capacity of an ecosystem and provides necessary services that safeguard food security (Oliver et al. 2015). Biodiverse ecosystems offer essential services such as pollination of crops, purification of water and air, climate regulation, and decomposition of organic matter. Such resilience ensures ecosystem function and productivity maintenance in any given landscape (Lefcheck et al. 2015). Biodiversity proposes great opportunities for scientific research, provides ecosystem services, minimizes environmental change, and improves ecological understanding (Soliveres et al. 2016).

10.2 Relationship Between Sustainable Land Management and Biodiversity

The association between SLM and biodiversity is a crucial concept for supporting food security, maintaining ecological balance, and stimulating sustainable development (Kirsop-Taylor 2022). Biodiversity is fundamental to SLM as it provides the genetic resources and ecosystem functions essential for resilient agricultural systems (Critchley and Radstake 2017). Ecosystem services such as pollination, nutrient cycling, and water regulation are crucial for maintaining biodiversity and supporting ecological balance (Kirsop-Taylor 2022). SLM practices enhance biodiversity by fostering healthy ecosystems, which improve soil fertility, water quality, and pest control. However, climate change affects biodiversity by altering habitats and species distribution, making it vital to adapt conservation strategies accordingly. Effective SLM policies can increase biodiversity conservation by promoting practices that protect and restore ecosystems, ensuring long-term ecological sustainability (Liniger et al. 2019). These interconnected elements highlight the importance of integrated land management approaches that prioritize both agricultural productivity and biodiversity conservation. Ecosystem resilience is the backbone for SLM practices and interventions against adverse climatic events such as flooding, droughts, and desertification. Furthermore, conventions such as the United Nations Convention to Combat Desertification and the Convention of Biological Diversity provide mechanisms for connecting SLM practices and biodiversity (Kirsop-Taylor 2022).

Sustainably managed forest ecosystems contribute to ecological stability, food security, livelihood, regulate and mitigate climate change, and enhance carbon sequestration, habitat restoration, flood and drought mitigation, and slope stability (IIED 2019).

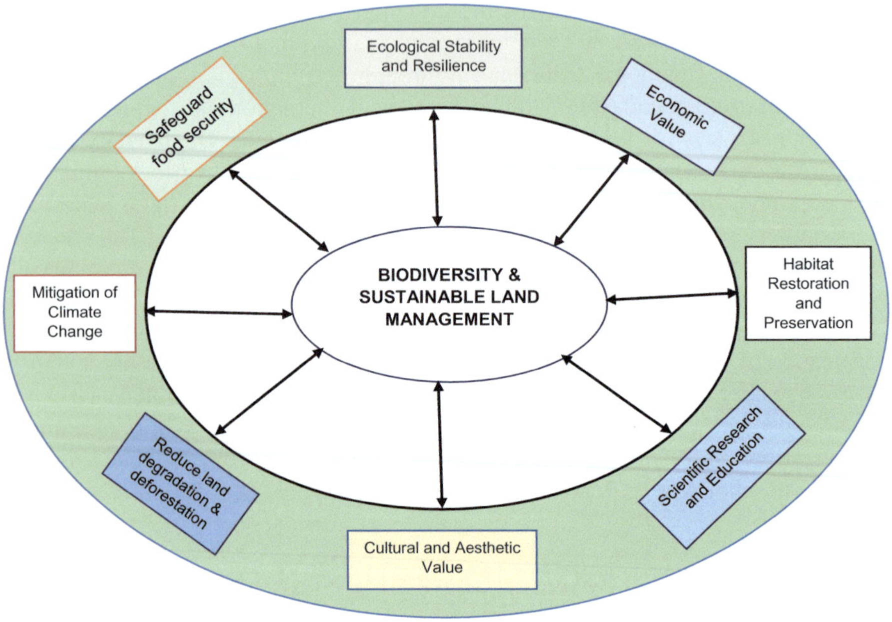

Fig. 10.1 Interaction between biodiversity and SLM

Sustainably managed ecosystems play a crucial role in nature-based solutions, enhancing biodiversity conservation through ecological restoration and the delivery of ecosystem services. This underscores the deep and intrinsic connection between sustainable land development and biodiversity across various dimensions (UNEP 2021b). The relationship between biodiversity and SLM highlights the interdependence between ecological health and human well-being. Consequently, promoting practices that protect biodiversity while ensuring sustainable land use is vital for achieving SDGs and mitigating the impacts of climate change globally (Fig. 10.1).

10.2.1 Connection Between Sustainable Development Goals and Sustainable Land Management

There is a strong relationship between SDGs, SLM, SDG activities, and SLM practices (Zdruli et al. 2016). While the SDGs contain 17 goals with 169 targets, SLM has four core principles: integrated land use planning; soil and water conservation; adaptive management and resilience; and participation and stakeholder management. While the SDGs are the international standard for achieving a sustainable future for current and future generations, SLM plays a critical role in supporting many of these goals by ensuring the efficient management and allocation of land resources (Bloomfield et al. 2017). SLM is considered as activities that integrate biodiversity conservation, soil and water management, and ecosystem maintenance

(Zhang et al. 2023). About 7 out of the 17 SDGs are interconnected with land resources. However, SDG 15 is solely dedicated to life on earth via land degradation, deforestation, biodiversity loss, and neutrality among others. SDG 15 targets 1, 2, and 3 are afforestation, land degradation, and land conservation.

SLM is essential in attaining the SDGs through the SDGs that enhance environmental protection, ecosystem function, integrated agricultural practices, and economic resilience (Lu et al. 2021; Ma et al. 2023). The efficient implementation of SLM practices is critical in achieving SDGs and aims at making the world a better place for the current and future generations (Abouelhamd 2020; Eppler et al. 2015). For instance, in achieving SDG8, SLM can result in more stable sustainable economic growth by making sure that longstanding land productivity supports livelihood activities and provides a cushion for emerging economic cyclical shocks (Abouelhamd 2020). Similarly, SLM has a significant role to play in achieving SDG 9 through the promotion of agricultural technologies that support sustainable land use management that enhances biodiversity increase (Abab et al. 2023). In addition, SDGs 10, 11, and 12 are all interconnected with land resource management (Meggiolaro et al. 2018). While SDG 10 advocates for reducing inequalities and improving land tenure security to ensure equitable resource allocation for all, SDG 11 focuses on integrating urban and peri-urban land resource management activities to support the sustainable development of urban areas, particularly regarding protecting natural habitats (Zhang et al. 2023). To minimize waste and promote the effective use of land resources for increased productivity, SDG 12 promotes the responsible production and consumption of agricultural and forestry-related practices (Shivakoti 2019). SDG 15 calls for actions that reduce land degradation, biodiversity loss, deforestation, restoration, prevent desertification, and ensure the sustainable conservation of flora and fauna associated with SDG 15 goal (Abouelhamd 2020). The SDG 15 targets incorporate SLM practices directly to achieve this goal (Munroe and Müller 2020).

10.3 Sustainable Land Practice and Its Impact on Food Security

About 1.6 billion people around the world depend on biodiversity and forests for food and livelihood sustenance (Borah et al. 2020; Newton et al. 2020). This makes biodiversity a critical element for safeguarding food security, especially for forest-dwelling communities around the world (Broegaard et al. 2017; Borah et al. 2020). Biodiversity contributes to food security through diverse pathways that support the livelihood activities of the local population of tropical countries. It is estimated that about six million tons of bush meats are extracted from the forests of Africa and Asia annually (Baudron et al. 2019; Overton et al. 2017; Wunder et al. 2014; Borah et al. 2020). According to Bharucha and Pretty (2010), the forests of Asia and Africa produce more than 800 species of wild variety of fruits. Some of these wild fruits are honey, bush yam, fungi, vegetables, mushrooms, bush meat, fish, and insects among others (Borah et al. 2020).

A sustainably managed conservation land is critical for flora and fauna welfare and abundance. These plants and animals form the food chain of healthy and sustainably managed land resources capable of storing water and food for living organisms (FAO 2019). According to the Food and Agriculture Organization (FAO) (2019) assertion, biodiversity is vital for food security safeguarding globally. A sustainably managed land practice can give rise to diverse plant and animal species that are used as food supplements and other diets needed for survival. In addition, highly conserved protected areas and lands serve as the main source of non-timber, forest products such as mushrooms, snails, fruits, honey, and so on. Different types of land use support various food security initiatives around the world (Dasgupta 2021). In many forest-edge communities in developing countries, biodiversity supports local food production activities and serves as a supplement to maintain a balanced diet, while, in remote communities around the world, majority of flora and fauna are good sources of food and a perfect substitution for food commodities. Over the years, conserving biodiversity has helped mitigate food security threats and enhance the sustainability of ecosystem services essential for food production. Sustainably managed land enhances biodiversity abundance and, by extension, ecosystem services that support food production, including the gathering of non-timber forest products. Ecosystems have the ability to support many living organisms that play a vital role in making sure that food production is enhanced. In contrast, unsustainable land use pattern in any ecosystem impedes the ability of biodiversity to provide food and other food supplements. The maintenance of crop diversity in any agricultural field helps support food productivity at local and national levels while at the same time enhancing biodiversity conservation (Montoya et al. 2020; Li et al. 2021).

10.3.1 Biodiversity Helps Resist Climate Change

While climate change affects both biodiversity and SLM practices, increasing biodiversity, on the other hand, helps to resist climate change's impact on managed and conserved ecologies (Convention on Biological Diversity (CBD) 2021). The adverse effect of climate change typically fuels biodiversity loss in fragile ecologies (Muluneh 2021). A well-preserved biodiversity conservation area has the potential to help mitigate climate change within a community. Furthermore, the study found that a great number of biological species can endure certain climate change uncertainty and risk (Muluneh 2021). Greater biodiversity is a catalyst for reducing climate change impacts and vulnerabilities in forested communities. As such, protected area strengthening alongside good sustainable management of biological resources can help limit climate change impact and risk in these communities (Soni and Ansari 2017).

10.4 Sustainable Farming or Agricultural Practices Contribution to Biodiversity

Sustainable agricultural practices like agroforestry help sustain biodiversity by integrating trees and shrubs into agricultural landscapes, thereby increasing habitat diversity and providing safe corridors for wildlife movement (Boinot et al. 2022). Recent studies have found that silvoarable agroforestry systems increase species diversity and richness compared to other cropland systems (Boinot et al. 2022). The study concludes that agroforestry has a weaker impact on biodiversity as compared to land use systems. An effective agroforestry system supports SLM by incorporating fast-growing trees and providing quick canopy cover. Agroforestry systems influence species composition, abundance, and ecological function in a given agricultural landscape (Blowes et al. 2019; Cardinael et al. 2019; Boinot et al. 2019; Varah 2015).

Diverse agricultural communities enhance biodiversity by providing varied food sources and habitats. Planting a mix of crop species, rather than relying on monocultures, supports soil health and suppresses harmful pathogens. This results in greater yields and resilience (Erisman et al. 2016). Different crops contribute to biodiversity in various ways within agricultural landscapes. Agricultural systems enhance biodiversity through mechanisms such as pollination, disease regulation, increased productivity, and improved plant resilience. Increased biodiversity can improve a range of functions in agricultural systems, including productivity, pollination, pest and disease regulation, and resilience to stress (Li et al. 2021). Different crops contribute varying nutrients to the soil, which in turn affects food productivity and environmental protection (Wittwer et al. 2021).

Minimizing chemical usage in agriculture plays a pivotal role in SLM. Specifically, by decreasing the reliance on fertilizers and pesticides, we can positively impact species conservation and overall biodiversity. Numerous studies have highlighted the effectiveness of integrated production and pest management strategies in mitigating environmental degradation and preserving biodiversity. Notably, the application of pesticides and fertilizers within farming systems significantly influences the abundance of both flora and fauna, as well as overall food productivity. Consequently, a reduction in chemical substances within agricultural practices not only promotes soil health but also supports diverse species populations.

10.5 Sustainable Land Management Contributes to Habitat Restoration and Preservation

SLM contributes to habitat restoration and stimulates an increase in biodiversity. Habitat restoration entails the maintenance of ecosystems that have been subjected to degradation or destruction by either natural or anthropogenic means (UNEP 2021a, b). Habitat preservation includes the protection of natural landscapes from degradation, damage, or destruction to upkeep biodiversity and maintain ecological integrity (Atkinson et al. 2022). SLM practices help to prevent habitat destruction and degradation by promoting land use methods that maintain natural habitats. Habitat restoration and preservation are critical for maintaining Earth's ecological balance and ensuring the survival of countless species, including humans. Habitat restoration and preservation are essential parts of ecological conservation intended to uphold biodiversity, enhance ecosystem services, and guarantee the maintainable use of natural resources (Atkinson and Bonser 2020). Ecosystem restoration supports various biodiversity of which some species are categorized as threatened and endangered (Atkinson et al. 2022).

Ecological restoration of habitat enhances biodiversity protection through the provision of relevant resources that ensure habitat suitability and foster biodiversity increase (Atkinson et al. 2022). Similarly, habitat preservation ensures the abundance of wildlife in any given ecosystem through the protection of keystone wildlife species (Cook 2023; Atkinson and Bonser 2020). Basically, ecological restoration stimulates biodiversity abundance and diversity as compared to degraded ecosystems or habitats (Gann et al. 2019). Understanding how wildlife species change over time due to habitat destruction or restoration is important for assessing biodiversity abundance, movement, and distribution in any given area. SLM is crucial for maintaining the health of ecosystems and providing adequate services and functions that stimulate the abundance of biodiversity. Globally, habitat restoration efforts have included restoring fragile or degraded lands back to their natural form to enable the thriving of various wildlife species.

10.6 Wetland Conservation Role in Biodiversity Sustainability

Wetland management plays a critical role in biodiversity conservation through several key mechanisms such as flood regulation, habitat protection, wildlife corridor, and coral reef protection (Kirsop-Taylor 2022). In particular, sustainable wetland water management plays a crucial role in the conservation of biodiversity as it directly influences the health of ecosystems and the wildlife species that depend on them (UN 2021). A conserved wetland ecosystem serves as an important habitat for different types of species, including birds, fish, amphibians, reptiles, and invertebrates (OECD 2018). An efficient and sustainable management of wetlands ensures the maintenance and restoration of keystone species (Kirsop-Taylor 2022). Wetlands around the world serve as natural water filters that sieve stabilize banks, sediments, pollutants, and nutrients and provide habitat for aquatic life. A healthy wetland ecosystem serves as a hotspot for diverse aquatic life, improves water quality life, and prevents contamination and pollution of these important corridors. Wetlands also regulate flooding and store flood water that supports biodiversity, and a sustainably managed wetland not only protects and maintains biodiversity but also enhances proper hydrological functions that support diverse ecosystems and provisioning services. Sustainably conserving wetlands help us in the preservation of diverse aquatic and terrestrial species that are threatened in other ecosystems (WWF 2020). Conserving and managing wetlands is essential for preserving the unique species diversity that thrives in these environments. Wetlands are resilient ecosystems that support a wide range of aquatic life and help mitigate environmental challenges and changes over time. They often harbor unique species and genetic varieties that are specifically adapted to wetland conditions. Protecting and managing wetlands is crucial for preserving genetic diversity, which enhances the resilience of ecosystems to environmental changes and stressors. In complex ecosystem situations, wetlands serve as a reliable habitat for threatened wildlife species survival (Soliveres et al. 2016). Endangered and threatened wildlife species mostly find refuge in wetlands ecologies because these ecosystems are considered safer than others. Sustainable management of wetlands involves protecting and restoring these ecosystems by maintaining native vegetation and water quality. Strategies such as effective water management, community engagement, biodiversity conservation, and policies to conserve them are required to prevent their degradation.

10.7 Sustainable Integrated Landscape Management Impact on Biodiversity

Sustainable integrated land management supports comprehensive methods for the preservation of biodiversity through social, economic, and ecological approaches. Biodiversity conservation is considered an integral component of land use development due to its positive influence on biodiversity abundance and distribution within any ecosystem, addressing social, ecological, and economic issues (IPBES 2019a, 2019b; UNEP 2021a, b). Multifunctional landscapes support multiple ecosystem functions and services that are vital to biodiversity increase in the twenty-first century. A study by Zanzanaini et al. (2017) found that sustainable landscape management provides ecosystem services such as soil protection, food security, water regulation, and climate regulation that promote biodiversity abundance and distribution. SLM can significantly contribute to the conservation of biodiversity, safeguarding the health and functionality of ecosystems (IPBES 2019a, b).

SLM significantly supports biodiversity conservation and distribution by responsibly handling soil and water resources, mitigating climate change impacts, promoting ecosystem connectivity, developing relevant policies, adopting sustainable agricultural practices, and engaging local communities in the management process (Rasmussen et al. 2017; Fayiah and Fayiah 2022). SLM approaches help in enhancing ecosystem function services, maintaining biodiversity population, and ensuring the connectivity of wildlife habitat. The incorporation of biodiversity conservation efforts into integrated ecosystem management strategies is critical for the sustenance, distribution, diversity, and abundance of biodiversity in any given landscape (Borah et al. 2020). The sustainable conservation of biodiversity plays a key and central role in protecting wildlife habitats, enhancing ecosystem services provision, supporting cultural and traditional beliefs, meeting the socio-economic needs of forest edge communities, alleviating rural poverty, and strengthening the local economies of remote communities (IUCN 2012).

10.8 Land Tenure System Impact on Biodiversity Abundance

The concept of land tenure refers to the various institutional arrangements governing access, ownership, usage rights, and utilization privileges over land. These arrangements can apply to individuals, communities, clans, societies, or specific regions. The nature of land tenure systems across diverse ecosystems globally has a significant correlation with either the augmentation or reduction of biodiversity (Kasimbazi 2017; Mukupa 2011). Globally, there are four renowned land tenure systems being practiced by nations on Earth. These tenure systems include private land ownership, communal ownership, state ownership, and open access, among others. Private land ownership gives individuals the right to own a piece of land and do as they wish with it. Communal land ownership gives communities the right to manage land resources for the benefit of the communities instead of individuals. State land tenure system, on the other hand, gives the government or countries the right to manage, allocate, and control the use of these lands as per the land policies of that state. The open-access land tenure system is barely practiced by individuals, communities, or states. An inappropriate land tenure system significantly reduces land productive capability and by extension biodiversity abundance (UN-HABITAT 2008). In addition, unsustainable land tenure systems affect the socio-economic, ecological, livelihood, and biodiversity increase in any landscape. An unsustainable land tenure system mostly results in land degradation and degradation leading to biodiversity loss and decline (World Bank 2014).

Land tenure system impacts on biodiversity are multi-faceted and complex and tend to influence the abundance of biodiversity either negatively or positively. Such land tenure systems seek to integrate effective land use, respect for norms and traditions of local people, and clear land ownership display, enhance economic benefits, regulate access, and most often stimulate and enhance biodiversity abundance and positive outcomes (Woinarski et al. 2013). A controversial land tenure system may tend to have little or no impact on biodiversity conservation due to the inconsistency of tenure and conservation area insecurity. Societies with mixed or irregular land tenure most often result in habitat exploitation, degradation, wildlife extinction, and constant biodiversity decline. Biodiversity increase and distribution greatly depend on sustainable land tenure features such as land ownership rights, utilization rights, and land use rights among others. Studies have found that the land tenure and biodiversity relationship is complex and dynamic and is influenced by societies, regions, and ecologies. The relationship between biodiversity and the land tenure system is multifaceted and involves economic, ecological, legal, and social features that depend on biodiversity abundance or decline (Fig. 10.2).

10.9 Sustainable Land Management and Biodiversity Threats

There exists a significant connection between SLM and biodiversity conservation across various ecosystems, but they are currently confronted with substantial threats. An ecosystem provides multiple services for humans in the form of water, food,

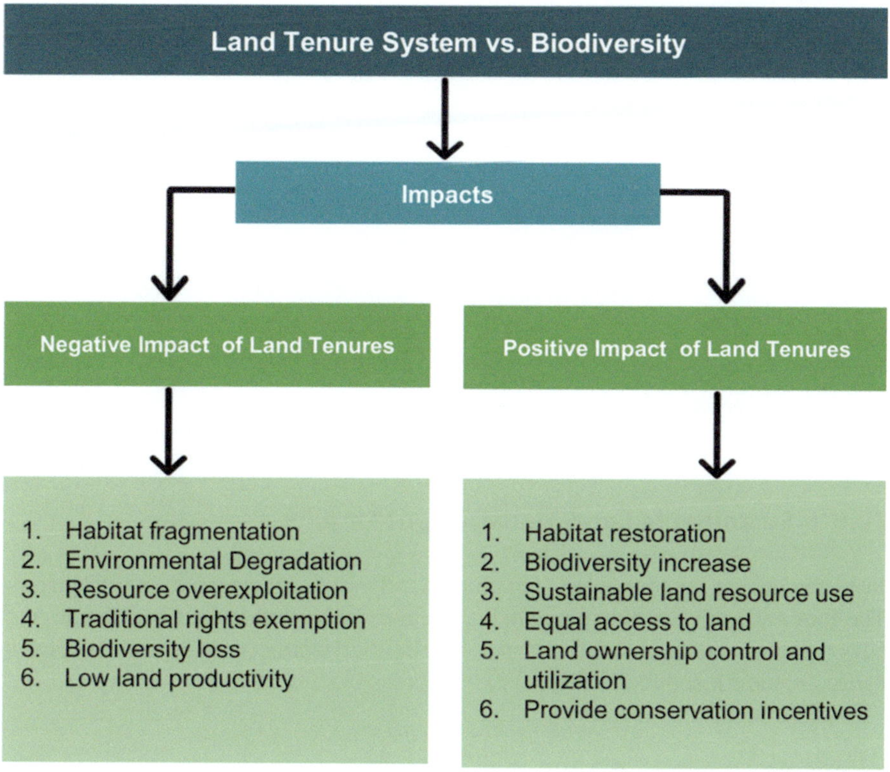

Fig. 10.2 Land tenure system impact on biodiversity

energy, biodiversity support, habitat sustenance, climate regulation, medicinal offset, recreational facilities, and wildlife protection (Lenzen et al. 2012). These threats affect both flora and fauna as well as the productive potential of the ecosystem. In addition, these threats expose ecosystems to food insecurity, environmental degradation, biodiversity loss, exposure of lands to climate change impacts, destruction of ecological habitats, pollution, and health of ecosystems. However, both biodiversity and SLM face numerous threats that affect the sustainable management of these practices (IPBES 2019a, b). Nonetheless, these threats are mostly country or region-specific (Lenzen et al. 2012). These threats have undermined efforts made to conserve biodiversity and have resulted in land degradation and biodiversity loss in various landscapes and ecologies around the world (Fayiah and Fayiah 2022). Therefore, biodiversity conservation and SLM are critical for maintaining environmental health and productive potential. Figure 10.3 presents some threats facing ecosystem services and productive functions of the ecosystem.

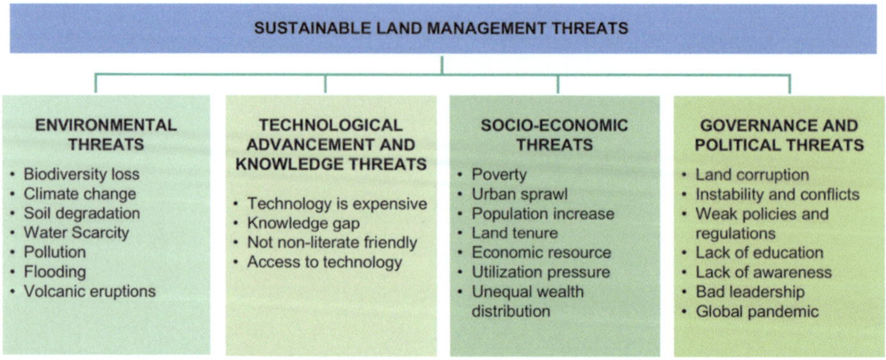

Fig. 10.3 Sustainable land management threats

10.9.1 Sustainable Land Management Threats

SLM threats are numerous but are summarized into four categories in this chapter. The four categories of threats include (1) environmental threats, (2) technological advancement and knowledge threats, (3) socio-economic threats, and (4) governance and political threats (Fig. 10.3).

10.9.2 Biodiversity Conservation Threats

Globally, biodiversity conservation faces a range of threats, both human-made and natural. These threats span from deforestation to inadequate policies and legislation in various countries. However, biodiversity threats differ by continent, region, country, and society. Some biodiversity threats are, however, interconnected and similar across regions (IUCN 2021). For instance, in Africa, deforestation, shifting cultivation, charcoal burning, illegal logging, and mining are common threats to biodiversity conservation. The loss or extinction of some important biodiversity flora and fauna has negative implications on human activities and survival (Bellard et al. 2022). Threats, such as climate change, are preventing ecosystems from providing adequate services and functions needed for the productivity of the environment. The threats against biodiversity are more severe in developing countries than in developed nations (Lenzen et al. 2012). This could be attributed to weak policies, conventional farming practices, poor economies, poverty, and complex land tenure systems (Leisher et al. 2022) (Table 10.1).

Table 10.1 Threats facing biodiversity conservation

	Threats	Effects	Sources
1	Deforestation	Habitat destruction for flora and fauna	Fayiah and Fayiah (2022)
2	Illegal logging	Ecosystem alteration and destruction	Bellard et al. (2022)
3	Slash & burn agriculture	Habitat loss and fragmentation	IPBES (2019a, b)
4	Wildfire outbreak	Loss of flora and fauna and habitat loss	Agduma et al. (2023)
5	Illegal artisanal fishing	Harm and reduce marine biodiversity	Abakah and Owusu (2023)
6	Illegal mining	Negatively affect ecosystem services	Agduma et al. (2023)
7	Climate change	Alter precipitation and land resilience	Bellard et al. (2022)
8	Droughts	Reduce land productivity	Penning et al. (2023)
9	Invasive species	Disrupt ecosystem services provision	Harfoot et al. (2021)
10	Overgrazing	Habitat degradation and vegetation loss	Agduma et al. (2023)
11	Urban sprawl	Reduce habitat and ecologies	Simkin et al. (2022)
12	Pollution	Degrade soil quality and damage habitat	Bellard et al. (2022)
13	Soil erosions	Reduce soil fertility	Tilman et al. (2017)
14	Technology	Chemicals make land unproductive	Pimm et al. 2015)
15	War and conflicts	Destruction of vegetation	Leisher et al. (2022)
16	Charcoal burning	Loss of forests and ecosystem services	Leisher et al. (2022)
17	Fuel wood collection	Degrade ecosystem and loss of species	Leisher et al. (2022)

10.10 Forest Management Effects on Biodiversity

Forest management practices play a critical role in biodiversity conservation and distribution across any ecological region. These practices are capable of minimizing ecosystem degradation, deforestation, and habitat fragmentation and enhancing ecosystem service provision. Practices such as selective logging, shelterwood cutting, and community forestry ensure the sustainable utilization of the forests and their resources and the preservation of biodiversity. Managed tropical forests are considered a complex biotic ecosystem that is characterized by vegetation and serve as a reservoir of flora and fauna of various lifeforms such as arthropods, soil microbes, and Lepidoptera (Asbeck et al. 2021). It is estimated that tropical forests alone host 50% of the planet's biodiversity (Loo 2009). Managed tropical forests around the world are believed to host a number of undiscovered flora and fauna species of great scientific and societal value. Sustainable forest management contributes to biodiversity well-being and ecosystem productivity through multifaceted service provision for both wildlife and humans (Wang et al. 2019; Raihan 2023; Taye et al. 2021). Managing forests sustainably increases ecosystem productivity and by extension habitat restoration. Biodiversity in a sustainably managed forest greatly relies on habitat continuity, tree species, and ecosystem function. Mature or old-growth natural forests are more productive in terms of biodiversity than secondary forests or plantations (EEA 2016). Nevertheless, both natural and plantation forests have great conservation value as compared to other ecologies. Tree species diversity has a great influence on biodiversity conservation due to its ability to attract diverse flora and fauna that constitute an ecosystem that provides various services (Bauhus et al. 2017). Studies have found that managed tropical forests are considered a hotspot for biodiversity and host different plants and animals of great conservation value and status (Loo 2009). On the other hand, a recent study found that biodiversity conservation helps in the timely propagation of some tree species in the forest (Dinca and Zhiyanski 2023). It is important to know that not all tropical forests are equally diverse in terms of flora and fauna and that could be attributed to soil fertility and rainfall patterns among other factors (Loo 2009).

In this twenty-first century, managed forests are key strategies that tackle biodiversity loss, and their decline can increase carbon stocks significantly. Managed forests have direct and indirect effects on biodiversity and significant effects on ecosystem function and productivity. Biodiversity-oriented managed forests provide a wide range of ecosystem services and functions that are compatible with biodiversity increase and distribution (Bohn and Huth 2017; Paillet et al. 2010; Villard and Jonsson 2009; Nascimbene et al. 2014). Poorly managed forests, on the other hand, threaten ecosystem service provision and biodiversity increase and are at the heart of the biodiversity crisis. Studies have also suggested that managed community forests help prevent forest resource degradation and support biodiversity conservation. Sustainably managed forests improve biodiversity conditions for all categories of wildlife as compared to other degraded ecologies or biomes (Acharya 2004).

10.11 Strategies to Overcome Threats Facing Sustainable Land Management

Globally, SLM problems are multifaceted in nature. Ecosystem degradation is influenced by various factors such as topography, rainfall pattern and intensity, land use type, population and urban sprawl, and socio-economic pressure among others (Haregeweyn et al. 2023). As such, to mitigate SLM problems, comprehensive strategies are required to promote SLM practices. These strategies could range from smart agricultural practices to integrated land use management and strengthened land policies. These strategies have the potential to mitigate land use problems and restore degraded ecosystems (Haregeweyn et al. 2023) For instance, the adoption of climate-smart agriculture and sustainable agroforestry and agroecology practices can reverse land degradation and enhance land productivity.

To mitigate these threats, a holistic approach comprising political strategies, socio-economic, environmental, and integrated policy framework will ensure the sustainability of land resources (Fig. 10.4). Overcoming SLM threats requires

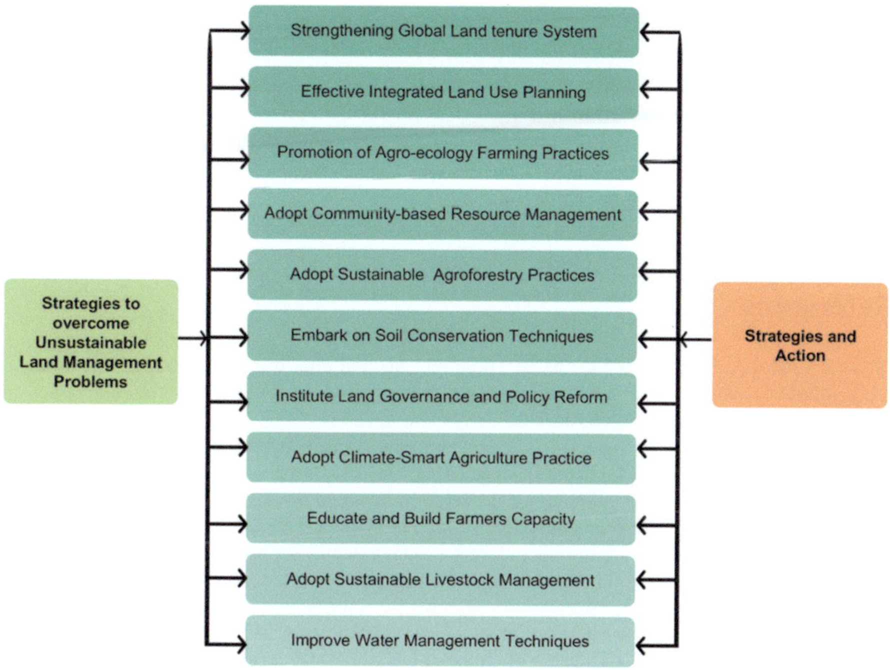

Fig. 10.4 SLM strategies

multifaceted approaches that combine political, academic, research, and private sector involvement to combat land degradation at all levels.

10.12 Conclusion

There is a strong correlation between SLM and biodiversity increase and distribution in any given ecosystem. A sustainably managed land resource enhances productivity, while a poorly managed land resource leads to biodiversity decline and loss. While SDGs are the international benchmarks set to find a sustainable future and path for the current and future generation, SLM, on the other hand, plays a critical role in supporting many of these goals through efficient land resources management and allocation irrespective of social status. SLM not only supports biodiversity increase but also enhances ecosystem services, functions, and productivity. As compared to SLM, biodiversity plays a fundamental role in the livelihood and survival activities of forest edge communities. Biodiversity conservation is considered an integral component of land use development for social, ecological, economic, and social issues that positively influence biodiversity abundance. Healthy wetlands ecosystem serves as important habitats for different types of species, including birds, fish, amphibians, reptiles, and invertebrates. Land tenure systems in different ecosystems around the world are directly linked to biodiversity increase or decline.

An inappropriate land tenure system significantly reduces land productive capability and by extension biodiversity abundance. Biodiversity conservation is exposed to various threats, either man-made or natural. Loss or extinction of some important flora and fauna has negative implications on human survival. It is recommended that land resources be sustainably managed for the enhancement of ecosystem services, functions, biodiversity increase, and livelihood improvement for the current and future generations.

References

Abab SA, Senbeta F, Negash TT (2023) The effect of land tenure institutional factors on small landholders' sustainable land management investment: evidence from the highlands of Ethiopia. Sustainability 2023(15):9150

Abakah S, Owusu V (2023) Impacts of illegal fishing and ocean dependence on the livelihoods of coastal fisherfolk in Ghana. Ghana J Geogr *15*(1):198–225. https://doi.org/10.4314/gjg.v15i1.9

Abouelhamd I (2020) Analytical study of the relation between the Sustainable land management indicators (SLM) and the Sustainable Development Goals (SDGs), January, 16

Acharya KP (2004) Does community forests management supports biodiversity conservation? Evidences from two community forests from the Mid Hills of Nepal. J For Livelihood 4(1):44–54

Agduma AR, Garcia FG, Cabasan MT, Pimentel J, Ele RJ, Rubio M, Murray S, Hilario-Husain BA, Dela Cruz KC, Abdullah S, Balase SM, Tanalgo KC (2023) Overview of priorities, threats, and challenges to biodiversity conservation in the southern Philippines. Reg Sustain 4(2):203–213. https://doi.org/10.1016/j.regsus.2023.05.003

Asbeck T, Sabatini F, Augustynczik ALD, Basile M, Helbach J, Jonker M, Knuff A, Bauhus J (2021) Biodiversity response to forest management intensity, carbon stocks and net primary production in temperate montane forests. Sci Rep 11(1):1–11. https://doi.org/10.1038/s41598-020-80499-4

Atkinson J, Bonser SP (2020) "Active" and "passive" ecological restoration strategies in meta-analysis. Restor Ecol 28:1032–1035

Atkinson J, Brudvig LA, Mallen-Cooper M, Nakagawa S, Moles AT, Bonser SP (2022) Terrestrial ecosystem restoration increases biodiversity and reduces its variability, but not to reference levels: a global meta-analysis. Ecol Lett. 25(7):1725–1737. https://doi.org/10.1111/ele.14025

Baudron F, Tomscha SA, Powell B, Groot JCJ, Gergel SE, Sunderland T (2019) Testing the various pathways linking forest cover to dietary diversity in tropical landscapes. Front Sustain Food Syst 3:97. https://doi.org/10.3389/fsufs.2019.00097

Bauhus J, Kouki J, Paillet Y, Asbeck T, Marchetti M (2017) How does the forest-based bioeconomy impact forest biodiversity? In: Winkel G (ed) Towards a sustainable european forest-based bioeconomy – assessment and the way forward. What science can tell us 8. European Forest Institute, Joensuu

Bellard C, Marino C, Courchamp F (2022) Ranking threats to biodiversity and why it doesn't matter. Nat Commun 13(1):10–13. https://doi.org/10.1038/s41467-022-30339-y

Bharucha Z, Pretty J (2010) The roles and values of wild foods in agricultural systems. Philos Trans R Soc B Biol Sci 365:2913–2926

Bloomfield G, Bucht K, Martínez-Hernández JC, Ramírez-Soto AF, Sheseña-Hernández I, Lucio-Palacio CR, Ruelas Inzunza E (2017) Capacity building to advance the United Nations sustainable development goals: an overview of tools and approaches related to sustainable land management. J Sustain For 37(2):157–177. https://doi.org/10.1080/10549811.2017.1359097

Blowes SA, Supp SR, Antão LH, Bates AE, Bruelheide H, Chase JM, Moyes F, Magurran AE, McGill BJ, Myers-Smith IH, Winter M, Bjorkman AD, Bowler DE, Byrnes JE, Gonzalez A,

Hines J, Isbell F, Jones HP, Navarro LM, Thompson PL, Vellend M, Waldock CA, Dornelas M (2019) The geography of biodiversity change in marine and terrestrial assemblages. Science. 2019(366):339–345
Bohn FJ, Huth A (2017) The importance of forest structure to biodiversity–productivity relationships. R Soc Open Sci 4:160521
Boinot S, Fried G, Storkey J, Metcalfe H, Barkaoui K, Lauri PÉ, Mézière D (2019) Alley cropping agroforestry systems: reservoirs for weeds or refugia for plant diversity? Agric Ecosyst Environ. 284:106584. https://doi.org/10.1016/j.agee.2019.106584
Boinot S, Barkaoui K, Mézière D, Lauri PE, Sarthou JP, Alignier A (2022) Research on agroforestry systems and biodiversity conservation: what can we conclude so far and what should we improve? BMC Ecol Evol 22(1):1–6. https://doi.org/10.1186/s12862-022-01977-z
Borah JR, Laumonier Y, Rc Bayala E, Djoudi H, Gumbo D, Moombe KB, Yuliani EL, Zida M (2020) The role of biodiversity in integrated landscape approaches. Available online at: https://www.cifor-icraf.org/publications/pdf_files/Books/2020-COLANDS-03.pdf. Accessed June 2024
Broegaard RB, Rasmussen LV, Dawson N, Mertz O, Vongvisouk T, Grogan K (2017) Wild food collection and nutrition under commercial agriculture expansion in agriculture-forest landscapes. For Policy Econ 84:92–101
Cardinael R, Hoeffner K, Chenu C, Chevallier T, Béral C, Dewisme A, Cluzeau D (2019) Spatial variation of earthworm communities and soil organic carbon in temperate agroforestry. Biol Fertil Soils. 55(2):171–183. https://doi.org/10.1007/s00374-018-1332-3
Convention on Biological Diversity (CBD) (2021) Website, Climate Change and Biodiversity. https://www.cbd.int/climate/
Cook A (2023) What is habitat restoration and why is it important. Available online at: https://defenders.org/blog/2023/06/what-habitat-restoration-and-why-it-important. Accessed 3 June 2024.
Critchley W, Radstake F (2017) Sustainable land management in Asia: introducing the landscape approach. Asian Development Bank
Dasgupta P (2021) The economics of biodiversity: the dasgupta review. Abridged Version. HM Treasury, London
Dinca L, Zhiyanski M (2023) Forest management and biodiversity conservation: introduction to the special issue. Diversity 2023(15):1078. https://doi.org/10.3390/d15101078
EEA (2016) European Forest ecosystems – state and trends. EEA Report No. 5/2016
Eppler U, Fritsche UR, Iriarte L (2015) Global sustainable land use concept and examples for systemic indicators, January 42
Erisman JW, van Eekeren N, de Wit J, Koopmans C, Cuijpers W, Oerlemans N, Koks BJ (2016) Agriculture and biodiversity: a better balance benefits both. AIMS Agric Food 1(2):157–174
FAO (2019) The State of the world's biodiversity for food and agriculture. In: Bélanger J, Pilling D (eds) FAO Commission on Genetic resources for food and agriculture assessments. Rome, p 572. http://www.fao.org/3/CA3129EN/CA3129EN.pdf
Fayiah M, Fayiah MS (2022) Challenges of biodiversity conservation in africa: a case study of Sierra Leone. In: Chibueze Izah S (ed) Biodiversity in Africa: potentials, threats and conservation. Sustainable development and biodiversity, vol 29. Springer, Singapore. https://doi.org/10.1007/978-981-19-3326-4_23
Gann GD, McDonald T, Walder B, Aronson J, Nelson CR, Jonson J et al (2019) International principles and standards for the practice of ecological restoration. Second edition. Restor Ecol 27:S1–S46
Grumbine RE, Xu J (2021) Five steps to inject transformative change into the post-2020 global biodiversity framework. Bio-Science 71(6):637–646. https://doi.org/10.1093/biosci/biab013
Haregeweyn N, Tsunekawa A, Tsubo M, Fenta AA, Ebabu K, Vanmaercke M, Borrelli P, Panagos P, Berihun ML, Langendoen EJ, Nigussie Z, Setargie TA, Maurice BN, Minichil T, Elias A, Sun J, Poesen J (2023) Progress and challenges in sustainable land management initiatives: a global review. Sci Total Environ *858*(May 2022):160027. https://doi.org/10.1016/j.scitotenv.2022.160027

Harfoot MBJ, Johnston A, Balmford A, Burgess ND, Butchart SHM, Dias MP, Hazin C, Hilton-Taylor C, Hoffmann M, Isaac NJB, Iversen LL, Outhwaite CL, Visconti P, Geldmann J (2021) Using the IUCN Red List to map threats to terrestrial vertebrates at global scale. Nat Ecol Evol *5*(11):1510–1519. https://doi.org/10.1038/s41559-021-01542-9

IIED (2019) Biodiversity loss is a development issue: a rapid review of evidence, Issue Paper, April 2019. https://pubs.iied.org/17636iied

IPBES (2019a) Summary for policymakers of the global assessment report on biodiversity and ecosystem services (summary for policy makers). https://doi.org/10.5281/zenodo.3553579

IPBES (2019b) Media release, nature's dangerous decline 'Unprecedented'; Species Extinction Rates 'Accelerating', May 2019. https://ipbes.net/news/Media-Release-Global-Assessment

IUCN (2012) Livelihoods and landscapes strategy. Results and reflections. International Union for Conservation of Nature (IUCN), Gland. https://portals.iucn.org/library/sites/library/files/documents/2012–055.pdf

IUCN (2021) The IUCN Red List of Threatened Species. Version 2021–2. https://www.iucnredlist.org. Downloaded on [12 November 2021]. Locke et al. 2021: A nature-positive world: the global goal for nature. https://library.wcs.org/doi/ctl/view/mid/33065/pubid/DMX3974900000.aspx

Izquierdo-Tort S, Alatorre A, Arroyo-Gerala P, Shapiro-Garza E, Naime J, Dupras J (*2023*) Exploring local perceptions and drivers of engagement in biodiversity monitoring among participants in payments for ecosystem services schemes in southeastern Mexico. Conserv Biol 2024:1–17. https://doi.org/10.1111/cobi.14282

Kasimbazi PE (2017) GLOBAL for improved land management land tenure. Global Land Outlook Working Paper, 28

Kirsop-Taylor N (2022) Biodiversity and sustainable development governance. In: Handbook on the governance of sustainable development, vol 9, pp 48–64. https://doi.org/10.4337/9781789904321.00011

Lefcheck JS, Byrnes JEK, Isbell F, Gamfeldt L, Griffin JN, Eisenhauer N, Hensel MJS, Hector A, Cardinale BJ, Duffy JE (2015) Biodiversity enhances ecosystem multifunctionality across trophic levels and habitats. Nat Commun 6:6936

Leisher C, Robinson N, Brown M, Kujirakwinja D, Schmitz MC, Wieland M, Wilkie D (2022) Ranking the direct threats to biodiversity in sub-Saharan Africa. Biodivers Conserv 31(4):1329–1343. https://doi.org/10.1007/s10531-022-02394-w

Lenzen M, Moran D, Kanemoto K, Foran B, Lobefaro L, Geschke A (2012) International trade drives biodiversity threats in developing nations. Nature. 486(7401):109–112. https://doi.org/10.1038/nature11145

Li L, Werf VD, Zhang F (2021) Crop diversity and sustainable agriculture: mechanisms, designs and applications long. Front Agric Sci Eng 8(3):359–361. https://doi.org/10.15302/J-FASE-2021417

Liniger H, Harari N, van Lynden G, Fleiner R, de Leeuw J, Bai Z, Critchley W (2019) Achieving land degradation neutrality: the role of SLM knowledge in evidence-based decision-making. Environ Sci Policy 94:123–134

Loo JA (2009) The role of forests in the preservation of biodiversity. For For Plants *3*:1–25

Lu X, Zhang Y, Lin C, Wu F (2021) Analysis and comprehensive evaluation of sustainable land use in China: based on sustainable development goals framework. J Clean Prod 310:127205

Ma Y, Zheng M, Zheng X, Huang Y, Xu F, Wang X, Liu J, Lv Y, Liu W (2023) Land use efficiency assessment under sustainable development goals: a systematic review. Land 12:894. https://doi.org/10.3390/land1204089

Meggiolaro L, Sato R, Sorensen N (2018) Land Governance and the Sustainable Development Goals (SDGs): consolidating and harmonizing monitoring initiatives. In: Paper prepared for presentation at the "2018 World Bank conference on land and poverty". The World Bank, Washington DC

Montoya D, Gaba S, de Mazancourt C, Bretagnolle V, Loreau M (2020) Reconciling biodiversity conservation, food production and farmers' demand in agricultural landscapes. Ecol Model 416:108889

Mukupa W (2011) Land administration to support sustainable development. FIG working week 2011 bridging the gap between cultures, Marrakech, Morocco, 18–22 May 2011

Muluneh MG (2021) Impact of climate change on biodiversity and food security: a global perspective – a review article. Agric Food Secur 10:36. https://doi.org/10.1186/s40066-021-00318-5
Munroe DK, Müller D (2020) Land-system science to support achieving the sustainable development goals. J Land Use Sci 15(4):477–481. https://doi.org/10.1080/1747423X.2020.1783085
Nascimbene J, Fontana V, Spitale DA (2014) Multi-taxon approach reveals the effect of management intensity on biodiversity in Alpine larch grasslands. Sci Total Environ 487:110–116
Newton P, Kinzer AT, Miller DC, Oldekop JA, Agrawal A (2020) The number and spatial distribution of forest-proximate people globally. One Earth 3:363–370
OECD (2018) Mainstreaming biodiversity for sustainable development. Policy Highlights. https://www.oecd.org/environment/resources/Policy-Highlights-Mainstreaming-Biodiversity-for-Sustainable-Development.pdf
Oliver TH, Heard MS, Isaac NJB, Roy DB, Procter D, Eigenbrod F, Freckleton R, Hector A, Orme CDL, Petchey OL (2015) Biodiversity and resilience of ecosystem functions. Trends Ecol Evol 30:673–684
Overton J, Davis S, Nguluka L, Nsende E, Sompa B, Simukonda C, Lindsey P (2017) The Illegal bushmeat trade in the greater Kafue Ecosystem, Zambia: Drivers Impacts and Solutions. FAO. http://www.fao.org/3/abc610e.pdfand.
Paillet Y, Bergès L, Hjältén J, Odor P, Avon C, Bernhardt-Römermann M, Bijlsma RJ, De Bruyn L, Fuhr M, Grandin U, Kanka R, Lundin L, Luque S, Magura T, Matesanz S, Mészáros I, Sebastià MT, Schmidt W, Standovár T, Tóthmérész B, Uotila A, Valladares F, Vellak K, Virtanen R (2010) Biodiversity differences between managed and unmanaged forests: meta-analysis of species richness in Europe. Conserv Biol. 24(1):101–112. https://doi.org/10.1111/j.1523-1739.2009.01399.x
Penning E, Peñailillo Burgos R, Mens M, Dahm R, de Bruijn K (2023) Nature-based solutions for floods AND droughts AND biodiversity: do we have sufficient proof of their functioning? Cambridge Prisms Water 1. https://doi.org/10.1017/wat.2023.12
Pimm SL, Alibhai S, Bergl R, Dehgan A, Giri C, Jewell Z, Joppa L, Kays R, Loarie S (2015) Emerging technologies to conserve biodiversity. Trends Ecol Evol 30(11):685–696. https://doi.org/10.1016/j.tree.2015.08.008
Raihan A (2023) A review on the integrative approach for economic valuation of forest ecosystem services. J Environ Sci Econ 2:1–18
Rasmussen LV, Watkins C, Agrawal A (2017) Forest contributions to livelihoods in changing agriculture-forest landscapes. For Policy Econ 84:1–8
Shivakoti BR (2019) Using the Sustainable Development Goals to address unsustainable land transformations in the Asia-Pacific region. Asia-Pac Landsc Transform Solut Sustain 1972(Un 1972):205–222
Simkin RD, Seto KC, McDonald RI, Jetz W (2022) Biodiversity impacts and conservation implications of urban land expansion projected to 2050. Proc Natl Acad Sci U S A 119(12):1–10. https://doi.org/10.1073/pnas.2117297119
Soliveres S, van der Plas F, Manning P, Prati D, Gossner MM, Renner SC, Alt F, Arndt H, Baumgartner V, Binkenstein J et al (2016) Biodiversity at multiple trophic levels is needed for ecosystem multifunctionality. Nature 536:456–459
Soni DK, Ansari F (2017) Climate change and biodiversity; impacts, vulnerability and mitigation in Indian perspective: a review. J Appl Nat Sci 9(1):632–638. https://doi.org/10.31018/jans.v9i1.1243
Taye FA, Folkersen MV, Fleming CM, Buckwell A, Mackey B, Diwakar KC, Saint Ange C (2021) The economic values of global forest ecosystem services: a meta-analysis. Ecol Econ 189:107145
Tilman D, Clark M, Williams DR, Kimmel K, Polasky S, Packer C (2017) Future threats to biodiversity and pathways to their prevention. Nature 546:73–81
UN (2021) The Sustainable Development Goals Report 2021. https://unstats.un.org/sdgs/report/2021/
UNEP (2021a) Becoming #GenerationRestoration: Ecosystem restoration for people, nature and climate. https://www.unep.org/resources/ecosystem-restoration-people-nature-climate

UNEP (2021b) Making peace with nature: a scientific blueprint to tackle the climate, biodiversity and pollution emergencies. Nairobi. https://www.unep.org/resources/making-peace-nature

UN-HABITAT (2008) Secure land rights for all. UN-Habitat, Nairobi World Resources Institute (2016), Focus on Land in Africa, Washington, DC

Varah A (2015) Can agroforestry reconcile conflicting demands for productivity, biodiversity conservation and delivery of ecosystem services? Ph.D. thesis, 312 pp

Villard M-A, Jonsson BG (2009) Tolerance of focal species to forest management intensity as a guide in the development of conservation targets. For Ecol Manag 258:S142–S145

Wang L, Zheng H, Wen Z, Liu L, Robinson BE, Li R, Kong L (2019) Ecosystem service synergies/trade-offs informing the supply-demand match of ecosystem services: framework and application. Ecosyst Serv 37:100939

Wittwer RA, Bender SF, Hartman K, Hydbom S, Lima RAA, Loaiza V, Nemecek T, Oehl F, Olsson PA, Petchey O, Prechsl UE, Schlaeppi K, Scholten T, Seitz S, Six J, Van Der Heijden MGA (2021) Organic and conservation agriculture promote ecosystem multifunctionality. Sci Adv 7(34):Article eabg6995. https://doi.org/10.1126/sciadv.abg6995

Woinarski JCZ, Green J, Fisher A, Ensbey M, Mackey B (2013) The effectiveness of conservation reserves: land tenure impacts upon biodiversity across extensive natural landscapes in the tropical savannahs of the Northern Territory, Australia. Land 2(1):20–36. https://doi.org/10.3390/land2010020

World Bank (2014) Agricultural land redistribution and land administration in Sub-Saharan Africa case studies of recent reforms, Washington

Wunder S, Börner J, Shively G, Wyman M (2014) Safety nets, gap filling and forests: a global-comparative perspective. World Dev 64:S29–S42

WWF (2020) Living Planet Report 2020. In: Almond REA, Grooten M, Petersen T (eds) Bending the curve of biodiversity loss. WWF, Gland. https://f.hubspotusercontent20.net/hubfs/4783129/LPR/PDFs/ENGLISH-FULL.pdf

Zanzanaini C, Trần BT, Singh C, Hart A, Milder J, DeClerck F (2017) Integrated landscape initiatives for agriculture, livelihoods and ecosystem conservation: an assessment of experiences from South and Southeast Asia. Landsc Urban Plan 165:11–21

Zdruli P, Ziadat F, Nerilli E, D'Agostino D, Lahmer F, Bunning S (2016) Sustainable development of land resources. Mediterra 2016 – zero waste in the Mediterranean: natural resources, food and knowledge, October, 91–111

Zhang L, Kuang B, Yang B (2023) Sustainable land use and management. Sustainability 15(23):16259. https://doi.org/10.3390/su152316259

Soil Stewardship: Integrating Eco-holistic Approaches for Land Degradation Neutrality and Sustainable Development Goals

11

Shreyansh Singh, Sanjay Singh, and Trisha Roy

Abstract

The chapter explores the concept of soil stewardship, emphasizing its integral role in achieving land degradation neutrality (LDN) and advancing sustainable development goals. It highlights the critical need for sustainable soil management to address soil degradation driven by factors such as erosion, pollution, and climate change. The chapter defines soil stewardship, outlines its principles, and situates its historical evolution from traditional agricultural practices to modern eco-holistic approaches. Key strategies discussed include agroecology, permaculture, regenerative agriculture, and integrated soil fertility management, each illustrated with global case studies showcasing their potential for restoring soil health and enhancing agricultural productivity. The chapter also explores policy frameworks, implementation strategies, and the importance of monitoring systems for achieving LDN. Case studies, such as China's Loess Plateau Project and Ethiopia's Sustainable Land Management Program, provide practical insights into successful soil stewardship initiatives. Challenges such as limited awareness, governance barriers, and economic constraints are addressed, with recommendations for policy integration, capacity building, and innovative practices. By integrating ecological, agronomic, and socio-economic interventions, the chapter advocates for robust policies and collective action to ensure sustainable land management, soil health restoration, and resilience for future generations.

S. Singh
Forest Research Institute, Dehradun, Uttarakhand, India
e-mail: shreyanshsingh@outlook.com

S. Singh
Centre of Excellence on Sustainable Land Management (COE-SLM),
Indian Council of Forestry Research and Education, Dehradun, Uttarakhand, India

T. Roy
ICAR-Indian Institute of Soil and Water Commission, Dehradun, Uttarakhand, India

S. Singh, M. K. Gautam (eds.), *Biodiversity Conservation and Land Degradation Neutrality*, https://doi.org/10.1007/978-981-96-7210-3_11

Keywords

Soil stewardship · Land degradation neutrality (LDN) · Eco-holistic approaches · Sustainable development goals (SDGs) · Soil health restoration

11.1 Introduction

11.1.1 Background

Soil health is fundamental to agricultural productivity, ecosystem stability, and overall environmental sustainability. Healthy soils support plant growth that regulate water, filter pollutants, and store carbon, playing a crucial role in mitigating climate change (FAO 2017a, b, c). However, soil degradation caused by factors such as erosion, deforestation, overgrazing, and improper agricultural practices threatens these essential functions. Globally, it is estimated that 33% of soils are moderately to highly degraded (FAO 2015a, b, c). Every year around 24 billion tons of fertile soil is lost from agricultural land affecting the sustainability of the production systems (Hannam 2022). The soil resource is finite in nature and contributes to more than 90% of the world's food production. The maintenance of soil resources in its present form and averting further degradation is a major challenge.

Land degradation not only affects biodiversity and ecosystem services but also has significant socio-economic impacts. It undermines food security, exacerbates poverty, and drives migration and conflict (UNCCD 2017; Singh et al. 2024). Recognizing the critical need to address these issues, international frameworks such as the United Nations Convention to Combat Desertification (UNCCD) and the Sustainable Development Goals (SDGs) emphasize the importance of achieving land degradation neutrality (LDN) and promoting sustainable land management practices (UN 2015b; Kumar et al. 2024). The LDN targets offer a clear roadmap to how soils across different ecosystems can be preserved and protected to restrict further degradation and perpetuate sustainability as a concept (Kumar et al. 2021). However, the identification of the potential target areas for achieving LDN is a major prerequisite for any nation. Along with this, soil governance and policies/legislations to regulate the use of soil are major factors that need to be put forth by nations/countries to ensure proper soil management. "*Soil stewardship*" in simple terms is the art of "careful and responsible management of soil" which is the key focus of the chapter.

This chapter aims to explore soil stewardship through eco-holistic approaches, emphasizing their role in achieving LDN and supporting the broader objectives of the SDGs. It will dig into the principles and practices of soil stewardship, review current challenges and innovative strategies, and present case studies that highlight successful implementations. By integrating scientific insights with practical applications, this chapter seeks to provide a comprehensive guide for stakeholders committed to fostering sustainable land management.

11.2 Understanding Soil Stewardship

11.2.1 Definition and Principles

Soil stewardship refers to the responsible management and conservation of soil resources to ensure their sustainability and productivity for present and future generations. It encompasses a range of practices aimed at maintaining or enhancing soil health, thereby supporting agricultural productivity, biodiversity, and ecosystem services (Lal 2020). Core principles of soil stewardship include the following:

- Sustainability: Ensuring that soil use does not compromise the ability of future generations to meet their needs.
- Biodiversity Conservation: Promoting practices that enhance soil biodiversity, which is critical for nutrient cycling, pest control, and resilience against environmental stresses (Barrios 2007).
- Integrated Management: Combining traditional knowledge with modern science to develop holistic soil management practices that consider the ecological, economic, and social dimensions (Altieri 1999).

The term soil is often broadened as a "*Source of Infinite Life.*" Soil forms the very basis of all civilizations and life forms. Thus, the process of soil conservation and management is keenly knitted with society. The concept of soil stewardship can be put into practice only through the participation of society. Delinking the conservation measures from natural and agricultural socio-ecosystems and focusing on soil erosion management and fertility management cannot bring about soil stewardship. "*Citizen Science*" or the involvement of people can help achieve this goal.

11.2.2 Historical Context

The concept of soil stewardship is rooted in ancient agricultural practices. Traditional societies developed complex soil management techniques that allowed them to sustain agricultural productivity over centuries. For instance, the terracing systems of the Andean civilizations and the raised fields of the Mayan civilization are prime examples of sustainable soil management (Denevan 2001). In the modern era, the Green Revolution of the mid-twentieth century brought significant increases in agricultural productivity but often at the expense of soil health due to the overuse of chemical fertilizers and pesticides (Pretty 2008). This led to a renewed interest in sustainable soil management practices, integrating both traditional wisdom and modern scientific advancements.

11.2.3 Current Challenges

The core principle of soil management revolves around nurturing society's well-being. Despite clear warnings from modern agricultural practices, the conventional business model persists, endangering the survival of future generations. Lal (2024) argues for a paradigm shift, emphasizing the perception of soil as a living entity to foster a collective sense of responsibility and better protect this vital natural resource (de la Bellacasa 2019). Several pressing challenges confront soil stewardship today. Soil erosion, driven by water and wind, stands as a foremost threat, diminishing soil fertility and agricultural productivity (Pimentel et al. 1995). Pollution from industrial activities, improper waste disposal, and extensive agrochemical use have resulted in widespread soil contamination, posing significant risks to human health and ecosystems (McLaughlin and Singh 1999). Climate change exacerbates these issues through more frequent extreme weather events, altered precipitation patterns, and rising temperatures, further degrading soil quality (Lal 2004). Meanwhile, urbanization, deforestation, and agricultural expansion continue to disrupt soil structure and function, exacerbating degradation (Foley et al. 2005; Rani and Kumar 2024).

To tackle these challenges effectively, a transition toward eco-holistic approaches is imperative. These approaches integrate scientific knowledge with traditional and community-based practices to restore and maintain soil health, thereby ensuring the sustainability of agricultural systems and the ecosystems they support. In addition to these challenges, anthropogenic activities such as the Russia–Ukraine war have also significantly impacted soil management. Solokha et al. (2023) highlighted widespread contamination of heavy metals such as Cu, Cr, Ni, Mn, Fe, and Pb in shelled areas compared to non-shelled areas, potentially jeopardizing groundwater and surface water resources alike.

11.3 Eco-holistic Approaches (Fig. 11.1)

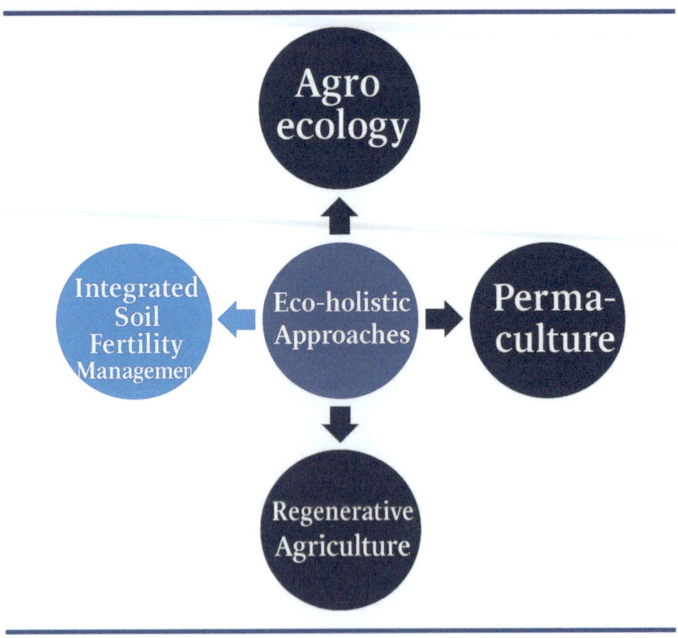

Fig. 11.1 Eco-holistic approaches linking agroecology, permaculture, regenerative agriculture, and soil fertility management. Eco-holistic approaches linking agroecology, permaculture, regenerative agriculture, and soil fertility management. Source: Created by Author

11.3.1 Agroecology

Agroecology is an integrative approach that applies ecological principles to agricultural systems to enhance sustainability, resilience, and biodiversity. It focuses on the interactions between plants, animals, humans, and the environment within agricultural systems. Key principles of agroecology include diversification of crops, integration of livestock, use of organic inputs, and conservation of natural habitats (Altieri 2002). These principles aim to create balanced and productive farming systems that are environmentally sound and socially just. Agroecological practices can improve soil health by increasing organic matter content, enhancing nutrient cycling, and promoting beneficial soil organisms (Gliessman 2007). It ensures the maintenance of both above- and belowground biodiversity and promotes the holistic and sustainable growth of the agroecosystems.

Agroecology promotes a variety of practices such as crop diversification and rotation, agroforestry, intercropping, the use of cover crops, and green manures. Biological pest control is favored over chemical inputs to maintain ecological balance. A prominent example of agroecology in practice is the "Campesino a Campesino" movement in Latin America. This farmer-to-farmer network emphasizes knowledge sharing and sustainable agricultural practices. Originating in Central America in the 1970s, it has spread across the region, leading to significant improvements in soil health, crop yields, and resilience to climate change. Farmers participating in the movement have reported reduced reliance on chemical fertilizers and pesticides, increased use of organic matter, and enhanced biodiversity on their farms, resulting in more resilient and productive agricultural systems (Holt 2006).

11.3.2 Permaculture

Permaculture is a design system for creating sustainable human environments that mimic the patterns and relationships found in nature. It integrates land, resources, people, and the environment through mutually beneficial synergies. Permaculture principles include observing and interacting with nature, catching and storing energy, producing no waste, and using small and slow solutions (Mollison 1988). Permaculture designs often incorporate elements such as swales, contour farming, and polycultures to improve water retention, reduce erosion, and enhance soil fertility. The design principles of permaculture involve integrating perennials and polycultures, implementing water management techniques such as swales and ponds, and building soil through composting and mulching. Zaytuna Farm in Australia, established by permaculture expert Geoff Lawton, serves as a successful case study. The farm, which was initially set on degraded land, has been transformed into a lush, productive landscape through the application of permaculture principles. Key practices include the creation of swales to capture and store water, extensive use of compost and mulch to build soil organic matter, and the integration of diverse plant and animal species to create a self-sustaining ecosystem. Over time, these practices have significantly improved soil structure, fertility, and moisture retention, demonstrating the potential of permaculture to rehabilitate degraded landscapes and promote sustainable agriculture (Lawton 2013).

11.3.3 Regenerative Agriculture

Regenerative agriculture focuses on restoring and enhancing the health and vitality of agricultural soils. It emphasizes practices that regenerate topsoil, increase biodiversity, improve water cycles, and strengthen the health of ecosystems. Core techniques include no-till farming, cover cropping, rotational grazing, and composting (Gosnell et al. 2019). These practices aim to sequester carbon, boost soil fertility, and improve crop resilience to climate variability. One of the most well-known success stories in regenerative agriculture is Gabe Brown's farm in North Dakota, USA. Over two decades, Brown implemented regenerative practices such as no-till farming, diverse cover crops, and holistic planned grazing. By avoiding tillage, Brown reduced soil disturbance and preserved soil structure. Cover crops helped to protect the soil from erosion, add organic matter, and enhance soil biology. Rotational grazing of livestock on pastures improved plant growth and soil health. These practices collectively transformed Brown's farm from degraded, low-productivity land into a highly productive and resilient system. The farm now boasts rich, dark soils with high organic matter content, improved water infiltration, and increased biodiversity, illustrating the potential of regenerative agriculture to restore soil health and productivity (Brown 2018).

11.3.4 Integrated Soil Fertility Management

Integrated soil fertility management (ISFM) combines the use of organic and inorganic soil amendments to optimize soil fertility and crop productivity. ISFM promotes the efficient use of fertilizers, organic inputs, and improved crop varieties adapted to local conditions. This approach aims to enhance nutrient availability, improve soil structure, and increase microbial activity (Vanlauwe et al. 2015). In practice, ISFM involves the strategic combination of compost, manure, crop residues, and mineral fertilizers tailored to specific soil and crop needs. An exemplary case is found in smallholder farming systems in sub-Saharan Africa. Here, ISFM practices have been implemented to address soil fertility constraints and improve food security. Farmers using ISFM techniques apply organic inputs like compost and manure to boost soil organic matter and nutrient content. They complement these with the judicious use of mineral fertilizers to address specific nutrient deficiencies. Improved seed varieties and crop rotations are also integral components of the system. Studies have shown that ISFM can significantly increase crop yields, enhance soil nutrient levels, and improve overall soil health, thereby contributing to sustainable agricultural intensification in resource-limited settings (Vanlauwe et al. 2015).

11.4 Strategies for Achieving Land Degradation Neutrality

11.4.1 Policy Frameworks

Achieving LDN requires robust policy frameworks that integrate environmental, agricultural, and social objectives. Effective policies must promote sustainable land management practices, incentivize conservation efforts, and ensure the participation of local communities. The UNCCD provides a global framework for LDN, emphasizing the need for national action programs and international cooperation (UNCCD 2017). Countries are encouraged to develop and implement policies that support sustainable land use, restoration of degraded lands, and prevention of further degradation. Policies should also promote research, education, and capacity building to equip stakeholders with the knowledge and skills needed for effective soil stewardship.

11.4.2 Implementation Strategies

Successful implementation of LDN strategies involves a combination of technical, institutional, and financial measures. Technically, it requires the adoption of best practices in land management, such as conservation agriculture, agroforestry, and ISFM. These practices are designed to enhance soil health, increase biodiversity, and improve water management (Lal 2015). Institutional measures include establishing clear land tenure rights, which are crucial for encouraging landholders to invest in sustainable practices. Strengthening local governance structures and fostering collaboration among stakeholders, including government agencies, non-governmental organizations, and local communities, are also key components.

Financially, achieving LDN necessitates adequate funding and investment. This can be facilitated through public and private sector partnerships, international aid, and innovative financing mechanisms such as payment for ecosystem services (PES) schemes. PES programs provide financial incentives to landholders for maintaining or enhancing ecosystem services, such as carbon sequestration, watershed protection, and biodiversity conservation. For example, Costa Rica's PES program has been successful in reducing deforestation and promoting reforestation by compensating landowners for preserving forest cover (Pagiola 2008).

11.4.3 Monitoring and Evaluation

Monitoring and evaluation are critical for assessing progress toward LDN and ensuring the effectiveness of implemented strategies. This involves establishing baseline data, setting measurable targets, and using indicators to track changes in land and soil health over time (Cowie et al. 2018). Remote sensing technologies, geographic information systems, and participatory monitoring approaches can be utilized to collect and analyze data. Effective monitoring systems should also incorporate feedback mechanisms to allow for adaptive management, enabling stakeholders to adjust practices based on observed outcomes and changing conditions.

The LDN Target Setting Programme, launched by the UNCCD, provides a framework for countries to set and achieve their LDN targets. The program supports countries in assessing land degradation, identifying priority areas for intervention, and developing action plans. Through this program, countries such as Senegal and Turkey have successfully integrated LDN targets into their national policies and land management strategies, demonstrating the potential for coordinated global action (UNCCD 2020).

11.4.4 Case Study

Several countries have made significant strides toward achieving LDN through comprehensive and innovative approaches. In China, the Loess Plateau Watershed Rehabilitation Project is a notable example (Xing et al. 2023). This project, implemented in one of the world's most severely eroded areas, employed a combination of reforestation, terracing, and sustainable agricultural practices. Over two decades, these efforts transformed the landscape, significantly reducing soil erosion, increasing vegetation cover, and improving agricultural productivity. The project also improved local livelihoods by increasing income from agricultural activities and creating new employment opportunities (World Bank 2003).

Another successful example is Ethiopia's Sustainable Land Management Program (SLMP), which integrates community-based approaches to land restoration. The program focuses on soil and water conservation, agroforestry, and the establishment of community watershed committees. By involving local communities in planning and implementation, the SLMP has enhanced ownership and sustainability of restoration efforts. As a result, the program has led to improved soil fertility, increased crop yields, and enhanced resilience to climate change (Gebremedhin et al. 2020).

These case studies highlight the importance of adopting integrated and participatory approaches to achieve LDN. By combining technical innovations with strong institutional frameworks and financial support, countries can effectively combat land degradation and promote sustainable development.

11.5 Sustainable Development Goals and Soil Stewardship

Soil health is essential for achieving the SDGs set by the United Nations. It plays a crucial role in food security, water filtration, climate regulation, and biodiversity, thereby directly affecting several SDGs, such as Zero Hunger (SDG 2), Clean Water and Sanitation (SDG 6), Climate Action (SDG 13), Life on Land (SDG 15), and Partnerships for the Goals (SDG 17) (UN 2015a, b).

11.5.1 Linking Soil Health to the SDGs

To achieve Zero Hunger (SDG 2), sustainable soil management is essential as it ensures long-term agricultural productivity and food security. Healthy soils provide essential nutrients for crops, resulting in higher yields and better food quality. Practices such as crop rotation, organic farming, and the use of biofertilizers improve soil fertility and structure, reducing the dependency on chemical inputs (FAO 2017a, b, c). For example, in India, the adoption of organic farming practices has led to enhanced soil health, increased crop yields, and higher farmer incomes, significantly boosting food security in rural areas (IFOAM 2016).

Soil health is closely connected to Clean Water and Sanitation (SDG 6). Healthy soils naturally filter pollutants and improve water quality. Techniques such as maintaining vegetation cover, reducing tillage, and using organic amendments help preserve soil structure and water infiltration capacity, minimizing runoff and erosion (FAO 2015a, b, c). The Tigris and Euphrates River Basin project highlights the importance of soil conservation practices in enhancing water quality. Integrated land and water management strategies, including soil erosion control and reforestation, have significantly improved water quality and availability for downstream communities (UNEP 2016).

Healthy soils significantly contribute to Climate Action (SDG 13) by acting as major carbon sinks. Practices such as no-till farming, cover cropping, and agroforestry can increase soil organic carbon sequestration, helping mitigate climate change (Kanta et al. 2024). The 4Per1000 initiative, introduced at COP21, aims to boost global soil organic matter stocks by 0.4% annually, demonstrating the potential of soil stewardship in climate mitigation efforts (Minasny et al. 2017). France's adoption of this initiative through extensive conservation agriculture and agroecology practices has shown promising results in increasing soil carbon storage and reducing greenhouse gas emissions (Rumpel et al. 2020).

Life on Land (SDG 15) emphasizes the sustainable management of forests, combating desertification, and preventing biodiversity loss. Soil stewardship practices contribute to this goal by improving soil health and promoting biodiversity both above and below ground. Methods such as integrated pest management, agroforestry, and habitat restoration help maintain ecological balance and support diverse ecosystems (FAO 2015a, b, c). In Kenya's Kakamega Forest, community-led agroforestry projects have successfully restored degraded land, increased biodiversity, and improved livelihoods, demonstrating the link between soil health and biodiversity conservation (Kitalyi et al. 2013).

Achieving the SDGs necessitates strong partnerships and collaborative efforts (SDG 17). Effective soil stewardship involves cooperation among governments, non-governmental organizations, local communities, researchers, and the private sector. Multi-stakeholder platforms and international collaborations, like the Global Soil Partnership, promote knowledge sharing, capacity building, and the development of best practices in soil management (UN 2015a, b). The Global Soil Partnership, coordinated by the Food and Agriculture Organization (FAO), promotes sustainable soil management worldwide through partnerships and collaborative projects, encouraging innovation and spreading successful practices across countries and regions (FAO 2017a, b, c).

11.6 Case Study: The Loess Plateau Watershed Rehabilitation Project, China

The Loess Plateau in northern China faced severe environmental degradation due to erosion-prone loess soil and unsustainable land practices. This degradation not only harmed local livelihoods but also exacerbated flooding downstream in the Yellow River. Initiated in 1994 with support from the World Bank and the Chinese government, the Loess Plateau Watershed Rehabilitation Project aimed to restore ecosystems, enhance soil and water conservation, increase vegetation cover, and promote sustainable agriculture. Strategies included extensive reforestation, terracing to reduce erosion, adoption of sustainable farming techniques, and community engagement to empower residents in decision-making (Chen et al. 2007).

The project achieved notable success, significantly increasing forest cover, reducing soil erosion, and improving water quality. Sustainable practices boosted agricultural productivity, diversified crops, and enhanced food security. Socially, the project reduced poverty, improved living standards, and fostered inclusive development through community organizations. Lessons learned emphasize long-term commitment, community involvement, and adaptive management, urging continued efforts to sustain gains, expand successful practices, and tackle evolving challenges such as climate change and urbanization (Chen et al. 2007).

11.7 Challenges and Potential Barriers

11.7.1 Soil Degradation and Land Use Pressures

One of the primary challenges facing soil stewardship efforts is the pervasive issue of soil degradation, exacerbated by intensive land use practices and increasing population pressure (FAO 2015a, b, c). Soil degradation manifests in various forms, including erosion, salinization, compaction, and loss of organic matter, leading to reduced soil fertility, diminished agricultural productivity, and degradation of ecosystem services (Lal 2015). The expansion of agriculture, urbanization, and industrial activities often leads to the conversion of natural habitats and ecosystems, further exacerbating soil degradation and fragmentation (UNEP 2018).

11.7.2 Limited Awareness and Capacity

A significant barrier to effective soil stewardship is the limited awareness and capacity among stakeholders, including farmers, policymakers, and the general public, regarding the importance of soil health and sustainable land management practices

(FAO 2017a, b, c). Many farmers continue to rely on conventional agricultural practices that degrade soil quality and undermine long-term productivity, often due to a lack of access to information, technical support, and financial resources (Giller et al. 2015). Similarly, policymakers may prioritize short-term economic gains over long-term environmental sustainability, leading to inadequate investments in soil conservation and land restoration efforts (Lipper et al. 2014).

11.7.3 Institutional and Governance Challenges

Institutional and governance challenges pose significant barriers to effective soil stewardship and sustainable land management. Weak institutional capacity, inadequate enforcement of regulations, and lack of coordination among government agencies often hinder the implementation of soil conservation measures and the enforcement of land use policies (FAO 2016a, b). Additionally, competing interests, power imbalances, and vested interests may impede efforts to promote sustainable land management practices and secure tenure rights for smallholder farmers and marginalized communities (Deininger and Byerlee 2011).

11.7.4 Economic and Market Dynamics

Economic and market dynamics can also present challenges to soil stewardship initiatives. The prevalence of market-driven agricultural systems, characterized by intensive monoculture production, input-dependent farming practices, and globalized supply chains, often incentivizes short-term profit maximization at the expense of soil health and environmental sustainability (Foley et al. 2011). Furthermore, the externalization of environmental costs and the lack of incentives for ecosystem services provision may discourage investment in soil conservation and sustainable land management practices (Swinton et al. 2007).

11.8 Recommendations for Policy and Practice

11.8.1 Policy Recommendations

Policy interventions are essential for creating an enabling environment for sustainable land management and soil stewardship. Based on the findings and insights presented in this chapter, several policy recommendations can be proposed. Governments should prioritize the integration of soil health considerations into national development policies, agricultural strategies, and environmental agendas. This can be achieved through the development and implementation of comprehensive soil conservation and land management policies that incentivize sustainable practices and provide support for soil rehabilitation efforts (FAO 2015a, b, c). Policymakers should explore innovative financing mechanisms to incentivize sustainable land management practices and soil conservation efforts. This could include the establishment of PES programs, tax incentives for adopting sustainable farming practices, and subsidies for investments in soil conservation measures (Pagiola 2008). Additionally, governments should invest in capacity-building initiatives to empower farmers, land managers, and local communities with the knowledge and skills needed for sustainable land management. This could involve the development of extension services, training programs, and knowledge-sharing platforms to disseminate best practices and promote peer-to-peer learning (FAO 2017a, b, c).

11.8.2 Practice Recommendations

In addition to policy interventions, on-the-ground practices are crucial for implementing sustainable land management and soil stewardship. Based on the case studies and success stories discussed in this chapter, several practice recommendations can be made. Farmers and land managers should prioritize the adoption of agroecological practices that promote soil health, biodiversity, and resilience. This includes agroforestry, crop rotation, intercropping, and organic farming techniques that minimize soil disturbance, enhance nutrient cycling, and improve soil structure (Gliessman 2007). Landowners and agricultural producers should invest in soil conservation measures to prevent erosion, reduce soil degradation, and improve soil fertility. This involves constructing terraces, bunds, and contour barriers to control water runoff, as well as implementing cover cropping and mulching techniques to protect soil from erosion (Lal 2015). Additionally, local authorities and land use planners should prioritize sustainable land use planning approaches that balance competing land uses, protect natural ecosystems, and promote landscape resilience. This includes zoning regulations, land use incentives, and participatory planning processes that engage stakeholders in decision-making and ensure the equitable distribution of land resources (FAO 2016a, b).

11.9 Conclusion and Future Directions

In this chapter, we explored the importance of soil stewardship in achieving LDN and advancing the SDGs. We reviewed various eco-holistic approaches, policy frameworks, implementation strategies, and case studies, highlighting their effects on soil health, agricultural productivity, and environmental sustainability. Key insights from our investigation reveal the necessity of integrating ecological, agronomic, and socio-economic interventions to address the complexities of land degradation and promote sustainable land management. Active participation and engagement of local communities are essential for the success and sustainability of soil stewardship initiatives. Empowering communities and enhancing their capacity for sustainable land management are crucial for achieving long-term results. Robust policy frameworks, supported by political will and financial backing, are required to encourage sustainable land management practices, stimulate investments in soil health, and foster collaboration among various stakeholders. Furthermore, a comprehensive monitoring and evaluation system is vital for tracking progress toward LDN, assessing soil health metrics, and identifying areas needing improvement. Adaptive management strategies incorporating feedback mechanisms and continuous learning are critical for ongoing enhancement.

Looking to the future, several priorities and pathways emerge to advance soil stewardship and achieve LDN. Scaling up successful initiatives and expanding proven practices are vital for maximizing impact and promoting widespread adoption of sustainable land management strategies. Ongoing investment in research, innovation, and technological advancements is essential to deepen our understanding of soil health, develop new tools and methodologies for soil conservation, and address emerging challenges such as climate change and land use changes. Strengthening collaboration and partnerships among governments, international organizations, academia, civil society, and the private sector is crucial for mobilizing resources, sharing expertise, and fostering collective action toward LDN. Integrating soil health considerations into broader policy agendas, development frameworks, and investment plans is necessary for mainstreaming soil stewardship and ensuring its long-term sustainability.

In conclusion, soil stewardship is fundamental in achieving LDN, promoting sustainable agriculture, and advancing the SDGs. By adopting eco-holistic methodologies, facilitating community engagement, and reinforcing policy support, we can effectively combat land degradation, restore soil health, and build resilient landscapes for future generations' prosperity.

References

Altieri MA (1999) The ecological role of biodiversity in agroecosystems. Agric Ecosyst Environ 74(1–3):19–31. https://doi.org/10.1016/S0167-8809(99)00028-6

Altieri MA (2002) Agroecology: the science of sustainable agriculture. CRC Press, Boca Raton

Barrios E (2007) Soil biota, ecosystem services and land productivity. Ecol Econ 64(2):269–285. https://doi.org/10.1016/j.ecolecon.2007.03.004

Brown G (2018) Dirt to soil: one family's journey into regenerative agriculture. Chelsea Green Publishing

Chen L, Wei W, Fu B, Lü Y (2007) Soil and water conservation on the Loess Plateau in China: review and perspective. Prog Phys Geogr 31(4):389–403. https://doi.org/10.1177/0309133307081290 (Original work published 2007)

Cowie AL, Orr BJ, Castillo Sanchez VM et al (2018) Land in balance: the scientific conceptual framework for land degradation neutrality. Environ Sci Pol 79:25–35. https://doi.org/10.1016/j.envsci.2017.10.011

de la Bellacasa MP (2019) Re-animating soils: transforming human-soil affections through science, culture, and community. Sociol Rev 67(2):391–407

Deininger K, Byerlee D (2011) Rising global interest in farmland: can it yield sustainable and equitable benefits? The World Bank 47475. https://doi.org/10.1596/1813-9450-47475

Denevan WM (2001) Cultivated landscapes of native Amazonia and the Andes: triumphs and tragedies. Oxford University Press

FAO (2015a) Soil and water: a larger perspective for soil health. Retrieved from [FAO]. http://www.fao.org/documents/card/en/c/4a9d3d14-e35e-4a5e-9f2b-4c42228d1bcf/

FAO (2015b) Status of the world's soil resources. Food and Agriculture Organization of the United Nations. Retrieved from [FAO]. http://www.fao.org/3/i5199e/i5199e.pdf

FAO (2015c) Sustainable soil management: A strategy for the provision of ecosystem services. Retrieved from [FAO]. http://www.fao.org/3/a-i5199e.pdf

FAO (2016a) Land-use planning for sustainable development. Retrieved from [FAO]. http://www.fao.org/3/a-i5210e.pdf

FAO (2016b) Land-use planning for sustainable development. Retrieved from [FAO]. https://www.fao.org/in-action/forest-landscape-restorationmechanism/knowledge-base/land-use-planning/en/

FAO (2017a) Soil and water: a larger perspective for soil health. Retrieved from [FAO]. https://openknowledge.fao.org/3/cc0900en/cc0900en.pdf on June 12, 2024, 11:10 AM

FAO (2017b) Soil organic carbon: the hidden potential. Retrieved from [FAO]. https://openknowledge.fao.org/server/api/core/bitstreams/584567aa-d89d-4d9a-82d8-1fad58266364/content on June 14, 2024, 10:16 AM

FAO (2017c) The future of food and agriculture: trends and challenges. Food and Agriculture Organization of the United Nations. Retrieved from [FAO]. http://www.fao.org/3/a-i6583e.pdf

Foley JA, DeFries R, Asner GP et al (2005) Global consequences of land use. Science 309(5734):570–574. https://doi.org/10.1126/science.1111772

Foley JA, Ramankutty N, Brauman KA et al (2011) Solutions for a cultivated planet. Nature 478(7369):337–342. https://doi.org/10.1038/nature10452

Gebremedhin B, Tesfaye K, Alemu D (2020) The impact of sustainable land management on household income and crop production in the highlands of Ethiopia. Agric Econ 51(6):879–894. https://doi.org/10.1111/agec.12588

Giller KE, Witter E, Corbeels M et al (2015) Conservation agriculture and smallholder farming in Africa: the heretics' view. Field Crop Res 174:17–23. https://doi.org/10.1016/j.fcr.2014.04.014

Gliessman SR (2007) Agroecology: the ecology of sustainable food systems. CRC Press, Boca Raton

Gosnell H, Gill N, Voyer M (2019) Transformational adaptation on the farm: processes of change and persistence in transitions to 'climate-smart' regenerative agriculture. Glob Environ Chang 59:101965. https://doi.org/10.1016/j.gloenvcha.2019.101965

Hannam I (2022) Soil governance and land degradation neutrality. Soil Security 6:100030

Holt-Giménez E (2006) Campesino a Campesino: voices from Latin America's farmer to farmer movement for sustainable agriculture. Food First Books

IFOAM (2016) The world of organic agriculture. Retrieved from [IFOAM]. https://www.ifoam.bio/sites/default/files/world_of_organic_agriculture_2016_web.pdf

Kanta C, Kumar A, Chauhan A, Singh H, Sharma IP (2024) The interplay between plant functional traits and climate change. In: Plant functional traits for improving productivity. Springer Nature, pp 41–58

Kitalyi A, Mwendia S, Maiyo L et al (2013) Community-led agroforestry practices improve livelihoods and biodiversity in Kakamega Forest, Kenya. Agrofor Syst 87(3):599–611. https://doi.org/10.1007/s10457-012-9574-1

Kumar A, Singh S, Chand HB, Kewat AK, Singh KP (2021) Road towards restoring land degradation in India: a UNCCD perspective. RJCES 9(4):27–38

Kumar A, Singh S, Kumar R, Kumar M, Rani K, James J (2024) Balancing conservation and development for achieving land degradation neutrality. Trends Curr Biol 2(4):17–24

Lal R (2004) Soil carbon sequestration impacts on global climate change and food security. Science 304(5677):1623–1627. https://doi.org/10.1126/science.1097396

Lal R (2015) Restoring soil quality to mitigate soil degradation. Sustain For 7(5):5875–5895. https://doi.org/10.3390/su7055875

Lal R (2020) Regenerative agriculture for food and climate. J Soil Water Conserv 75(5):123A–124A. https://doi.org/10.2489/jswc.2020.0620A

Lal R (2024) Soil, soul, spirituality, and stewardship. J Soil Water Conserv 79(1):10A–14A

Lawton G (2013) Permaculture design course. Zaytuna Farm
Lipper L, Thornton P, Campbell BM et al (2014) Climate-smart agriculture for food security. Nat Clim Chang 4(12):1068–1072. https://doi.org/10.1038/nclimate2437
McLaughlin MJ, Singh BR (1999) Cadmium in soils and plants. Plant Soil 199(2):127–138. https://doi.org/10.1023/A:1004342211533
Minasny B et al (2017) Soil carbon 4 per mille. Geoderma 292:59–86. https://doi.org/10.1016/j.geoderma.2017.01.002
Mollison B (1988) Permaculture: a designer's manual. Tagari Publications
Pagiola S (2008) Payments for environmental services in Costa Rica. Ecol Econ 65(4):712–724. https://doi.org/10.1016/j.ecolecon.2007.07.033
Pimentel D, Harvey C, Resosudarmo P et al (1995) Environmental and economic costs of soil erosion and conservation benefits. Science 267(5201):1117–1123. https://doi.org/10.1126/science.267.5201.1117
Pretty J (2008) Agricultural sustainability: concepts, principles and evidence. Philos Trans R Soc B 363(1491):447–465. https://doi.org/10.1098/rstb.2007.2163
Rani K, Kumar A (2024) Biotechnological innovations in urban forestry: adapting and mitigating climate change and environmental pollution. In: Singh H (ed) Urban forests, climate change and environmental pollution. Springer, Cham, p 573
Rumpel C, Chabbi A, Garcia-Franco N (2020) The impact of conservation agriculture and agroecology on soil carbon sequestration. Agronomy 10(9):1–22
Singh S, Kumar A, Kumar A (2024) Harmony in extraction: navigating mining, land degradation and ecological restoration for sustainable futures. In: Forests and climate change: biological perspectives on impact, adaptation, and mitigation strategies. Springer Nature, pp 513–534
Solokha M, Pereira P, Symochko L, Vynokurova N, Demyanyuk O, Sementsova K et al (2023) Russian-Ukrainian war impacts on the environment. Evidence from the field on soil properties and remote sensing. Sci Total Environ 902:166122
Swinton SM, Lupi F, Robertson GP et al (2007) Ecosystem services and agriculture: cultivating agricultural ecosystems for diverse benefits. Ecol Econ 64(2):245–252
UN (2015a) Transforming our world: the 2030 agenda for sustainable development. Retrieved from [UN]. https://sustainabledevelopment.un.org/post2015/transformingourworld
UN (2015b) Transforming our world: the 2030 agenda for sustainable development. United Nations. Retrieved from [UN]. https://sustainabledevelopment.un.org/post2015/transformingourworld/publication
UNCCD (2017) Global land outlook. United Nations convention to combat desertification. Retrieved from [UNCCD]. https://www.unccd.int/sites/default/files/documents/2017-09/GLO_Full_Report_low_res.pdf
UNCCD (2020) Land degradation neutrality target setting Programme: country reports. United Nations convention to combat desertification. Retrieved from [UNCCD]. https://www.unccd.int/actions/ldn-target-setting-programme
UNEP (2016) The Mesopotamian Marshlands: demise and the pathway to recovery. Retrieved from [UNEP]. https://wedocs.unep.org/handle/20.500.11822/7747
UNEP (2018) Global environmental outlook. Retrieved from [UNEP]. https://www.unep.org/resources/global-environment-outlook-6
Vanlauwe B, Wendt J, Giller KE (2015) A fourth principle is required to define conservation agriculture in sub-Saharan Africa: the appropriate use of fertilizer to enhance crop productivity. Field Crop Res 155:10–13. https://doi.org/10.1016/j.fcr.2013.10.002
World Bank (2003) Implementation completion report on a loan in the amount of US$149.2 million to the People's Republic of China for a loess plateau watershed rehabilitation project. Retrieved from [World Bank]. http://documents.worldbank.org/curated/en/391321468743924773/pdf/multi0page.pdf
Xing J, Zhang J, Wang J, Li M, Nie S, Qian M (2023) Ecological restoration in the loess plateau, China necessitates targeted management strategy: evidence from the Beiluo River basin. In forests (vol. 14, issue 9, p. 1753). MDPI AG. https://doi.org/10.3390/f14091753

Combating Land Degradation and Desertification Through Forestry Interventions

12

Sanjay Singh, Abhishek Kumar, and Mukesh K. Gautam

Abstract

Land degradation poses a signification challenge globally. In India, 97.85 million hectares of land is affected by various drivers of degradation. Water erosion (11.0%), vegetation degradation (9.15%), and wind erosion (5.46%) are identified as the primary drivers. Identifying land degradation as a pressing challenge impacting ecosystems, biodiversity, livelihoods, and food security, India has launched multiple initiatives, including afforestation, watershed management, and agroforestry programs, to combat degradation and achieve land degradation neutrality (LDN). This chapter examines the status of land degradation in India, its commitments under international frameworks, and the efforts being made to restore degraded landscapes. A multi-tier landscape-level planning approach is proposed to align regional interventions with national and global targets. Additionally, the convergence of resources and monitoring mechanisms is emphasized to ensure effective implementation and tracking of progress. This study underscores the importance of integrated efforts in addressing land degradation and advancing sustainable development in India.

S. Singh (✉)
Centre of Excellence on Sustainable Land Management, Indian Council of Forestry Research and Education, Dehradun, Uttarakhand, India
e-mail: sanjaysn@icfre.org

A. Kumar
Department of Biotechnology, Graphic Era (Deemed to be University), Dehradun, Uttarakhand, India

M. K. Gautam
Biology Department, Medgar Evers College, City University of New York, NY, USA

S. Singh, M. K. Gautam (eds.), *Biodiversity Conservation and Land Degradation Neutrality*, https://doi.org/10.1007/978-981-96-7210-3_12

Keywords

Ecosystem restoration · Land degradation · Sustainable land management · Land degradation neutrality · Multi-tier landscape approach

12.1 Status of Land Degradation in India

The most recent mapping (2018–2019) by the Space Applications Centre Indian Space Research Organisation (ISRO) shows that 97.9 million hectares (Mha) (29.8% of India's total geographic area) are undergoing land degradation. Nine state accounts for 80% of the total land degradation in the country, while 13 states have more than 30% of their area degraded. Among these, the highest degraded land was recorded in Rajasthan (22% of the total degraded land) followed by Maharashtra (15%), Gujarat (10%), Ladakh (7%), Karnataka (7%), Jharkhand (6%), Odisha (5%), Madhya Pradesh (4%), and Telangana (4%), while the other states together account for only 20% of the total degraded land (SAC 2021).

Between 2003–2005 and 2018–2019, India saw an increase of approximately 3.32 Mha of degraded land. There are 14 states where the degraded land has increased by over 10% between the two reporting periods (Sengupta 2021). Mizoram experienced the highest increase in land degradation percentage of becoming between the two reporting periods (87%), followed by Tripura (37%), Arunachal Pradesh (47%), Punjab (80%), and Assam (46%). Additionally, more than 10% of land was degraded in Nagaland, Delhi, Jammu and Kashmir, Meghalaya, Uttarakhand, Haryana, Kerala, Bihar, and Himachal Pradesh (Table 12.1).

The Desertification and Land Degradation Atlas of India (2021) identifies the main drivers of degradation as water erosion (11.0%), vegetation degradation (9.15%), and wind erosion (5.46%)/net degradation of around an additional 3.32 Mha approximately (Fig. 12.1). It is evident that the area affected by human settlement, mass movements, frosts, waterlogging, wind erosion, and degradation of vegetation has increased over the past decades. On the other hand, the area defined as a rocky barren area has remained the same; however, the region affected by wind erosion and salinity has decreased. The pattern indicates the dynamic shifts in the driver of land degradation (SAC 2021).

12.2 India's Commitment to Achieve Land Degradation Neutrality

Globally land degradation is identified as one of the major threats to ecosystem, biodiversity, livelihood, food security, and overall well-being of people. The Government of India has launched various initiatives in the form of rural and watershed development programs for combating land degradation since the 1980s. Among these efforts, Soil Conservation in the Catchment of River Valley Project (1962–1963), the Reclamation and Development of Alkali and Acid Soils during the

Table 12.1 Total area under degradation of the states

S. no	States	TGA (km^2)	Total area under degradation (km^2)			TGA under degradation (%)		
			2003–2005	2013–2015	2018–2019	2003–2005	2013–2015	2018–2019
1	Andhra Pradesh	162,989	22,677.28	22,987.58	23,780.42	14.16	14.35	14.84
2	Arunachal Pradesh	83,743	1366.86	1539.33	2006.83	1.63	1.84	2.4
3	Assam	78,438	5722.15	7165.96	8345.3	7.3	9.14	10.64
4	Bihar	94,171	6595.39	6948.09	7465.86	7.0	7.38	7.93
5	Chhattisgarh	135,194	21,763.88	22,111.53	23,065.31	16.1	16.36	17.06
6	Delhi	1483	735.14	898.68	915.43	49.57	60.6	61.73
7	Goa	3702	1864.58	1929.73	1948.77	50.37	52.13	52.64
8	Gujarat	196,024	100,774.6	102,616.4	102,480.6	51.35	52.29	52.22
9	Haryana	44,212	3145.83	3389.64	3641.54	7.12	7.67	8.24
10	Himachal Pradesh	55,673	21,413.66	23,942.4	24,003	38.46	43.01	43.11
11	Jammu and Kashmir	222,236	9667.95	10,647.21	11,295.03	17.86	19.67	20.86
12	Jharkhand	79,706	54,186.57	54,987.26	54,822.6	67.97	68.98	68.77
13	Karnataka	191,791	69,409.43	69,510	69,598.47	36.19	36.24	36.29
14	Kerala	38,863	3705.12	3795.87	4222.99	9.54	9.77	10.87
15	Ladakh	308,252	65,786.53	69,115.73	71,119.68	39.14	41.12	42.31
16	Madhya Pradesh	307,690	37,718.53	38,043.15	38,597.35	12.24	12.34	12.52
17	Maharashtra	22,327	133,486	138,259.4	143,060.3	43.38	44.93	46.49
18	Manipur	22,429	5930.93	6019.59	6125.66	26.56	26.96	27.44
19	Meghalaya	21,081	4788.25	4948.8	5575.76	21.35	22.06	24.86
20	Mizoram	16,579	958.73	1874.53	2758.27	4.55	8.89	13.08
21	Nagaland	155,707	6423.04	7866.78	8289.43	38.74	47.45	50
22	Orissa	50,362	53,219.03	53,041.14	53,590.14	34.18	34.06	34.42
23	Punjab	342,239	931.15	1446.53	1679.89	1.85	2.87	3.34
24	Rajasthan	7096	216,256	215,265.1	212,376.7	63.19	62.9	62.06
25	Sikkim	130,058	784.82	787.49	846.1	11.06	11.1	11.92
26	Tamil Nadu	112,079	15,166.6	15,438.98	15,999.81	11.66	11.87	12.3

(continued)

Table 12.1 (continued)

S. no	States	TGA (km^2)	Total area under degradation (km^2)			TGA under degradation (%)		
			2003–2005	2013–2015	2018–2019	2003–2005	2013–2015	2018–2019
27	Telangana	10,486	36,584.82	35,988.56	36,385.08	31.86	31.34	31.68
28	Tripura	53,483	3273.02	4371.28	4473.78	31.21	41.69	42.66
29	Uttar Pradesh	240,928	18,352.63	15,289.97	15,496.08	7.62	6.35	6.43
30	Uttarakhand	88,752	5812.41	6482.53	6738.94	10.87	12.12	12.6
31	West Bengal	9490	16821.81	17339.31	17843.45	18.95	19.54	20.1
	Total	*3,287,263*	*945,323*	*964,049*	*978,549*	*28.76*	*29.32*	*29.77*

Source: SAC (2021)

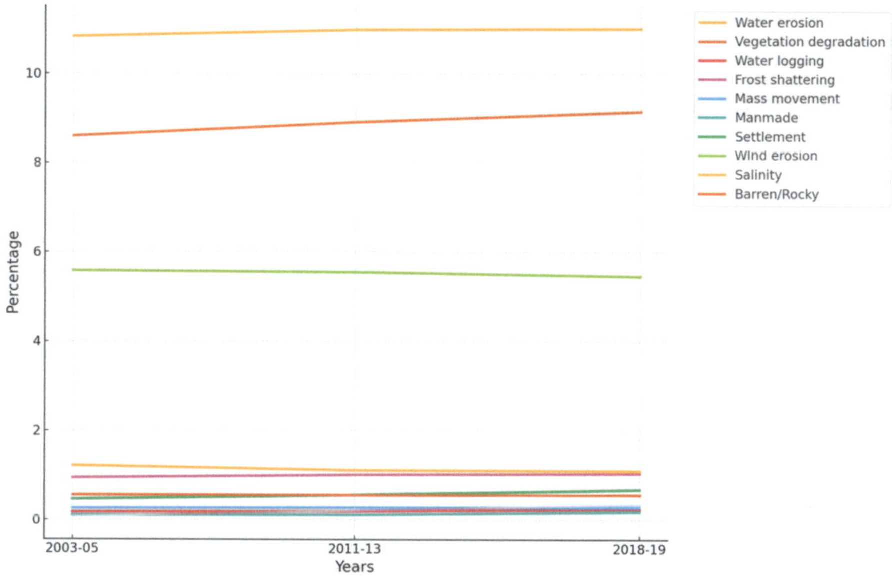

Fig. 12.1 Percentage of total degraded land affected by the various drivers of degradation. (Source: SAC (2021))

Seventh Five-Year Plan, and the Watershed Development Project for Shifting Cultivation Areas (1987–1988) are notable early efforts of Government of India (Singh et al. 2023).

Some of the projects like the Drought-Prone Area Program (1973–1974), the National Watershed Development Project for Rainfed Areas (1985–1986), the Desert Development Program for 1977–1978, and the Integrated Watershed Development Program for 1989 were merged into the Integrated Watershed Management Program in 2008 and are now part of the Pradhan-Mantri Krishi Sinchai Yojana (Prime Minister's agricultural irrigation scheme) under the Ministry of Agriculture and Farmers Welfare. In addition to the above, programs such as the Mahatma Gandhi National Rural Employment Guarantee Act (MGNREGA) and the Green India Mission have further contributed to combating land degradation (Singh et al. 2023).

The sectoral policies of the Government of India, viz., the National Water Policy (2012), the National Forest Policy (1988), the National Agriculture Policy (2000), the National Policy for Farmers (2007), and the National Agroforestry Policy (2014), address various aspects of land degradation, further strengthening the commitment of the Government of India in addressing land degradation (Singh et al. 2023).

India recognizes the importance of land degradation neutrality (LDN) as land degradation has been a focus of the three Rio Conventions for more than 20 years and restoration programs. Each convention has its own goals and commitments, yet they are interconnected. United Nations Convention to Combat Desertification

Table 12.2 International commitments and their linkages to LDN

International commitments	Link with LDN
UNCCD	Since 2015, LDN has spearheaded the progress of the SDGs and the implementation of conventions
UNFCCC	FLR initiatives to provide outcome-based compensation for mitigation efforts related to forests and land use
CBD	Contributions from FLR and LDN are connected to the Kumming Montreal Biodiversity Framework
Bonn challenge	The Bonn challenge's FLR technique can be used as a foundation for cooperatively accomplishing national goals and agreements
SDG	SDG goals 1, 2, 5, 6, 7, 8, 12, and 13 are all connected to LDN (SDG goal 15.3)

(UNCCD) envisions attaining LDN by 2030. This complements and accelerates land use-related United Nations Framework Convention on Climate Change (UNFCCC) and CBD aims. Land management and restoration are vital to the CBD and are addressed in the Kumming Montreal Biodiversity Framework (restore ecosystems) (Ekardt et al. 2023). Land management and restoration promote climate change adaptation by reducing emissions and improving carbon sequestration (Table 12.2).

Under the Bonn challenge, India was committed to restoring more than 13 Mha of deforested land by 2020 and an additional 8 Mha by 2030 (Borah et al. 2018). This target was further revised and increased by an additional 6 Mha at the 14th meeting of the Conference of Parties (COP) of the United Nations Convention to Combat Desertification, in 2019 (Singh et al. 2023). Under the Paris Agreement, at the 21st meeting of COP of the UNFCCC, India outlined the creation of an additional carbon sink of 2.5–3 billion tons of CO_2 through additional forest and tree cover (an increase of about 680–817 million tons of carbon stock) (Bisht and Singhal 2021).

12.3 Scope of Achieving LDN in India and National Action Plan

According to the Desertification and Land Degradation Atlas (2021) by SAC, ISRO, 97.9 Mha are experiencing desertification or land degradation. SAC data per land use class report that 60% of all land degradation occurs on unirrigated agricultural and forest land (SAC 2021). Forest Survey of India (FSI)'s reports identify 63 Mha of potential areas for restoration across different categories. The estimation establishes that the largest potential of creating additional carbon sinks lies in the restoration of forests that have been impaired in the last 15–20 years, along with activities such as restoration of open forests, afforestation on wastelands, and agroforestry (Ashutosh et al. 2021).

Ten categories of activities have been considered by the FSI for the potential increase in carbon sink and forest cover in the country. These include restoration of impaired forests, restoration of open forests, afforestation on wastelands, agroforestry, setting up of green corridors, undertaking plantations along roads and railways, undertaking plantation on railway sidings, along rivers and canals, and the creation of urban green spaces (Ashutosh et al. 2021).

12.3.1 Afforestation Achievements (2011–2012 to 2021–2022)

Conservation and forest restoration, biodiversity enhancement, and green cover expansion have been given high priority under various policies, acts, rules, and programs. Effective collaborations between both the National and State Governments have strengthened afforestation efforts through multiple plans and schemes. The afforestation efforts are being implemented under various plans and schemes. As a result, a total of 18.9 Mha area has been brought under afforestation from 2011–2012 to 2021–2022 by different implementing agencies and under different schemes or components of schemes (NAP 2023).

The current rate of afforestation and eco-restoration activities falls short of about 7 Mha from the stipulated target of 26 Mha (Table 12.3). To accommodate the shortfall, it is critical to choose additional areas from states that are prone to desertification or land degradation or have a considerably high area classified as open forests (NAP 2023).

Table 12.3 Afforestation achievements showing the afforested area (in million hectares) covered from 2011–2012 to 2021–2022

Year	Target	Target achieved
2011–2012	1.74	1.6
2012–2013	1.54	1.63
2013–2014	1.48	1.62
2014–2015	1.53	1.35
2015–2016	1.16	1.38
2016–2017	1.09	1.99
2017–2018	1.45	1.68
2018–2019	1.72	1.63
2019–2020	1.88	2.07
2020–2021	2.23	2.21
2021–2022	2.69	1.78
Total	20.74	18.94

Source: NAP (2023)

12.3.2 Government Schemes for Achieving LDN

Nationally defined contributions (NDCs) have emerged as India's best endeavor to balance the developmental imperatives of sustainable development and poverty eradication. In fact, many proactive eco-restoration activities are underway to meet the commitments, inherent in the approaches to the ideas of common but differentiated responsibilities, respective capabilities, and equity in terms of the National Action Plan, 2023. Nature-based solutions (NbSs) via an ecosystem-based approach stand paramount among the treatment measures. Four sectors might be employed to categorize government programs to achieve LDN (Singh et al. 2023), such as different sector-specific interventions that have been put in place to deal with environmental and agricultural issues. In the forestry sector, interventions include the Green India Mission, Eco-development forces, afforestation, the Compensatory Afforestation Fund Management and Planning Authority, the National Coastal Mission, and river rejuvenation projects. Significant initiatives in the agriculture sector are the National Mission on Sustainable Agriculture, the Rainfed Area Development Program, and the Soil Health Cards, a specific program for the distribution of the same. Watershed Management through the Integrated Watershed Program is included under MGNREGA and other state-specific programs under the Integrated Watershed Management Programme (IWMP). Other initiatives include the National Livestock Mission and its sub-mission on feed and fodder development, the National Green Highway Mission, Atal Bhujal Yojana, Namami Gange, and the Command Area Development and Water Management Program.

12.4 Land Degradation Neutrality Through Sustainable Land Management

LDN according to Parties to the Convention is defined as: "A state whereby the amount and quality of land resources, necessary to support ecosystem functions and services and enhance food security, remains stable or increases within specified temporal and spatial scales and ecosystems." LDN will help achieve the goals of CBD, UNFCC, and UNCCD and also the SDG goals. Sustainable Land Management (SLM) is an effective tool to achieve LDN. SLM practices can preserve and enhance ecosystem services in all land use systems while also supporting human well-being. Furthermore, SLM has great potential and adaptability to local conditions.

The United Nations defines SLM as "the use of land resources, including soils, water, animals and plants, for the production of goods to meet changing human needs, while simultaneously ensuring the long-term productive potential of these resources and the maintenance of their environmental functions." Adding to this, TerrAfrica (2005) further elaborates SLM as "the adoption of land use systems that through appropriate management practices enable land users to maximize the economic and social benefits from the land while maintaining or enhancing the ecological support functions of the land resources" (FAO 2023).

The essence of Sustainable Land Management (SLM) lies in its key components. These components are soil conservation techniques, water management strategies, agroecological approaches, and integrated landscape management (FAO n.d.).

12.5 Center of Excellence on Sustainable Land Management

To address the issue of land degradation and to promote South–South cooperation, the establishment of the Center of Excellence on Sustainable Land Management (CoE-SLM) was announced in 2019 during COP 14 of UNCCD, and in 2023, it was established at the Indian Council of Forestry Research and Education. In India and other developing countries that are signatories to the UNCCD, the Center aims to promote SLM practices, achieve LDN, and provide technical assistance at the national and subnational levels (Singh et al. 2023). The center aims to provide support for the development and implementation of strategies, policies, regulations, by-laws, and community oversight mechanisms to create a strong institutional framework for SLM for the global south countries. COE-SLM is dedicated to advancing and upscaling of SLM practices through awareness generation, enhancing the profile, and securing the recognition of SLM to ensure its integration into mainstream land use practices (Singh et al. 2023).

COE SLM aims to leverage on the existing knowledge by connecting actors with networks comprising international and national agencies, institutions, and individuals to facilitate collaboration and knowledge sharing. The center maintains repositories of resources, including tools and training materials, to provide stakeholders with easy access to information on best practices. The center is working to stimulate greater action, commitments, and investments in SLM, driving its adoption and success across diverse landscapes. As of March 2024, the center has documented 43 Agronomic best practices, 27 best vegetative practices, 36 structural best practices, and 6 other best practices for SLM, besides performing the capacity-building program at the national and international levels.

12.5.1 Multi-Tier Landscape Level Planning and Implementation

The CoE-SLM proposes a multi-tier landscape-level planning and implementation, where regional SLM interventions contribute to national goals and global targets (Fig. 12.2). This approach recognizes the diversity of landscapes, ecosystems, and services derived from the ecosystem. It advocates utilizing tailored solutions at the regional level, designed through the participation of various stakeholders, which includes communities, governments, and businesses. It further emphasizes on people's participation in identifying land degradation risks, opportunities for restoration, and promotion of sustainable land use practices. Based on the priority areas for interventions, suitable SLM practices can be customized to the prevailing social and ecological conditions that can be utilized for restoration (Singh et al. 2023).

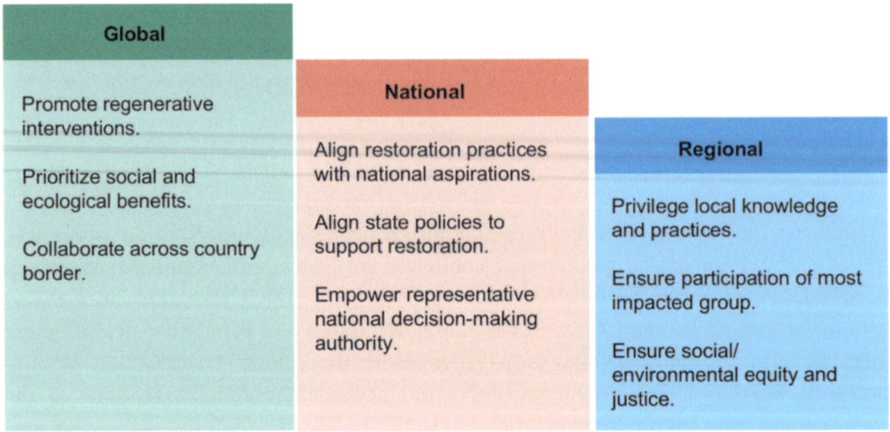

Fig. 12.2 Multi-tier landscape level planning and implementation. (Source: Singh et al. (2023))

While designing the regional intervention, it should be ensured that these initiatives are aligned with national LDN targets, thus ensuring coherence and synergy in efforts to combat land degradation. The approach further emphasizes on leveraging upon the available resources and expertise through partnership and collaboration to implement innovative solutions and advance towards achieving LDN while promoting biodiversity conservation, climate resilience, and inclusive development (Singh et al. 2023).

12.5.2 Landscape Approach for SLM

Though the application must be done on the regional level, the planning is ideal at the landscape level. The landscape approach for achieving LDN emphasizes on two main strategies for land management—reclamation/restoration of degraded land using NbSs and decentralized implementation models for proactive multi-stakeholder involvement (Global Mechanism of the UNCCD 2019). The landscape approach aims to promote self-reliance among forest-dependent communities through skill enhancement and livelihood diversification. Key components of the landscape approach include adaptive management, common concern entry points, multifunctionality, multiscale approaches, negotiated and transparent change logic, clarification of rights and responsibilities, participatory monitoring, resilience to variability and trends, and strengthening stakeholder capacity (Fig. 12.3) (Sayer et al. 2013).

This approach would align with the three major SDGs: SDG 2—end hunger—achieve food security and improved nutrition and promote sustainable agriculture, SDG 6—ensure availability and sustainable management of water and sanitation for all, and SDG 15—protect, restore, and promote sustainable use of terrestrial ecosystems, sustainably manage forests, combat desertification, and halt and reverse land

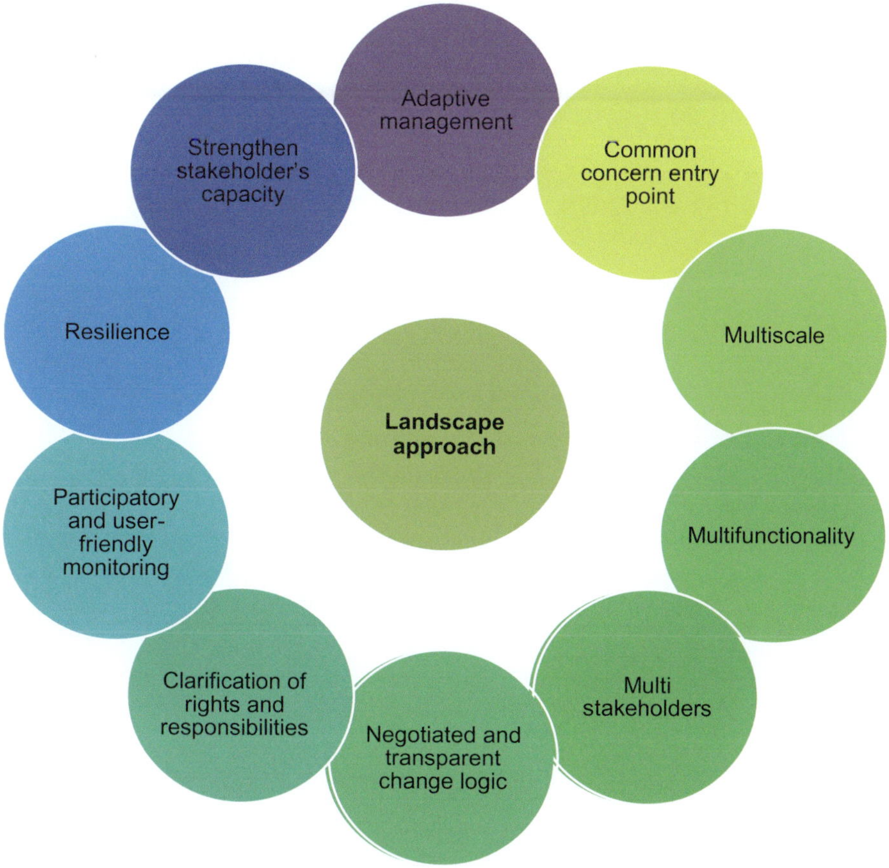

Fig. 12.3 Components of the landscape approach

degradation and halt biodiversity loss. This approach shall contribute to the success of the vision that all existing schemes together will contribute to fulfilling the NDC targets.

12.5.3 Convergence of Resources

There are many ongoing national and sectoral schemes aimed at the restoration of degraded landscapes supported by various stakeholders (Fig. 12.4). However, these schemes while addressing different aspects of land degradation, viz., afforestation, reforestation, soil moisture conservation, and integrated approach for overall socio-economic and ecological recovery, often run parallel to each other in the same areas, each with its own implementation and monitoring process (NAP 2023). Convergence of resources will not only cut the cost of the projects, but proper alignment of the processes will help increase the impact of the project manifold, providing a

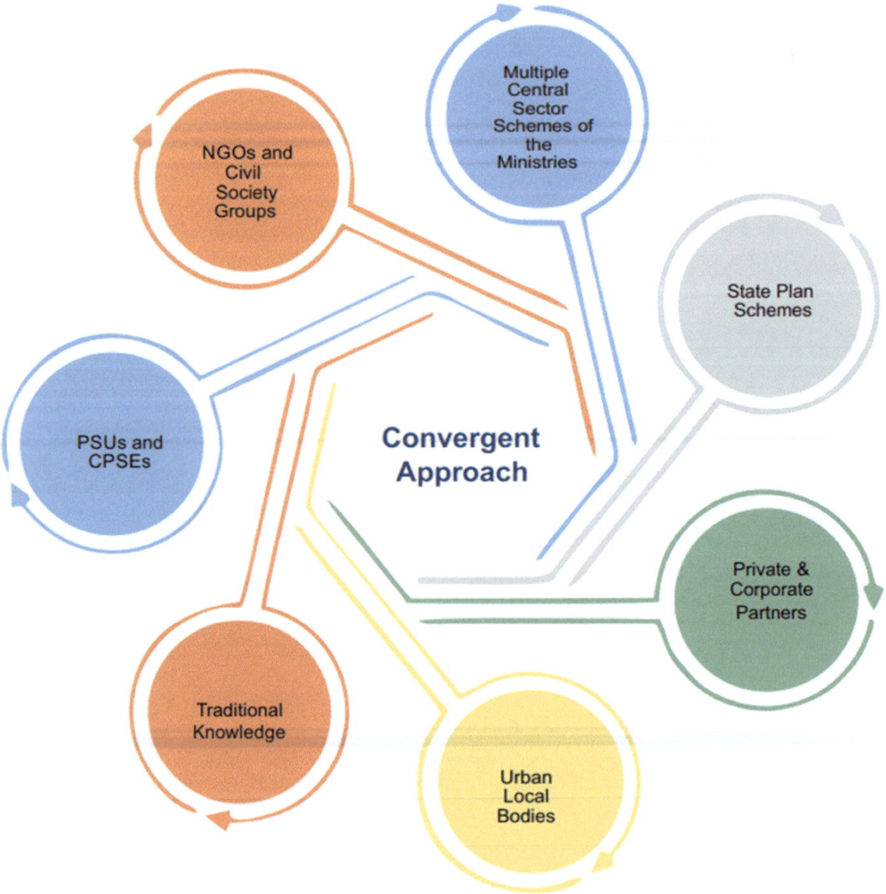

Fig. 12.4 Convergent approach for land restoration for achieving LDN

comprehensive socioeconomic and ecological recovery for the affected ecosystem (Singh et al. 2023).

Recognizing the challenges and enabling efficient planning and implementation of the eco-restoration initiatives in the country, the Ministry of Environment, Forests and Climate Change has developed the guidelines for the convergent approach for LDN. These guidelines aim to unify various national and sectoral schemes to provide impetus and focus in addressing the sustainable management of forest and natural resources available in the specific landscape and an effective planning and a wider selection of vulnerable and potentially vulnerable sites capable of supporting progressive and sustainable eco-restoration interventions (NAP 2023). Through the promotion of an integrated and convergent approach, effective planning and a broader selection of vulnerable and potentially vulnerable sites, capable of supporting progressive and sustainable eco-restoration interventions, can be facilitated. This strategic framework will facilitate restoration aimed at the resilience of the

CoE-SLM	GIS Monitoring Cell Central web based integrated management information system, powered by GIS to agglomerate and monitor implementation of LDN activities on real time basis. Working in tandem with FSI, NRSA and IIRS, centralized spatial database using LISS IV, Cartosat shall be developed with polygons of SLM project areas Monitoring (i) land use change, (ii) change in net primary productivity (iii) change in SOC in project area
Local monitoring	Self-monitoring by local community, implementing organization and forest department Social audits by JFMC's
Third party	Field review by external agency of randomly selected sites Monitoring (i) land use change, (ii) change in net primary productivity (iii) change in SOC in project area
Monitoring by implementing agencies	Monitoring to assess the impact and efficacy of interventions Monitoring of impact and sustainability, gender mainstreaming, environmental services, surface-runoff, groundwater levels etc.

Fig. 12.5 Layout of the proposed monitoring and evaluation plan

ecosystem in congruence with national and international commitments to achieving LDN and sustainable development goals.

12.5.4 Monitoring and Evaluation

For monitoring of the progress of LDN activities, a thorough monitoring structure at four different levels is proposed to be established at CoE-SLM (Fig. 12.5). In addition to on-the-ground self-monitoring by implementing agencies and communities, the Ministry of Environment, Forest and Climate Change will also use geographic information system-based real-time monitoring, and remote sensing with Global Positioning System (GPS) mapping of the project areas under various schemes and satellite imageries for monitoring at the output and outcome levels (Singh et al. 2023).

The social audit of the village-level operations will be carried out by Gram Sabha. This data will be made available to the general public. Using Geographic Information System (GIS) techniques, third-party monitoring and long-term monitoring of UNCCD indicators for favorable changes in land cover, higher land productivity, and higher carbon stocks will be conducted. As part of the monitoring process, a third party will keep an eye on the specified intervention regions on a regular basis (NAP 2023).

12.5.5 Elements of Outreach Strategy and Proposed Activities

As per the mandate of the CoE-SLM, the center will act as the nodal agency for training, capacity building, outreach, and dissemination activities. The CoE will adopt the multi-dimensional and multi-hierarchical approach for building the capacities of the stakeholders through the proactive involvement of diverse knowledge partners and premier institutes. The strategies for promoting the SLM for LDN approach and disseminating its products include regularly updated websites, brochures, newsletters, briefing notes, presentations, outreach sessions, press releases, and opinion pieces. Additionally, it emphasizes the distribution of reports, best practices, success stories, and training manuals to various stakeholders, along with analyzing stakeholder documents and surveying media coverage (Singh et al. 2023; NAP 2023).

12.6 Conclusion

If the world community intends to find remedies to these problems of desertification, food insecurity, climate change, and loss of biodiversity, among others, damaged forests and landscapes in drylands have to be restored now. This action plan is the culmination of lessons learned through numerous activities that have already been undertaken. Although this strategy is tailored to accommodate local and regional contexts, its ambition is to be globally replicable. It defines the key components of planning, implementing, and sustaining restoration projects that build social and ecological resilience and benefit the community. It is a strenuous effort to improve local governance and foster local leaders and advocates for restoration. That can be done, among other things, by increasing small- and medium-sized enterprises, community-based groups, local administrations, and forest producer organizations. Funding opportunities from the various funding mechanisms must be further explored to promote restoration and the implementation of these recommendations.

References

Ashutosh S, Negi SS, Retired IFS, Bhardwaj S, Ghosh S, Pawar S (2021) FSI technical information series. India's nationally determined contribution of creating an additional carbon sink of 2.5 to 3 billion tonnes of CO_2 eq through additional forest and tree cover: possibilities, scale and costs for formulating strategy. Forest Survey of India, Government of India, Dehradun

Bisht J, Singhal S (2021) India's road to net-zero. Thediplomat.com. https://thediplomat.com/2021/11/indias-road-to-net-zero/

Borah B, Bhattacharjee A, Ishwar NM (2018) Bonn challenge and India: progress on restoration efforts across states and landscapes. IUCN and MoEFCC, Government of India, New Delhi

Ekardt F, Günther P, Hagemann K, Garske B, Heyl K, Weyland R (2023) Legally binding and ambitious biodiversity protection under the CBD, the global biodiversity framework, and human rights law. Environ Sci Eur 35(1):80

FAO (2023) Sustainable land management. Food and Agriculture Organization of the United Nations

FAO (n.d.) SLM practices. Food and Agriculture Organization of the United Nations. www.fao.org

Global Mechanism of the UNCCD (2019) Land degradation neutrality transformative projects and programmes: operational guidance for country support. UNCCD Publication, Bonn

NAP (2023) National Action Plan (NAP) to combat desertification and land degradation through forestry interventions. Ministry of Environment Forest and Climate Change, Government of India

SAC (2021) Desertification and Land Degradation Atlas of India (Assessment and analysis of changes over 15 years based on remote sensing). Space Applications Centre, ISRO, Ahmedabad

Sayer J, Sunderland T, Ghazoul J, Pfund JL, Sheil D, Meijaard E et al (2013) Ten principles for a landscape approach to reconciling agriculture, conservation, and other competing land uses. Proc Natl Acad Sci 110(21):8349–8356

Sengupta R (2021) Land degradation in India hurts farmers and forest dwellers the most. Down to Earth. Center for Science and Environment. https://www.downtoearth.org.in/news/environment/land-degradationin-india-hurts-farmers-and-forest-dwellers-the-most-78701

Singh S, Giri K, Mishra G, Kumar M, Singh RK, Pandey S, Mullick M, Sharma R (2023) Pathways to achieve land degradation neutrality in India. Indian Council of Forestry Research and Education, Dehradun

TerrAfrica (2005) Improving country investment programming through advocacy, alliances, and alignment. In: UN expert group meeting on sustainable land management and agricultural practices in Africa: Bridging the gap between research and farmers (April 16–17, 2009). University of Gothenburg, Sweden

Community-Centric Approaches for Land Degradation Neutrality and Biodiversity Conservation

13

Rachana Shah, Rajan Prasad Paudel, Prawesh Poudel, Bikalpa Karki, Bhaskar Singh Karky, and Mohan Chand

Abstract

Land degradation is the reduction in the capability of the land to produce benefits from a particular land use under a specified form of land management. Over the last half-century, Chitwan has been one of the districts that have undergone the most significant changes in land use and land cover. It has one of the highest population growth rates in the country due to rapid urbanization and immigration. The eradication of malaria in the 1960s, highway construction, and the re-settlement policy of the government were the catalysts for change. By the end of the 1960s, 70% of the primary forest and grasslands had been converted into settlements. Chitwan continues to be at the center of conservation and development as global efforts to safeguard the remaining tiger, rhino, elephant, and bear populations must dovetail with a national road and infrastructure development vision. It raises the urgency of biodiversity conservation and land degradation neutrality (LDN) to be customized for the region.

This chapter presents how customizing specific sustainable interventions can contribute to the United Nations Conventions to Combat Desertification's goal of LDN to address and reverse this degradation trend. By taking examples of natural resource management through community forest groups, improved grassland management, and wetland restoration, the chapter will show how community-led sus-

R. Shah · R. P. Paudel · P. Poudel · B. Karki
National Trust for Nature Conservation, Biodiversity Conservation Center, Chitwan, Nepal

B. S. Karky (✉)
Asian Development Bank, Kathmandu, Nepal
e-mail: bhaskarsinghk@gmail.com

M. Chand
Cryospheric Society of Nepal, Kathmandu, Nepal

S. Singh, M. K. Gautam (eds.), *Biodiversity Conservation and Land Degradation Neutrality*, https://doi.org/10.1007/978-981-96-7210-3_13

tainable interventions have contributed to the restoration of degraded lands, avoiding new degradation while conserving the biodiversity and flow of ecosystem services.

Keywords

Chitwan National Park · Community-based conservation · Land degradation neutrality · Biodiversity conservation · Sustainable land management

13.1 Introduction

Land degradation is the reduction in the land's capability to produce benefits from a particular land use under a specified form of land management (FAO 1999). There are several interrelated components of land degradation, all of which may contribute to a decline in land production. The most important, according to Douglas (1994):

- "*Soil degradation*: decline in the productive capacity of the soil as a result of soil erosion and changes in the hydrological, biological, chemical, and physical properties of the soil.
- *Vegetation degradation*: a decline in natural biomass quantity and/or quality and a decrease in vegetative ground cover.
- *Water degradation*: decline in surface and groundwater resources quantity and/or quality.
- *Climate deterioration*: changes in the micro- and macro-climatic conditions that increase the risk of crop failure.
- *Losses to urban/industrial development*: decline in the total area of land used, or with potential, for agricultural production due to arable land being converted to urban, industrial, and infrastructure uses."

Deterioration of land productivity due to poor management is land degradation. Overuse, under-use, or neglect can lead to a decline in productivity and overall output of the land, leading to its degradation. Land degradation as a result of changes in land use and land cover has been a global agenda of concern (Nepal et al. 2020). The science-based definition endorsed by the United Nations Conventions to Combat Desertification (UNCCD) Conference of the Parties in 2015: "Land degradation neutrality is a state whereby the amount and quality of the land resources necessary to support ecosystem functions and services and enhance food security remain stable or increase within specified temporal and spatial scales and ecosystem" (UNCCD 2015). Land degradation neutrality (LDN) has emerged as a pivotal concept in sustainable development discourse, particularly in regions experiencing rapid land transformations like in Chitwan.

Over the past five decades, Chitwan has undergone profound transformations in its landscape, emblematic of the challenges of population growth and rapid vegetation degradation, mainly from urbanization and agriculture expansion. The history of resettlement in Chitwan district goes back as far as 1956 with the advent of the

Rapti Valley Development Program as a result of consecutive landslides in hilly regions of Nepal in 1951, 1953, and 1954, respectively (Agergaard 1999). Chitwan has experienced a surge in population due to urbanization and immigration, leading to significant alterations in land cover and land use patterns. The eradication of malaria in the 1960s, coupled with highway construction and government resettlement policies, catalyzed these changes, converting much of its primary forests into farmland (Chaudhary 2000). These transformations posed a challenge to safeguarding the immense biodiversity, including charismatic megafauna such as Bengal tigers, greater one-horned rhinoceroses, Asian elephants, and sloth bears. This juxtaposition of conservation and development underscores the pressing need for biodiversity conservation and LDN customized for the region such that it can halt the loss of productive land and restore degraded areas, thereby achieving a delicate equilibrium between land degradation and restoration.

This chapter presents how customizing specific sustainable interventions can contribute to the UNCCD's goal of land LDN to address and reverse this degradation trend. Taking examples of natural resource management through community forest (CF) groups, improved grassland management, and wetland restoration shows how community-led sustainable interventions have contributed to the restoration of degraded lands, avoiding new degradation while conserving the biodiversity and flow of ecosystem services (Fig. 13.1).

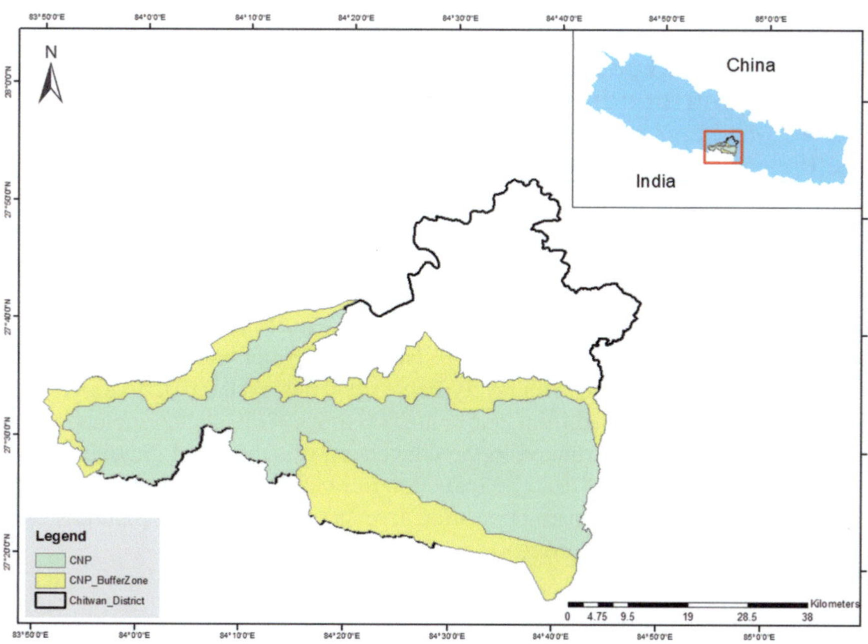

Fig. 13.1 Map of Chitwan district showing Chitwan National Park and buffer zone area

13.2 Ecosystem Degradation in Chitwan

Land use pressures in biodiversity hotspots like Chitwan are exacerbating as land management remains poorly regulated (Upreti et al. 2017). The landscape of Chitwan has undergone dramatic shifts over the past five decades, from loss to recovery, mainly driven by a complex interplay of socio-economic factors and environmental dynamics, along with the locals' awareness of the need to participate in conservation.

Until the early twentieth century, the Chitwan region was predominantly forested, sparsely populated, and inhabited by indigenous ethnic groups such as Tharus and Maithalis (Agergaard 1999; Muller-Boker 1999). However, initiatives to attract farmers from hill regions, malaria eradication programs, and resettlement schemes facilitated rapid deforestation and agricultural expansion (Paudel 2016). Intensive agriculture practices, reliant on forest resources, further accelerated land degradation, leading to the conversion of forests into farmlands and settlements. As much as 25% of forests in the Terai region were harvested between 1960 and 1990s during the party-less "Panchayat" regime, with much of the wood sold to India (Stapp et al. 2015). Increasing settlements made agriculture more important in the region, making Terai known as the fertile "bread basket" of Nepal. However, after 2010, there has been some evidence of improvement in forest cover. Studies evaluating land use and land cover dynamics provide insights into the evolving landscape of Chitwan (Table 13.1).

13.3 Land Use Land Cover Change in Chitwan and Surroundings

Land Use and Land Cover (LULC) maps spanning three distinct periods, namely, 2000, 2010, and 2020, were acquired from the International Center for Integrated Mountain Development (ICIMOD 2022) portal for analysis. These datasets were derived from Landsat imagery corresponding to the respective years and were extracted for the Chitwan district for decadal change in LULC. The tabulated LULC classification (Table 13.2) and accompanying map (Fig. 13.2) exhibit a continuous rise in built-up areas from 2000 to 2020. Notably, forested areas increased (4%) between 2000 and 2010, with a relatively stable trend observed in 2020. Conversely, the cropland area diminished from 26.5% in 2000 to 23.9% in 2020. Fluctuations in grassland areas may be attributed to the conversion of substantial grasslands into forests between 2000 and 2010. The reduction in riverbed area may also be linked to increased embankment construction along riversides, subsequently repurposed as either grassland or forested terrain.

Table 13.1 Major development events impacting wildlife habitat

Year	Event	Impact
Until 1950s	Chitwan as the hunting ground of Ranas	Restrictions on hunting and punishments for poachers
1957	Forest Nationalization Act	Established the government's ownership of all forested land, but its unintended impact was the accelerated deforestation of forests in and around settlements as it undermined community-level management practices.
1960	Eradication of Malaria and resettlement program by deforestation	Opening of Terai lands for hill people to migrate for agriculture, settlement, and business
1973	National Park and Wildlife Conservation Act	Chitwan National Park was established as the first national park of Nepal. Legal restrictions on land and resource use
1977	Construction of Muglin–Narayghat Highway	Contributed to forest fragmentation and movement of people
1980	Construction of Narayghat–Hetauda Highway and linking with Butwal	Contributed to forest fragmentation, encroachment, and expansion of settlements along the highway
1986	Annapurna Conservation Area Project establish	Initiation of community-based conservation and development in the mountains
1990	Expansion of industrial corridor	
1993	The Forest Act and National Park and Wildlife Conservation Act amendment	Recognized community rights to forest resources
1999	Buffer zone management guideline	Facilitated local community engagement in conservation and sharing of benefits from conservation
2005	End of 10-year Maoist Civil War	Improved security, socio-political stability
2020	Expansion of highways	Loss of trees and land disturbance

Table 13.2 Land use land cover classification of the Chitwan District in 2000, 2010, and 2020

LULC classes	2000		2010		2020	
	Area (km^2)	Area (%)	Area (km^2)	Area (%)	Area (km^2)	Area (%)
Water	40.7	1.82	43.4	1.94	29.7	1.33
Forest	1462.4	65.33	1551.7	69.32	1552.5	69.36
Riverbed	63.0	2.81	37.6	1.68	6.5	0.29
Built-up	3.5	0.15	5.0	0.22	22.3	1.00
Crop	592.4	26.47	543.3	24.27	534.3	23.87
Bare land	0.0	0.00	0.0	0.00	0.2	0.01
Grassland	76.4	3.41	57.5	2.57	92.9	4.15
Total	2238.4	100.00	2238.4	100.00	2238.4	100.00

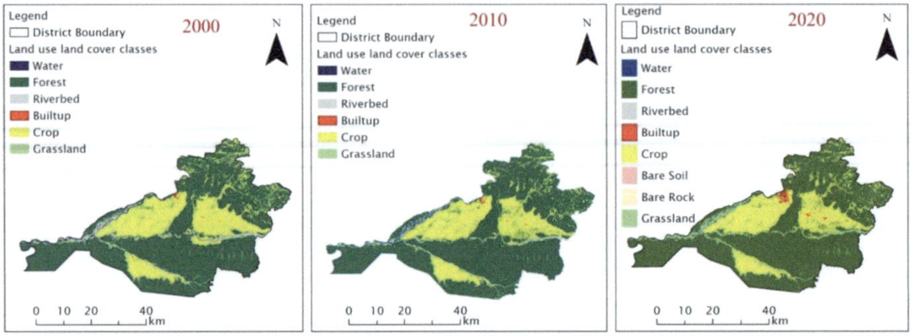

Fig. 13.2 Land use land cover map of Chitwan district of the year 2000, 2010, and 2020

13.4 Forest Cover Change in Buffer Zones of Chitwan National Park

Chitwan's evolving land use patterns reflect the complex interactions between human activities, environmental processes, and policy interventions. Until the 1950s, the Chitwan National Park (CNP) and the buffer zone around it were dense forests related to wildlife like tigers and elephants. The buffer zone was inhabited by a few ethnic Tharu households as well. However, by the 1970s, the area witnessed increased migration from the hills and mountains. Based on data from Land Resource Mapping Project (LRMP 1986) 1978 and Earth Resources Technology Satellite (LANDSAT) Thematic Mapper 1999, land use change showed a net gain of 67.28 km^2 in agricultural land and a loss of 62.2195 km^2 in forest area (Baidya et al. 2009). The built-up area was not prominent in 1978. However, by 1999, satellite imagery revealed a noticeable net increase of built-up area, mostly at the expense of agriculture and grasslands (Table 13.3). The dominance of forests as a primary land use type was replaced by agriculture during this period.

The conservation and community-based forest management efforts initiated in the 1990s have contributed to reforestation and habitat restoration in certain areas. The United Nations Collaborative Initiative on Reducing Emissions from Deforestation and Forest Degradation (2014) estimated that overall forest loss in Nepal dropped to 0% annually between 2005 and 2010. Analysis of normalized difference vegetation index using Landsat imagery for 1989, 2005, and 2013 showcases the changes in forest cover over time in the buffer zone areas of CNP (Table 13.4). Buffer zone in CNP lost 9.9% of total forest cover between 1989 and 2005 and regained 7.5% between 2005 and 2013; the net loss between 1989 and 2013 was 3.1% (Stapp et al. 2015). Total forest cover in the area has largely regenerated to the levels that existed in 1989. This indicates the stabilization of forest cover in and around Chitwan in recent decades.

The trend of forest cover restoration is reflected in the Chitwan and its surrounding areas. The area under the forest has remained more or less constant during the recent period throughout the Chitwan-Annapurna landscape (WWF 2013). Landsat

Table 13.3 Land use change in the buffer zone of Chitwan National Park in 1978–1999

Land use type	Net land use change (km^2)
Agriculture land	67.2829
Built-up area	0.8055
Forest	−62.2195
Grass land	−11.3839
Shrub land	7.6772
Water bodies	−2.1622

Table 13.4 Forest cover change in the buffer zone of Chitwan National Park

Details	1989	2005	2013	Overall
Mean forest area in buffer zone village (ha)	1754	1581	1700	−54
Forest cover change (%) between 1989–2005, 2005–2013, and 1989–2013	–	−9.9	+7.5	−3.1

Source: Stapp et al. (2015)

Table 13.5 Land use change in Chitwan-Annapurna landscape from 1990 to 2020

	Land cover type	2000	2010	2020
1	Water bodies	53.2	52.7	54.04
2	Barren area	60.8	56.1	48.62
3	Grassland	94.1	88.24	47.32
4	Riverine forest	60.03	52.16	53.25
5	Sal dominated forest	393.15	411.3	423.65
6	Mixed forest	899.1	1062.48	1235.9
7	Cropland	1102.8	923.7	773.35
8	Developed area	86.3	102.8	113.35
	Total	2749.48	2749.48	2749.48

Source: Adhikari et al. (2022)

images between 2000 and 2020 (Table 13.5) show an increase in Sal-dominated forests, mixed forests, and developed areas and a decrease in riverine forests, barren areas, croplands, and grasslands (Adhikari et al. 2022).

The finding that the forest area remained largely unchanged during the period does not mean that there was no forest loss or gain in the landscape. In some areas, forest areas have been substantially lost due to infrastructure development, resettlement, road construction, urban expansion, and agriculture (WWF 2013). It is also noteworthy to consider that the forest gained in recent decades may not replicate the forest lost in the areas as the characteristics of regenerated forest seen today are likely to vary significantly. The structure and composition of the forest and the diversity of flora and fauna may not reflect the original condition. The decrease in agricultural land and water-associated ecosystems, particularly the grassland and riverine forests, is concerning as these habitats are crucial to supporting the Chitwan food web. Many grassland-dependent species, such as the Bengal florican, Swamp francolin, hispid hare, and pigmy hog, have gone locally extinct from the grassland patches of Chitwan. These lands have undergone encroachment from woody vegetation as well as engulfed by the rapidly establishing invasive alien species such as Mikenia, Lantana, and Parthenium.

13.5 Drivers of Change

The drivers and underlying causes of Chitwan's land use and land cover change are multifaceted (ICIMOD 2017). Direct drivers are human activities that originate from human choice of land use, which directly impacts forest cover. The underlying causes of deforestation and forest degradation are structural (or systemic). These can be seen as a complex of social, political, economic, technological, and cultural variables constituting initial conditions in human–environmental relations (Geist and Lambin 2002). In addition, climate change can trigger and exacerbate these drivers, and underlying causes such as an increase in temperature and a rain delay can increase the incidences of forest fires. The cessation of forest loss in Nepal is likely due to a combination of these direct and underlying factors. Decentralized forest management institutions such as community forestry have played an important role in not just slowing but halting and possibly reversing forest loss and degradation in areas of the buffer zone of CNP (Table 13.6).

Forest and agricultural land is rapidly being converted to built-up areas, propelled by a nexus of politicians, land brokers, and real-estate actors, leading to the loss of agrobiodiversity, food insecurity, and conflicts between local communities and stakeholders involved in land deals. The conversion of forests into agricultural land, driven by the need for food security and livelihoods, contributes significantly to deforestation in Chitwan. Practices such as shifting cultivation, illegal timber extraction, fuelwood collection, and expanding roads and settlements lead to habitat fragmentation, further exacerbating forest cover change and loss of biodiversity. Deliberate fires used in some instances to get rid of the ground litter and promote grass for animals often escape the intended areas, causing catastrophic damage to the forest and biodiversity. Between 2012 and 2022, CNP experienced 1135 forest fire incidents among the highest for Nepal (DoFSC 2024). Weak enforcement of

Table 13.6 Direct drivers and underlying causes of land use change

Direct drivers	Underlying causes
1. Unsustainable forest resource harvesting (e.g., illegal logging and fuelwood collection, and sand and stone mining) 2. Forest encroachment (e.g., for road construction, agriculture expansion, and human settlement invasive species) 3. Forest fire	1. Demand for forest resources 2. (e.g., forest land for hydropower, road, and industry) 3. Forest-dependent livelihoods (e.g., for timber, fuel, fodder, and wild vegetables) 4. Socio-economic differences 5. Population growth 6. Technological limitations 7. Forest policy and its enforcement

forest conservation laws, inadequate governance mechanisms, and policy gaps undermine land management efforts.

Buffer zone CFs (BZCFs) in CNP highlight the transition from land degradation to restoration. It has garnered considerable success in promoting sustainable land management practices and balancing conservation and community development goals. Implementing buffer zone community forestry programs has played a crucial role in forest conservation, mitigating the impacts of population pressure and external disturbances on the ecosystems of Chitwan. Establishing CF user groups, such as the Bagmara BZCF, exemplifies local initiatives aimed at forest restoration and sustainable resource management. Analysis of spatial data reveals the recovery of forest cover, wildlife population, and habitat (Pokhrel 2018).

It is important to consider that the current land use land cover change detections have their own limitations. High-resolution images free of errors are not always available, and techniques to differentiate between different features are not always accurate. For example, it may be difficult to distinguish the bare soil and agricultural land in some instances, leading to misclassification of the features. Addressing forest cover change in Chitwan requires comprehensive strategies that integrate community participation, sustainable land management practices, and effective governance mechanisms.

13.6 Trends in Species

The CNP and its surrounding BZCFs are renowned for their rich biodiversity and iconic wildlife species. Over the years, monitoring efforts and research initiatives have provided valuable insights into the trends of wildlife populations, highlighting challenges and conservation successes in the region. CNP serves as a sanctuary for diverse wildlife species, including charismatic megafauna such as Bengal tigers, greater one-horned rhinoceroses, Asian elephants, and sloth bears. Additionally, the park is home to many other mammalian, avian, and reptilian species, contributing to its status as a United Nations Educational, Scientific and Cultural Organization (UNESCO) World Heritage Site and a global biodiversity hotspot. The BZCFs surrounding CNP play a crucial role in wildlife conservation by providing additional habitats and corridors for wildlife movement. Community-led initiatives aimed at sustainable forest management, habitat restoration, and biodiversity conservation contribute to the overall health and resilience of the ecosystem. By fostering coexistence between local communities and wildlife, BZCFs enhance the ecological connectivity and integrity of the landscape, benefiting both people and wildlife alike. The trends of wildlife species in CNP and its BZCFs underscore the importance of continued conservation efforts and collaborative management approaches. By addressing threats such as habitat loss, poaching, and human–wildlife conflict, while promoting sustainable land use practices and community engagement, Chitwan can ensure the long-term survival of its iconic wildlife species and maintain its status as a global biodiversity hotspot.

Greater One-Horned Rhinoceroses CNP is renowned for its successful conservation of the greater one-horned rhinoceros, with the park serving as a stronghold for this endangered species. Population surveys and monitoring programs have indicated a steady increase in rhinoceros numbers. Historically, the one-horned rhinoceros (*Rhinoceros unicornis*) faced severe threats such as poaching and habitat loss, dramatically declining their numbers. Within two decades, from the 1950s to the 1960s, the rhino population decreased from 800 to just 100 (Thapa et al. 2013). The main reason behind the decline was extensive habitat destruction (~70% of forests) due to human settlement expansion into the Rhino habitats after the malaria eradication program at Chitwan (Caughley 1969; Laurie 1978). However, concerted conservation efforts, including strict anti-poaching measures, habitat restoration, community involvement, and translocation programs, have led to a remarkable recovery of the rhinoceros population in CNP. Over the years, the population of one-horned rhinoceros in CNP has shown a positive trend, with steady growth in numbers. The last rhino census revealed a remarkable increase of more than 90 rhinos in 6-year intervals (NTNC 2021). This result reflects effective protection measures and habitat management strategies implemented within the park and its buffer zones. Community participation in rhinoceros conservation through initiatives such as community-based anti-poaching patrols has further contributed to the species' recovery.

Bengal Tigers The tiger population in CNP has been the subject of extensive research and conservation efforts. Despite facing threats such as habitat loss, poaching, and human–wildlife conflict, the tiger population has shown resilience data, with population estimates indicating a stable or slightly increasing trend in recent years.

Nepal has succeeded in the "Tiger doubling mission" as the number of tigers increased from 121 to 355 (DNPWC and DFSC 2022). The tiger population of CNP has also gradually increased from 91 in 2009 (WWF 2018) to 128 in 2022 (DNPWC and DFSC 2022). Despite various worries, the increased tiger population has not overshot the estimated carrying capacity of 136 individuals (Shah et al. 2024). Conservation measures, including anti-poaching patrols, habitat restoration, and community engagement, have played a crucial role in safeguarding tiger populations in the park.

Asian Elephants The population of Asian elephants in CNP is smaller than that of tigers and rhino populations. The estimated population of elephants in the Chitwan Parsa complex was 25–30 in 2008 (DNPWC 2008), which increased to 45–53 in 2020 (Ram and Acharya 2020). Habitat fragmentation of the corridor used by elephants led to the population being forced to live in smaller forest patches (Carter et al. 2012). Conservation initiatives focused on mitigating human–elephant conflicts, preserving elephant corridors, and promoting habitat connectivity have been instrumental in ensuring the survival of these keystone species in the region.

Sloth Bears Sloth bears are an iconic but less studied species of CNP. Their actual population status is unknown. However, a recent ecological study (Paudel et al. 2022) and the cases of sighting bears in the Park suggest their stable or possible increase in their population (Lamichhane et al. 2016). Also, these bears are captured in the camera traps deployed to study tigers (Dhakal ct al. 2014). Thc cfforts to monitor and conserve the bear species are equally important. Similarly, efforts to address human–wildlife conflicts and enhance habitat quality in the CNP have yielded positive outcomes for sloth bears, contributing to the conservation of these elusive creatures.

Avian and Reptilian Species CNP is renowned for its avian diversity, with over 500 species of birds recorded within its boundaries. From endangered vultures to colorful migratory birds, the park provides critical habitat for many avian species. Similarly, reptilian diversity, including species such as gharials, mugger crocodiles, and various snake species, adds to the ecological richness of the park.

13.7 Status of Wetlands

The wetlands of CNP play a vital role in maintaining the ecological balance and supporting diverse wildlife populations. These habitats, characterized by their unique hydrological and vegetative features, serve as critical breeding grounds, foraging areas, and corridors for numerous species, contributing to the park's rich biodiversity. However, the status of wetlands and grasslands in CNP has been subject to various anthropogenic pressures and environmental changes, necessitating concerted conservation efforts.

The wetlands within CNP encompass rivers, lakes, ponds, and marshes, providing essential habitats for aquatic species, migratory birds, and amphibians. These wetlands serve as breeding sites for fish and amphibians, support diverse vegetation communities, and regulate local hydrology, contributing to the ecosystem's overall health.

At least 58 different-sized wetlands are in the park, of which 38 are inside the core areas and the remaining 20 are in the buffer zone (Jha and Sharma 2018). Only few wetlands are in excellent condition and safe from threats. In contrast, majority of wetlands are largely affected by siltation, invasive species, conversion to other lands, mainly grassland, drying up, pollution, fishing, and other extraction problems (Bhatta et al. 2018). Outside protected areas, agricultural runoff mixed with pesticides from the nearby farmlands to wetlands is another major threat, for example, in Beeshazari Lake (Kafle and Savillo 2009).

Conservation initiatives aimed at protecting and restoring wetland habitats in CNP are underway. These efforts include habitat restoration projects, water quality monitoring programs, and community-based conservation initiatives. Collaboration between government agencies, local communities, and non-governmental organizations (NGOs) has led to the establishment of protected areas, wetland management plans, and sustainable resource utilization practices aimed at safeguarding wetland ecosystems.

13.8 Status of Grassland

The grasslands of CNP are characterized by diverse grass species and support a myriad of herbivores, including deer, rhinoceroses, and wild boars. These grasslands serve as important grazing areas, nesting sites for ground-nesting birds, and feeding grounds for numerous mammalian species, contributing to the park's overall ecological balance.

Mapping of grassland habitats in Chitwan found 425 grassland patches with a total area of 10497.2 ha (6.24% of the total area) in CNP and its buffer zone (CNP 2016). In the core area, 8955.2 ha is covered by grasslands, which equals 9.6% of the total core area. Similarly, in the buffer zone, 1541.9 ha (2.06% of the total Buffer Zone area) is covered by grasslands. Invasion by alien species, woody vegetation, and livestock grazing threaten grassland habitats (CNP. 2016).

Efforts to conserve and manage grassland habitats in CNP focus on habitat restoration, fire management, and sustainable land-use practices. Community-based conservation initiatives, such as rotational grazing schemes and grassland restoration projects, engage local communities in sustainable resource management and biodiversity conservation. Additionally, research and monitoring programs assess the health and status of grassland ecosystems, guiding conservation interventions and adaptive management strategies.

In conclusion, the status of wetlands and grasslands in CNP reflects both the ecological significance of these habitats and the challenges they face due to anthropogenic pressures and environmental changes. By implementing targeted conservation measures, promoting community participation, and fostering sustainable land-use practices, CNP can ensure the long-term conservation of its wetland and grassland ecosystems, thereby preserving the park's biodiversity and ecological integrity for future generations.

13.9 Measures Taken for Biodiversity Conservation and Land Degradation Neutrality

In response to the critical need for biodiversity conservation and LDN, various measures have been implemented in Chitwan, Nepal. These efforts aim to mitigate the adverse effects of land use changes, promote sustainable development practices, and safeguard the region's unique biodiversity. The initiatives encompass a range of strategies, including community-based conservation, scientific research, wildlife-friendly infrastructure development, and policy interventions.

13.9.1 Community-Based Conservation Initiatives

CF Groups In a global context, CF has covered approximately 732 million hectares of land, or 18.2% of the world's forest area in the past four decades. This management practice has significantly mitigated many serious environmental and social problems, including climate change and poverty alleviation (Gilmour 2016). In Nepal, CF is defined as the national forests that are handed over to the local user group under Section 18 of the Forest Act, 2019, to conserve and use the forest and forest products independently and sustainably by fixing their prices according to the approved operational plan, CF, and it was started in the late 1970s (Kanel and Kandel 2004); since then, CF user groups have emerged as key stakeholders in natural resource management and biodiversity conservation in Nepal. According to the Federation of Community Forest Users Nepal, there are 22,415 CFUGs with 29,00,000 households protecting 23,00,000 CFs. These user groups engage local communities in sustainable forest management practices, afforestation efforts, and wildlife habitat restoration.

More than 48,758 households are involved in 75 CF user groups to manage 15,205 ha, the buffer zone forest of the CNP. Outside the buffer zone of CNP, 51,287 households are involved under 93 CFUGs to manage 23,100 ha of forest land (Dhungana et al. 2024). These user groups have been evolving and operating for the past three decades. These grassroots organizations engage local communities in sustainable forest management practices, afforestation efforts, and habitat restoration. Examples include the Bagmara BZCF, where participatory forest management has regenerated degraded habitats and increased wildlife populations.

Buffer Zone Community Forestry Program BZCF is a concept of management, involving the forest areas and the forest users around the protected areas with the aim of sustainable use of the natural resources and the conservation of biodiversity by the local people (Bhandari and Jianhua 2017). In the Chitwan district buffer zone communities implemented in collaboration with CNP authorities, the Buffer Zone Community Forestry Program empowers local communities to manage and conserve forest resources in the park's buffer zones. Research has shown that rural impoverished communities rely more heavily on forest resources because a significant portion of their income is derived from gathering them (Bhandari and Jianhua 2017). The program aims to mitigate human–wildlife conflicts and promote coexistence between people and wildlife through community-led initiatives, such as fire prevention, anti-poaching patrols, and habitat restoration.

13.9.2 Scientific Research and Monitoring

Ongoing biodiversity surveys and monitoring programs provide valuable insights into the status and trends of wildlife populations, vegetation communities, and ecosystem health in Chitwan. These research efforts inform conservation strategies, habitat management plans, and adaptive management practices aimed at preserving biodiversity and ecological integrity.

Remote sensing and geographic information system technology are employed to assess land cover changes, habitat fragmentation, and landscape dynamics in Chitwan. By analyzing satellite imagery and spatial data, researchers can identify priority areas for conservation, evaluate the effectiveness of conservation interventions, and detect land degradation trends over time.

13.9.3 Wildlife-Friendly Infrastructure Development

The construction of wildlife corridors and habitat connectivity networks facilitates wildlife species' movement between Chitwan's fragmented habitats. These corridors mitigate habitat fragmentation, reduce human–wildlife conflicts, and enhance landscape connectivity, promoting genetic exchange and biodiversity conservation.

Road construction projects in Chitwan prioritize ecologically sensitive design principles, such as wildlife underpasses, overpasses, and fencing, to minimize habitat fragmentation and mitigate the impacts of infrastructure development on wildlife populations. Road projects aim to maintain landscape connectivity and preserve critical wildlife habitats by incorporating wildlife-friendly infrastructure features.

13.9.4 Policy Interventions

CNP authorities formulate and implement protected area management plans to guide conservation efforts, habitat restoration initiatives, and sustainable tourism practices within the park. These management plans integrate biodiversity conservation objectives with community livelihoods, sustainable development goals, and land use regulations, fostering a balanced approach to conservation and development.

Legislative reforms and policy frameworks are enacted to strengthen environmental protection, regulate land use practices, and enhance biodiversity conservation efforts in Chitwan. Government policies, such as land tenure reform, habitat protection laws, and environmental impact assessments, aim to mitigate land degradation, preserve critical habitats, and promote sustainable land management practices.

The multifaceted measures implemented to address biodiversity conservation and LDN in Chitwan reflect a holistic approach to sustainable development and environmental stewardship. By engaging local communities, leveraging scientific expertise, promoting wildlife-friendly infrastructure, and enacting supportive policies, Chitwan endeavors to preserve its rich biodiversity, restore degraded landscapes, and achieve LDN for the benefit of present and future generations.

13.10 Major Challenges

Despite concerted conservation efforts, Chitwan faces significant challenges in achieving biodiversity conservation and LDN. Addressing these challenges requires collaborative partnerships, innovative solutions, and adaptive management strategies to safeguard the region's ecological integrity and promote sustainable development. Some of the key challenges include the following.

13.10.1 Human–Wildlife Conflicts

Challenge: Encroachment of wildlife habitats and increasing human–wildlife conflicts pose threats to both local communities and wildlife populations in Chitwan. Conflict incidents, such as crop raiding and predation on livestock, result in economic losses and negative perceptions toward conservation efforts.

Mitigation Efforts: To mitigate human–wildlife conflicts, initiatives such as community-based conflict resolution mechanisms, establishment of wildlife corridors, and deployment of innovative deterrent measures (e.g., chili smoke bombs and beehive fences) are implemented. These efforts aim to foster coexistence between people and wildlife while minimizing conflict incidents.

13.10.2 Habitat Degradation and Fragmentation

Challenge: Habitat degradation, fragmentation, and loss due to land conversion, infrastructure development, and unsustainable land use practices threaten the integrity of ecosystems in Chitwan. Deforestation, illegal logging, and habitat encroachment exacerbate habitat fragmentation, reducing wildlife habitat availability and connectivity.

Mitigation Efforts: Conservation initiatives focus on habitat restoration, afforestation, and reforestation programs to mitigate habitat degradation and fragmentation. Buffer zone community forestry programs, habitat restoration projects, and rewilding initiatives aim to restore degraded habitats, enhance landscape connectivity, and promote ecological resilience.

13.10.3 Illegal Wildlife Trade and Poaching

Challenge: Illegal wildlife trade, poaching, and trafficking of endangered species remain pervasive threats to biodiversity conservation in Chitwan. Poaching activities target iconic species such as tigers, rhinos, and elephants, driven by demand for wildlife products and traditional medicine.

Mitigation Efforts: Enhanced law enforcement efforts, wildlife crime monitoring, and anti-poaching patrols are implemented to combat illegal wildlife trade and poaching activities. Collaboration between law enforcement agencies, conservation organizations, and local communities strengthens wildlife protection measures and deters illegal activities.

13.10.4 Climate Change Impacts

Challenge: Climate change exacerbates existing conservation challenges in Chitwan, including altered precipitation patterns, increased frequency of extreme weather events, and habitat shifts. These climate-induced changes threaten biodiversity, disrupt ecosystem dynamics, and compromise the resilience of ecosystems.

Mitigation Efforts: Adaptation strategies, such as climate-smart land management practices, resilient ecosystem restoration, and habitat corridors for species migration, are implemented to mitigate the impacts of climate change on biodiversity. Integrated approaches that combine climate change adaptation with biodiversity conservation enhance ecosystem resilience and promote long-term sustainability.

13.10.5 Socio-Economic Pressures

Challenge: Rapid population growth, urbanization, and socio-economic pressures significantly stress natural resources and ecosystems in Chitwan. Increased demand for land, resources, and livelihood opportunities exacerbates land use conflicts, deforestation, and environmental degradation.

Mitigation Efforts: Sustainable development initiatives enhance community livelihoods, promote alternative income generation activities, and foster sustainable land use practices. Community-based ecotourism, eco-friendly agriculture, and capacity-building programs empower local communities to participate in conservation efforts while improving their socio-economic well-being.

Addressing Chitwan's major challenges requires a multi-faceted approach that integrates conservation, sustainable development, and community engagement. By implementing targeted mitigation efforts, fostering collaborative partnerships, and fostering adaptive management practices, Chitwan strives to overcome these challenges and achieve its biodiversity conservation and LDN goals for the benefit of present and future generations.

13.11 Conclusion

Considerable land use changes have occurred in Chitwan as a consequence of years of human settlements. Similarly, changes in land tenure and access to land resources have further contributed to the degradation and fragmentation of lands. Lack of adequate income generating opportunities, inadequacy of the agriculture to support prosperity aspirations have exacerbated pressure on land resources.

Chitwan stands at the forefront of biodiversity conservation and LDN efforts, grappling with a myriad of challenges while striving for sustainable development. Over the past decades, the district has witnessed profound land use changes, rapid urbanization, and encroachment upon natural habitats, placing immense pressure on its ecosystems and wildlife populations. Despite these challenges, Chitwan has demonstrated resilience and innovation in its conservation endeavors. Through collaborative partnerships between government agencies, NGOs, local communities, and research institutions, significant strides have been made in biodiversity conservation, habitat restoration, and sustainable land management.

Establishing community-based conservation initiatives, such as buffer zone community forestry programs and grassroots conservation projects, has empowered local communities to become stewards of their natural resources. These initiatives not only contribute to biodiversity conservation but also enhance community resilience and livelihoods. Furthermore, scientific research and technological advancements have provided invaluable insights into the dynamics of land use change, species trends, and ecosystem resilience. Chitwan continues to refine its conservation approaches and adapt to emerging challenges by harnessing the power of data-driven conservation strategies and evidence-based decision-making.

However, the journey toward biodiversity conservation and LDN is ongoing, and numerous challenges remain to be addressed. Human–wildlife conflicts, habitat degradation, illegal wildlife trade, impacts of climate change, and socio-economic pressures threaten the region's ecological integrity. To overcome these challenges, concerted efforts are needed to strengthen law enforcement measures, promote sustainable land use practices, enhance community engagement, and foster resilience to climate change. By embracing innovation, collaboration, and adaptive management, Chitwan can chart a path toward a more sustainable future where biodiversity thrives, ecosystems flourish, and communities prosper.

The conservation journey in Chitwan is a testament to the power of collective action and the enduring commitment to safeguarding our natural heritage for generations to come. As we reflect on the achievements and challenges encountered along this journey, let us remain steadfast in our resolve to protect and preserve the rich biodiversity and ecosystems that define the essence of Chitwan and inspire conservation efforts worldwide.

References

Adhikari JN, Bhattarai BP, Rokaya MB, Thapa TB (2022) Land use/land cover changes in the central part of the Chitwan Annapurna Landscape, Nepal. PeerJ 10:e13435. https://doi.org/10.7717/peerj.13435

Agergaard J (1999) Settlement and changing land use in the Chitwan district of Nepal. Geografisk Tidsskrift, SPEC. ISS. 1

Baidya NG, Bhuju DR, Kandel P (2009) Land use change in buffer zone of chitwan national park, Nepal between 1978 and 1999. Ecoprint: An International Journal of Ecology 16:79–86

Bhandari D, Jianhua Z (2017) Household Dependency on Buffer Zone Community Forest and its implication for Management of Chitwan National Park, Nepal. Int J Sci 3(03):68–80. https://doi.org/10.18483/ijsci.1214

Bhatta LD, Ning W, Udas E, Agrawal NK, Ranabhat S, Basnet D (2018) Wetlands in the Himalaya: securing services for livelihoods. International Centre for Integrated Mountain Development (ICIMOD)

Carter NH, Shrestha BK, Karki JB, Pradhan NMB, Liu J (2012) Coexistence between wildlife and humans at fine spatial scales. Proc Natl Acad Sci USA 109:15360–15365. https://doi.org/10.1073/pnas.1210490109

Caughley G (1969) Wildlife and recreation in the Trishuli watershed and other areas in Nepal. HMG/FAO/UNDP Trishuli Watershed Development Project, Project Report No 6, Kathmandu, Nepal

Chaudhary RP (2000) Forest conservation and environmental management in Nepal. Biodivers Conserv 9:1235–1260

CNP (2016) Grassland habitat mapping in Chitwan National Park. Chitwan National Park, Kasara, Chitwan

Dhakal M, Thapa MK, Jnawali SR et al (2014) Status of tigers and prey in Nepal. Department of National Parks and Wildlife Conservation, Kathmandu, Nepal

Dhungana N, Lee CH, Khadka C, Adhikari S, Pudasaini N, Ghimire P (2024) Evaluating community forest user groups (CFUGs)'performance in managing community forests: A case study in Central Nepal. Sustainability 16(11):4471

DNPWC (2008) The Elephant ConservationAction Plan for Nepal. Ministry of Forest and Soil Conservation. Department of National Parksand Wildlife Conservation, Kathmandu, Nepal. (18) (PDF) Current Status of Asian Elephant in Nepal. Available from: https://www.researchgate.net/publication/282602174_Current_Status_of_Asian_Elephant_in_Nepal. Accessed May 08 2024

DNPWC and DFSC (2022) Status of tigers and prey in Nepal 2022. Department of National Parks and Wildlife Conservation and Department of Forests and Soil Conservation. Ministry of Forests and Environment, Kathmandu, Nepal

DoFSC (2024) Forest Fire Detection and Monitoring System in Nepal

Douglas MG (1994) Sustainable Use of Agricultural Soils. A Review of the Prerequisites for Success or Failure. Development and Environment Reports No. 11, Group for Development and Environment, Institute for Geography, University of Berne, Switzerland

FAO (1999) Poverty alleviation and food security in Asia. In: Land degradation, Chapter 3. Food and Agricultural Organization of the United Nations (FAO)

Geist HJ, Lambin EF (2002) Proximate causes and underlying driving forces of tropical deforestation: Tropical forests are disappearing as the result of many pressures, both local and regional, acting in various combinations in different geographical locations. BioScience 52(2):143–150

Gilmour D (2016) Forty years of community-based forestry: a review of its extent and effectiveness. FAO Forestry Paper, 176

ICIMOD (2017) District REDD Action Plan Chitwan

ICIMOD (2022) Land cover of HKH region [Data set]. ICIMOD. https://doi.org/10.26066/RDS.1972511

Jha PK, Sharma CK (2018) Status of wetland Birds in Chitwan national parK, nepal. Tribhuvan University Journal 32(1):15–24

Kafle G, Savillo IT (2009) Present status of Ramsar Sites in Nepal. Int J Biodivers Conserv 1(5):146–150

Kanel KR, Kandel BR (2004) Community forestry in Nepal: achievements and challenges. J Fores Livelihood 4(1):55

Lamichhane BR, Gurung A, Poudel S, Pokheral CP, Subedi N (2016) Sloth bear sightings as a population index in Chitwan National Park, Nepal. Int Bear 25:1

Laurie WA (1978) The ecology and behavior of the greater one-horned rhinoceros. PhD Dissertation. University of Cambridge, Cambridge, UK

LRMP (1986) Summary Report. Land Resource Mapping Project, His Majesty's Government of Nepal and Government of Canada. Kenting Earth Sciences Limited

Muller-Boker U (1999) The Chitwan Tharus in southern Nepal: an ethnoecological approach. Franz Steiner Verlag Stuttgart, France

Nepal P, Khanal NR, Zhang Y, Paudel B, Liu L (2020) Land use policies in Nepal: an overview. Land Degrad Dev 31(16):2203. https://doi.org/10.1002/ldr.3621

NTNC (2021) 752 one-horned rhinos in Nepal determined by the National Rhino Count 2021, Khumaltar, Kathmandu, Nepal

Paudel RP (2016) Protected areas, people and tourism: political ecology of conservation in Nepal. J Fores Livelihood 14(1):13–27

Paudel RP, Kadariya R, Lamichhane BR, Subedi N, Sashika M, Shimozuru M, Tsubota T (2022) Habitat occupancy of sloth bear Melursus ursinus in Chitwan National Park, Nepal. Ecol Evol 12(3):e8699. https://doi.org/10.1002/ece3.8699

Pokhrel J (2018) Land use and land cover changes in Chitwan national park and adjacent areas from 1988 to 2017. A Thesis Submitted to the Faculty of the School of Natural Resources and the Environment. In Partial Fulfillment of the Requirements For the Degree of Master of Science with a major in natural resources. The University of Arizona, USA

Ram AK, Acharya H (2020) Status distribution and habitat use by Asian elephants in Nepal. A compendium of Conservation Bulletin published by Department of National Parks and Wildlife conservation, Nepal, 155–160

Shah SK, Karki JB, Bhatta B, Subedi N, Lamichhane BR, Amin R, et al (2024) Recovery of tigers (Panthera tigris): Assessing ecological carrying capacity in Bardia-Banke Complex, Nepal. Global Ecology and Conservation 56:e03326

Stapp JR, Lilicholm RJ, Upadhaya S, Johnson T (2015) Evaluating the impacts of forest management policies and community-level institutions in the Buffer Zone of Chitwan National Park, Nepal. J Sustain For 34(5):445–464. https://doi.org/10.1080/10549811.2015.1025080

Thapa K, Nepal S, Thapa G, Bhatta SR, Wikramanayake E (2013) Past, present and future conservation of the greater one-horned rhinoceros Rhinoceros unicornis in Nepal. Oryx 47:345–351

UNCCD (2015) Decision 3/COP.12 Integration of the Sustainable Development Goals and targets into the implementation of the United Nations Convention to Combat Desertification and the Intergovernmental Working Group report on land degradation neutrality

Upreti BR, Breu T, Ghale Y (2017) New challenges in land use in Nepal: reflections on the booming real-estate sector in Chitwan and Kathmandu Valley. Scott Geogr J 33(1). https://doi.org/10.1080/14702541.2017.1279680

WWF (2013) Chitwan Annapurna Landscape: Drivers of Deforestation and Forest Degradation. Baluawatar, Kathmandu, Nepal

WWF (2018) Nepal set to become first country to double wild tiger population. Baluwatar, Kathmandu, Nepal

Indigenous Knowledge and Community Involvement in Sustainable Ecosystem Management in Sri Lanka: Insights and Future Directions

14

Thilanka Gunaratne and Kamani Sylva

Abstract

Despite being renowned for its rich biodiversity and cultural heritage, Sri Lanka faces significant threats from development projects, agricultural expansion, and the overexploitation of natural resources, leading to unsustainable ecosystem management (EM). However, as in many parts of the world, local communities possess both documented and undocumented indigenous knowledge, acquired through real-life experiences, that is deeply connected to their environment. This information can be utilized to manage various land uses in the island in a sustainable manner. This chapter explores the significance of indigenous knowledge systems and the importance of community involvement in ensuring the preservation and stewardship of natural and man-made ecosystems in the country, examining key components and governance structures of sustainable EM systems, issues, innovative strategies, and governance frameworks. Furthermore, we emphasize the importance of inclusion, collaboration, and empowerment of local communities to achieve sustainable development goals (SDGs). The study is built around the model, Grassroots EM, which involves decentralization, cooperation, public engagement, and a holistic worldview to concurrently advance the environment, society, and economy known as the three pillars of sustainability. A conceptual framework based on the Strengths, Opportunities, Aspirations, and Results approach (SOAR) is proposed as a way forward for the effective management of ecosystems in the country, integrating indigenous knowledge systems.

T. Gunaratne (✉)
Department of Botany, Faculty of Science, University of Peradeniya, Peradeniya, Sri Lanka
e-mail: thilankag@pdn.ac.lk

K. Sylva
Department of Engineering Management, Faculty of Engineering, University of Peradeniya, Peradeniya, Sri Lanka

S. Singh, M. K. Gautam (eds.), *Biodiversity Conservation and Land Degradation Neutrality*, https://doi.org/10.1007/978-981-96-7210-3_14

Keywords

Indigenous knowledge · Land management · Sustainability

14.1 Introduction

Healthy terrestrial and aquatic ecosystems are vital components of the biosphere, possessing the resilience to maintain their structure, processes, and functions over time and across space, even in the face of increasing external pressures. The man-made ecosystems established by local communities have aided the well-being of these communities since ancient times, fostering harmonious relationships with nature. Sustainable ecosystem management (EM) is vital to face escalating environmental challenges such as biodiversity loss, pollution, overexploitation of natural resources, rapid spread of invasive species, and climate change. Indigenous people in Sri Lanka have demonstrated a high level of environmental consciousness and ecological knowledge, which is evident in their strategies for managing ecosystems sustainable (Dharmasena 1995; Mendis 2007). In the past, they have skillfully integrated productive systems like shifting cultivation, paddy fields, and home gardens with various landscape features.

Sri Lanka (latitudes 5°55′ and 9°51′ N and longitudes 79°41′ and 81°53′ E) is a continental tropical island in the Indian Ocean covering 65,610 km^2. Extensive geological faulting and erosion have led to a wide range of topographical features, including three major elevation zones: a coastal belt, undulating middle plains, and central highlands with a peak elevation of 2524 m (Kekulandala et al. 2021). The island has a unique ridge-to-reef connection through waterways originating in the central highlands. Considering the mean annual rainfall received by different parts of the island, three distinct climatic zones are identified: wet zone (<2500 mm/year without pronounced dry periods), dry zone (<1750 mm/year with a distinct dry season from May to September) and the intermediate zone (1750–2500 mm/year with a short and less prominent dry season) (Punyawardena 2010). Furthermore, its location in the tropical region, diverse topography, climate, and geology have fostered the evolution of various natural ecosystems, ranging from tropical montane forests in the highest mountains to coral reefs at sea level. Although the island harbors highly biodiverse ecosystems and cultural heritage sites of global significance, ever-increasing pressures from human population growth, including development projects, agriculture expansions, and unsustainable practices, have led to land degradation, threatening the environmental integrity and socio-economic stability of the country. Sri Lanka's traditional land use systems, enriched with indigenous knowledge and community participation, such as the Kandyan Forest Gardens (KFGs) and the Tank Cascade Systems (TCSs), have sustained local communities for centuries while promoting biodiversity, food security, and resilience to environmental shocks. These indigenous systems can be aligned directly or indirectly to achieve most of the sustainable development goals (SDGs) declared by the United Nations in 2015. Thus, it is crucial to understand the fundamental concepts encapsulated in indigenous knowledge systems and the challenges and opportunities in managing these systems to enable landscape-level decision-making for a sustainable future.

The work of Edward P. Weber on Grassroots Ecosystem Management (GREM) was taken as the basis of this discussion owing to the similarity of the concepts presented by Weber (2000) in identifying place-based EM rather than on the specific biophysical scale of management. The study was done to identify the characteristics related to decentralization, collaboration, citizen participation, and holistic worldview introduced by Weber (2000) in his new forefront of EM. The chapter presents a brief literature on sustainable EM. Thereafter, the case of Sri Lanka is presented as a place-specific EM system using indigenous knowledge and community participation. Furthermore, it highlights the features of GREM, decentralization, collaboration, citizen participation, and a holistic worldview relevant to the case to demonstrate place-based EM. Finally, a conceptual framework is proposed based on the SOAR (Strengths, Opportunities, Aspirations, and Results) approach.

14.2 Sustainable Ecosystem Management

Sustainability is the buzzword in the era of sustainable development, and most development approaches focus on human sustainability while aiming to enhance human well-being. Some are short-term focused, and some consider the ecosystem's long-term sustainability as a nutrition supplier for living beings. Ecosystem and EM have been researched for their crucial role in both anthropogenic and ecocentric approaches (De Lucia 2015; Xu et al. 2024). However, there is attention to ecocentrism as a long-term solution to all problems related to anthropological activities. In this context, sustainable EM is reviewed to determine the long-term sustainability of Earth for the benefit of all living organisms and abiotic components of ecosystems.

14.2.1 Ecosystem, Components, and Characteristics

Ecosystem is introduced as the organic and inorganic systems that overlap, interlock, and interact with one another (Tansley 1935). Some authors identify the ecosystem as a system with input and output (Agee and Johnson 1988; Sylva and Iyer-Raniga 2023). All ecosystems (natural and man-made) consist of biotic and abiotic components and interactions between them. The biotic or living components (plants, animals, microbes) play three key roles: producers, consumers, and decomposers, which facilitate the transfer of nutrients and energy throughout the ecosystem. The abiotic component in an ecosystem is represented by sunlight, temperature, water, soil, minerals, and so on, and they can set limits to the growth, survival, and diversity of organisms in an ecosystem. Moreover, the intricate biotic interactions between organisms, including predation, mutualism, parasitism, commensalism, and competition, are vital for sustaining and maintaining biodiversity in any ecosystem. Thus, the ecosystem connects biotic and abiotic components through nutrient and energy cycles to promote growth. Earth's four spheres interact with each other to sustain organisms in the ecosystems by providing the necessary inputs and absorbing the outputs. Human activities disrupt ecosystem balance by overusing natural resources and producing waste faster than circular systems can absorb or decompose (Inglezakis et al. 2016; Narasimhan 2009; Sylva and Iyer-Raniga 2023). While anthropogenic needs are fulfilled, managing the ecosystem balance for the well-being of all organisms has been identified as crucial for the sustainability of life on Earth.

14.2.2 Ecosystem Management and Its Importance

EM involves understanding the interactions between living organisms and their environment and focuses on a holistic approach to maintaining the balance between them. Furthermore, it focuses on anthropogenic activities impending the system. EM tries to control such activities through various means, such as biodiversity conservation, sustainable resource use, monitoring and assessment, stakeholder engagement, adaptive management, restoration, and rehabilitation. Most definitions of EM focus on human sustainability in terms of human needs and economic stability. For example, EM refers to "the use of an ecological approach that blends social, physical, economic, and biological needs and values to assure productive, healthy ecosystems" (Kaufmann et al. 1994), emphasizing human needs. According to Lackey (1998), the fundamental worldviews on EM are twofold: the biocentric approach, which prioritizes considering ecological health as a goal, and the anthropocentric approach, which focuses on benefits to humanity (De Lucia 2015). Since humans have a fundamental obligation to manage ecosystems, the convergence of biocentric and anthropocentric approaches will ensure the long term sustainability of the biosphere.

Edward P. Weber's work on GREM in the 1990s emphasizes human accountability as a crucial element in EM. His research primarily focuses on rural communities, where local economies are closely linked to natural resource harvesting (Weber 2000, 2003). This concept has been developing ever since to incorporate strategies such as community engagement, local knowledge and expertise, capacity building, empowerment and ownership, collaborative partnerships, and adaptive management. Moreover, accountability, bottom-up approaches, and collaboration are highlighted in every effort in EM related to a geographical space.

In more specific terms, GREM is organized based on geographic "place." The local areas are rural economies that rely on nature's resources through farming, forestry, commercial fishing, outdoor recreation, tourism, or ranching. The size and boundaries of a "place" can vary and are determined more by political agreements among those involved rather than scientific criteria. As a result, "place" is often defined as a valley and its surrounding landscape or as a watershed rather than strictly as an ecosystem. The term ecosystem in EM connotes the crosscutting, holistic, comprehensive approach to the notion of conservation and management that focuses on environmental protection, economic development, and community well-being rather than on the specific biophysical scale of management (Weber 2003). In other words, GREM participants are expected to manage valleys, watersheds, forests, or landscapes as an integrated system rather than in a fragmented, piecemeal manner (Weber 2003). GREM movement adopts a cross-cutting, win–win–win ecosystem approach to blend nature with economy and community (Weber 2000). It agrees with the definition of EM by the Forest Ecosystem Management Assessment Team (1993): "strategy or plan to manage ecosystems for all associated organisms, as opposed to a strategy or plan for managing individual species" (Thomas 1993). GREM focuses on the place and its uses as a habitat for living beings, whether human or not.

The Sri Lankan TCS of agricultural society is a classic example of vernacular settlements that manage space rather than a community group to fulfill its needs (Sylva and Sylva 2021; Sylva and Iyer-Raniga 2023). In these systems, human needs are fulfilled by maintaining ecosystem balance in areas under different land uses. Optimum use of landscape, water, and other natural resources, accountability, and local community participation are critical features of TCSs, which enable the preservation of ecological balance with attention to biodiversity and protection of natural ecosystems. Decentralization, collaboration, citizen participation, and a holistic worldview, introduced by Weber (2000) in "The New Vanguard for the Environment," exemplified in this case study for promoting sustainability with an emphasis on society, economy, and the environment.

14.3 Indigenous Knowledge and Community Participation in Ecosystem Management: Sri Lanka as a Case Study

Indigenous knowledge of EM refers to "how the residents of an area have come to understand their natural environment and how they organize this knowledge of environmental conditions, flora and fauna, cultural beliefs, practices, and history to enhance their lives" (Semali and Kincheloe 2002). Studies done by various researchers on ancient land management systems in Sri Lanka showcase the effectiveness of indigenous knowledge systems and community participation in sustaining ecosystems and promoting resilience in the face of environmental perturbations (Dharmasena 1995; Mendis 2007). For instance, Sri Lanka's traditional land use systems, such as the Kandyan Home Gardens or KFGs and the TCSs, have resulted from centuries of biological and cultural exchanges between humans and the rest of the environment (biotic, abiotic, and interactions between them). These systems have sustained local communities for centuries while providing many benefits, including provisioning, regulating, cultural, and supporting ecosystem services. This has resulted in food and water security, environmental conservation, and resilience to environmental shocks amidst climate change.

EM was central to the planning and practices of native agricultural societies in the country during the reign of the kings. The settlements were built around river basins and water bodies, and the villages were identified considering the natural topography of water sources. Water use for irrigation, flood control, drainage, and conservation of the environment, as well as flora and fauna, has been significant in Sri Lankan agricultural societies (Dharmasena 1995; Mendis 2007). However, modern measuring standards for development have considered indigenous knowledge and practices undeveloped and backward. Thus, the rich history of context-specific sustainable practices has been ignored and rarely considered in developing current practices (Dayaratne 2018; Mendis 2002; Sylva and Sylva 2021). Destroying ancient ecosystems by the imposition of new hydraulic engineering designs is a classic example of this ignorance (Mendis 2002).

14.3.1 History of Sustainable Ecosystem Management in Sri Lanka

Over time, Sri Lanka has undergone a unique and dynamic evolution in its laws, regulations, and institutions governing land and land use (Weerahewa et al. 2021). The traditional agricultural society in the dry climatic zone of Sri Lanka was built upon the foundational triangle of the tank (Wewa), paddy field (Keta), and temple (Dagaba), each serving a specific role: the temple managed both spiritual and administrative matters within village communities, the paddy field provided food security, and the tank was the essential heart of rural life (Köpke et al. 2019) (Fig. 14.1). The ancient villages were often located near the bank bunds, and the paddy fields supported the food requirement of the people engaged in shifting cultivation (Chena) and home gardens contributed to food security (Bandara 2010). Pre-colonial kings promoted self-sufficiency through a semi-feudal land distribution system, focusing on food crops and spice exports (Weerahewa et al. 2021). In pre-colonial Sri Lanka, the highlands were covered by montane and sub-montane forests, and grasslands had limited human interference. Most of the Chena cultivations were at elevations below 1000 m, only a few made into the higher elevations (Wickramagamage 2017).

However, during colonial rule, their focus shifted from self-sufficiency to earning foreign exchange via the commercial cultivation of cash crops such as coffee, tea, rubber, coconut, and spices. Coffee, the first plantation crop introduced to Sri Lanka, was responsible for the deforestation of a large part of the hydrologically most sensitive highlands in central Sri Lanka, where the destruction of forests was not confined to areas that were suitable for coffee but instead included steep slopes, which should have been kept under permanent tree cover (Wickramagamage 2017). Furthermore, land tenure laws were changed to facilitate the establishment of cash crops, directly affecting domestic rice production, necessitating the island to import rice (Weerahewa et al. 2021). During the post-colonial era, Sri Lanka's consecutive governments adopted varied regulations on land utilization and management to meet the political goals of the ruling party (Weerahewa et al. 2021). Currently, the lack of updated long-term research and result-based decision-making, along with the involvement of numerous institutions in land management, has led to the mismanagement of ecosystems during both the colonial and post-colonial periods. This

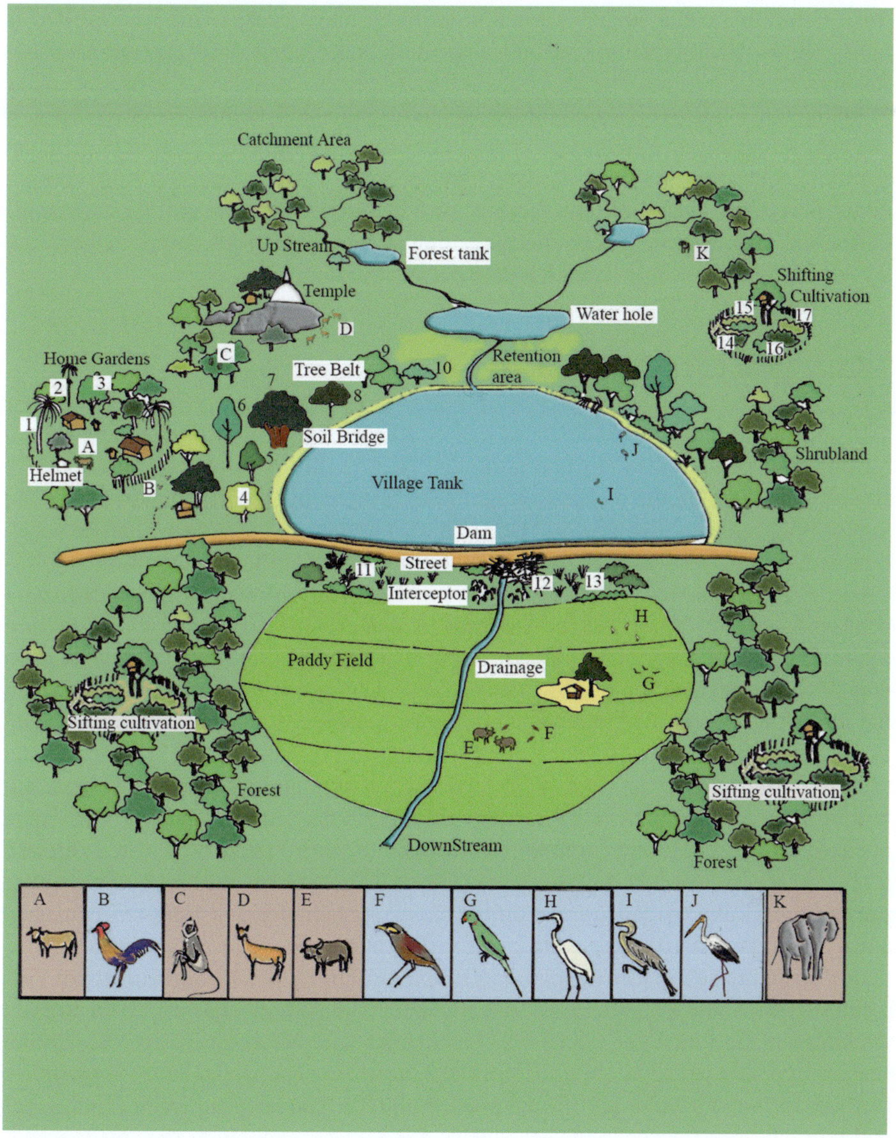

Fig. 14.1 Components of the TCS and adjacent land use categories in Sri Lanka. *Flora*: Home garden: 1—*Cocos nucifera* (Coconut), 2—*Areca catechu* (Arecanut), 3—*Artocarpus heterophyllus* (Jack Fruit), Tree belt: 4—*Azadirachta indica* (Neem), 5—*Cassia fistula* (Golden Flower Tree), 6—*Chloroxylon swietenia* (Ceylon Satinwood), 7—*Ficus* sp. (Fig), 8—*Manilkara hexandra* (Palu), 9—*Drypetes sepiaria* (Weera), 10—*Dichrostachys cinerea*, Interceptor: 11—*Calamus* spp. (Rattan), 12—*Bambusa* spp. (Bamboo), 13—*Cyperus pangorei* (Mat Grass), Shifting Cultivation: 14—*Eleusine coracana* (Finger Millet), 15—*Panicum miliaceum* (Common Millet), 16—*Cucumis melo* (Cooking Melon), and 17—*Cucurbita pepo* (Pumpkin). *Fauna*: A—*Bos indicus* (Cattle), B—*Gallus lafayetii* (Jungle Fowl), C—*Semnopithecus dussumieri* (Gray Langur), D—*Axis axis* (Spotted Deer), E—*Bubalus bubalis* (Domestic Water Buffalo), F—*Acridotheres tristis* (Common Mynah), G—*Psittacula krameri* (Rose-ringed Parakeet), H—*Egretta alba* (Great Egret), I—*Phalacrocorax carbo* (Great Cormorant), J—*Mycteria leucocephal* (Painted Stork), and K—*Elephas maximus maximus* (Wild Elephant). (© Dilmini Samarakoon/University of Peradeniya)

misuse and mismanagement of natural and man made ecosystems has resulted in their degradation, creating various environmental, social, and economic issues for the nation.

Indigenous knowledge embedded in traditional land management systems is strongly aligned with SDGs. Thus, it provides insights into the sustainable management of ecosystems through a holistic approach that involves learning from components, processes, functions, and services provided by ecosystems for the sustenance of life on Earth. Two such prominent sustainable man-made ecosystems in the country are the TCSs and KFGs, which date back more than 2,000 years.

14.3.2 Tank Cascade Systems

Two-thirds of Sri Lanka experiences a dry climate. Thus, irrigation remains crucial for agricultural activities (Köpke et al. 2019). Agriculture has spread into the dry zone by overcoming the water shortage, as evidenced by the distribution of ancient tanks (Wickramagamage 2017). The TCS, also known as the Wewa-Ellangava system, is a traditional water management system with historical and cultural significance. An Ellangava is a very low-gradient, sluggishly moving water path from a head-end summit (mudunna) of a micro- or meso-valley to a large reservoir or a secondary stream (oya) at down-end, filling a chain of tanks one after another in its way down, gently spilling over the tanks excess water into long flat plains in between the tanks and in that whole process the tanks accumulating, preserving, releasing and guiding water to downstream fields for irrigation (Tennakoon 2015). Dharmasena (2020) defines TCS as an ecosystem where water and land resources are organized within the micro catchments of the dry zone landscape, to sustainably meet basic needs to human and biotic communities through the integrated use of water, soil, air, and vegetation. Some TCSs date back over 2,000 years but remain operational at present, providing evidence for a unique case of resilience and long-term sustainability (Bandara 2010). These have evolved to irrigate paddy fields during the two paddy cropping seasons corresponding to the rainfall in the dry climatic zone: Yala (minor cultivation season) and Maha (major cultivation season). Yala season depends on the first inter-monsoonal rains from mid-March to early May. Maha season depends on high rainfall from the northeast monsoon from September/October to January/February (Punyawardena 2010). The need for more economical and rational use of water in the seasonally dry environment led to the development of the recycling or reuse strategy, where water from the upper parts of the cascade was used and reused several times before it reached the sea (Bandara 2010).

The rolling topography of the terrain in most areas creates a myriad of small watersheds, providing the necessary natural setting for the evolution of tank cascade-based agriculture (Bandara 2010). Tanks that capture rainwater vary in size and capacity from small village reservoirs to large regional tanks. Each cascade delineates a distinct small watershed or meso-catchment ranging in extent from 13 to 26 km^2 with an average of 20 km^2 (Paranage 2018). Panabokke (2002) have estimated a total of 15,499 tanks in the dry climatic zone, of which only 7,202 village tanks are currently functioning and 7,753 have been abandoned due to the location of meso-catchments in areas with poor hydrological conditions or in rocky and gravelly terrain referred to as rock-knob-plains, sharper relief of the meso-landform and the construction of dam embankments using easily dispersible sodic soils, which led to embankment breaches. TCSs guarantee uninterrupted cultivation in both the wet season (Maha) and in the dry season (Yala) in the dry climatic zone. Furthermore, the TCS has been recognized as a Globally Important Agricultural Heritage System by the Food and Agriculture Organization of the United Nations in 2017, to safeguard traditional agricultural systems of global importance, which harness the harmonious relationship between people and nature. The comprehensive studies on TCS highlight the environmental consciousness, indigenous wisdom, skill level, and community participation of locals to elucidate a solution for water issues in the dry zone of Sri Lanka. According to Dharmasena and Kadupitiya (2021), the TCSs greatly contribute to biodiversity as they create links between many ecosystems such as wetlands, seasonally wet and dry lands, paddy fields, uplands, forests, scrublands, tank beds, home gardens, rocky lands, and water streams. These sustainable EM systems have evolved through traditional water management practices, disaster forecasting systems, suitable cropping calendars, drought-resistant rice and other seed varieties, unique land tenure and land use systems, institutional mechanisms, cultural adjustments, and traditional values and beliefs, making them prime examples of social and technological adaptations to drought by indigenous communities in the dry climatic zone (Bandara 2010). Moreover, it is important to evaluate the contribution of TCSs to achieving the SDGs. TCSs impact 9 out of the 17 SDGs, including Goal 1 (No Poverty), Goal 2 (Zero Hunger), Goal 3 (Good Health and Well-being), Goal 4 (Quality Education), Goal 6 (Clean Water and Sanitation), Goal 8 (Decent Work and Economic Growth), Goal 13 (Climate Action), Goal 14 (Life Below Water), and Goal 15 (Life on Land), either directly or indirectly.

14.3.3 Components and Function

The water catchments protected by forests, man-made reservoirs, canals, bunds, and other structures of TCSs are interconnected and act as a single unit when capturing, storing, and distributing rainwater. Archaeological studies have revealed that the main parts of a cascade system were constructed in harmony with the local ecosystem (Köpke et al. 2019). Through experimenting under real-life scenarios, indigenous communities in the dry zone likely discovered various land use systems tailored to their landscape. For instance, they established paddy fields on richer alluvial humic gley soils, shifting cultivation on the slopes with reddish-brown soil in the upper segments of the catena, and infertile areas of the watersheds were left intact under rocks and forests (Bandara 2010).

The key components of a typical TCS include catchment areas protected by forests in upslopes surrounding reservoirs that collect rainwater and channel it into the tanks in lower slopes (Fig. 14.1). These tanks of different sizes and functions increase the volume of water available for storage by capturing rainwater runoff from surrounding hillsides and watersheds. The forest tank (Kalu Wewa in Sinhala), situated in the jungle above the village, was not used for land irrigation. Instead, it functioned as a sediment filtering tank and provided water to wild animals, reducing their likelihood of descending into cultivated areas in search of water (Bandara 2010; Dharmasena and Kadupitiya 2021). According to Dharmasena and Kadupitiya (2021), water purification is one of the most important ecological services provided by TCSs, which is achieved using the interceptor (Kattakaduwa in Sinhala), the reserved land below the tank bund and sediment filter (Perahana in Sinhala) developed under the tree belt. The interceptor with different microhabitats prevents the entering of salts and Ferric ions into the paddy field, and the sediment filter dominated by grasses, sedges (*Cyperus* spp., *Schoenoplectus grossus*), and reeds (*Typha angustifolia*) filters the sediment flow coming from upstream chena lands. The larger trees, such as *Terminalia arjuna* in the tree belt located in the upper inundation area of almost all TCSs, act as a wind barrier and reduce the occurrence of waves in the tank, consequently helping to prevent evaporation. Most of the plant species in the catchment forests and tree belt are represented by species of tropical dry-mixed evergreen forests (climax vegetation type in dry climatic zone), which have evolved corresponding to the arid climatic conditions in the region. Moreover, a water hole (Godawala in Sinhala) is used to trap sediments, and it provides water to wild animals, a strategy to avoid human–animal conflict. Iswetiya or Potawetiya is an upstream soil ridge constructed on either side of the tank bund to prevent entry of eroded soil from upper land slopes (Dharmasena and Kadupitiya 2021). Tank (Wewa in Sinhala) is used by villagers to store rainwater runoff during the wet season. This ensures water availability during the dry season to provide all the ecosystem services valued by the local communities. Canals (Elas in Sinhala) are earthen channels that connect different reservoirs within the TCS, allowing for the transfer of water from upstream to downstream areas. These distribute water to agricultural fields, homesteads, and other areas for irrigation or domestic water supply. Tank embankments (Bunds in Sinhala) are raised earthen structures built around the perimeter of reservoirs to contain water and prevent overflow. They minimize water loss due to seepage and ensure efficient utilization of available water resources. Spillways and sluice gates regulate the flow of water in and out of reservoirs, controlling water levels and preventing flooding during periods of heavy rainfall. Sluice gates are opened to release excess water during floods in the wet season and closed to retain water during the dry season, allowing for optimal irrigation scheduling in the dry climatic zone. Water distribution networks are composed of channels, pipes, and outlets that distribute water from canals and reservoirs to agricultural fields, homesteads, and other users. Kurulu Paluwa is a small section of unharvested paddy land allocated to support bird life and aid in the pest management of paddy fields. In ancient times, the water distribution network through the TCSs ensured equitable access to water resources, promoting efficient irrigation practices to maximize agricultural productivity and enhance the well-being of all organisms.

Maintenance structures such as sluices, weirs, and desilting outlets are built within the TCS to facilitate regular maintenance and repair activities. Although tanks were managed by local communities to suit their livelihoods during the pre-colonial period, during the British colonial period, community-led management structures were disregarded and largely disbanded as the government converted the governance of larger water storages into centrally managed structures (Kekulandala et al. 2021). In the tenth century and earlier, the maintenance of the vast and complex irrigation system in efficient working conditions would have constituted a very important responsibility of the state (Nicholas 1960). Furthermore, "pre-colonial Sri Lankan society retained a high degree of homogeneity through kinship and caste system that ensured a higher community spirit and self-reliance"; thus, the construction and management of the TCSs has been done through local community participation (Bandara 2010).

14.3.4 Ecosystem Services

According to Mendonce et al. (2022), TCSs provide various provisioning services, including water for paddy cultivation, traditional medicines, biopesticides, livestock fodder, roofing materials, opportunities for freshwater aquaculture, daily water needs, and diversification of income sources through vegetables, fruits, and ornamental flowers. It also regulates services such as flood and drought mitigation, silt trapping, water desalination, microclimate regulation, pollinator habitats, carbon sequestration, and biological pest control. Supporting services include nutrient cycling and habitats for wildlife such as wild elephants and aquatic fauna (birds, reptiles, fish, insects, etc.). Cultural services encompass sites for education, recreation (bathing, swimming), community meeting places, and tourism.

14.3.5 Kandyan Forest Gardens

The KFGs are man-made ecosystems that act as sustainably managed land parcels, contributing to conserving biodiversity outside protected areas. KFGs are distributed over wet and intermediate climatic zones of central Sri Lanka. These diverse smallholder agroforestry systems represent indigenous knowledge passed down from generation to generation over the past 2,000 years, cover an average land area of 0.4 ha, and exemplify socio-ecological systems characterized by deliberate human interactions (Landreth and Saito 2014). Moreover, KFGs reduce the pressure on natural forests by providing many provisioning services, including food, timber, and medicine to local communities.

According to Wickramasinghe (1995), the plant density of individuals in these gardens ranged between 2,240 and 3,926 per hectare; trees averaged 2,800 per hectare, endemic and naturalized trees make up nearly 40–45% of the garden, and almost 34% of the gardens contain livestock, primarily dairy cattle and poultry, with fewer buffaloes and goats. They have the same vertical structure as tropical rainforests, although the height of the strata is much lower than that of the natural forest (Fig. 14.2). Moreover, like in tropical rainforests, different types of life forms, such

Fig. 14.2 Canopy layer: 1—*Artocarpus heterophyllus* (Jack Fruit), 2—*Cocos nucifera* (Coconut), 3—*Areca catechu* (Arecanut), 4—*Tectona grandis* (Teak) Subcanopy layer: 5—*Durio zibethinus* (Duriyan), 6—*Mangifera indica* (Mango), 7—*Persea americana* (Avocado), 8—*Caryota urenus* (Fish Tail Palm) Understory tree layer: 9—*Murraya koenigii* (Curry Leaf Tree), 10—*Annona muricata* (Soursop), 11—*Carica papaya* (Papaya), 12—*Phyllanthus acidus* (Star Gooseberry), 13—*Syzygium samarangense* (Rose Apple), 14—Musa sp. (Banana), 15—*Flacourtia indica* (Indian Plum), 16—*Psidium cattleyanum* (Cattley Guava), 17—*Citrus aurantiifolia* (Lime), 18—*Sesbania grandiflora* (Vegetable Hummingbird) Understory shrub layer: 19—*Tabernaemontana divaricate* (Pinwheel Flower), 20—*Brunfelsia pauciflora* (Yesterday-today-and-tomorrow), 21—*Gardenia jasminoides* (Gardenia), 22—*Manihot esculenta* (Cassava), 23—*Pandanus amaryllifolius* (Pandan) Ground layer: 24—*Colocasia esculenta* (Taro), 25—*Phyllanthus myrtifolius* (Mousetail Plant), 26—Anthurium sp. (Laceleaf). Climbers: 27—*Piper nigum* (Peper), and 28—*Vanilla planifolia* (Vanilla). Fauna: A—*Macaca sinica* (Toque Macaque), B—*Psittacula krameri* (Rose-ringed Parakeet), C—*Funambulus palmarum* (Indian Palm Squirrel), D—*Turdoides affinis* (Yellow-billed Babbler), E—*Hystrix indica* (Indian Crested Porcupine), and F—*Sus scrofa* (Wild Boar). (© Dilmini Samarakoon/University of Peradeniya)

as trees, shrubs, climbers, lianas, and herbs, are found in them. The intricate vertical structure and floristic composition in KFGs create a variety of micro-habitats providing food sources and nesting sites for faunal species such as Indian crested porcupines, wild boar, Toque macaques, bats, birds and insects. As a result of human activities fragmenting natural ecosystems, presently KFGs function as biodiversity refugia within human-dominated landscapes. However, the increased frequency of visitation by wild animals to KFGs results in significant crop damage, leading to human-wildlife conflicts at present.

KFGs provide many ecosystem services, including provisioning services such as food, fuel, timber, medicine, pesticides, livestock fodder, fertilizer, and genetic resources. They regulate local climate, soil erosion, water purification, and pollination. Moreover, KFGs are centers of carbon sequestration, nutrient cycling and provide habitats for pollinators, seed dispersers, and predators, enhancing the ecological balance and sustainability. Their high aesthetic and spiritual value offers cultural services, including spiritual and recreational benefits, and thus can be used to promote ecotourism and overall community well-being through ecotherapy. Moreover, these gardens contribute to achieving 9 out of 17 SDGs, including SDG goal 1 (No Poverty), 2 (Zero Hunger), 3 (Good Health and Well-being), 4 (Quality Education), 6 (Clean Water and Sanitation), 8 (Decent Work and Economic Growth), 13 (Climate Action), 14 (Life Below Water), and 15 (Life on Land) thus contribute for the well-being of all organisms.

14.3.6 Citizen Participation and Collaboration

In ancient times, most of the activities related to the agriculture of a village have been looked after by the chief of the village "Vidane" (who gives orders), and according to the historical information, the village activities related to irrigation have been decided by the Vidane (Sylva and Sylva 2021). This person, also called the Vel-vidane, oversees the paddy cultivation and water management and has been elected by the farmers. He has been paid for his service with a quantity of paddy from the paddy cultivation related to the village tank (Dharmasena 2016). However, the irrigation system has been maintained collectively with the involvement of all villages. According to Dharmasena (2016), the main catchment area for the village tank has been protected and prohibited from being cleared for any purpose, and this has been done through community understanding.

Paddy has been the primary cultivation, and around 2,000 varieties of paddy have been cultivated in the past (Sylva and Sylva 2021). The decisions on the paddy type to cultivate, the managing and diverting of water from the paddy lands to other systems, which plot will be used by whom, and which plot to be reserved for cultivation for charity and foraging of the birds (commonly cultivated) have been monitored and administered by the chief of the field, Vidane (Sylva and Sylva 2021). The villages protected the village micro-landscapes to ensure sustainability of TCSs (Dharmasena 2016). They have developed self-formulated rules to sustain society. Dharmasena and Kadupitiya (2021) explain some of them as having controls on water when the amount of rainwater is limited, selecting the types of paddy for each season, protecting biodiversity, including trees, filling the lakes with rainwater, a simple life, teamwork, storing paddy for off-season, not to make life dependent on what is not available, but to protect knowledge as wealth, and to reduce stress by use of community music.

14.3.7 Decentralization and Holistic Worldview

Management of water sources has been done using a top-down method, from rivers to streams to village tanks in the TCSs or Ellangava system (Dharmasena and Kadupitiya 2021). However, maintenance of TCSs has been done through community education and an understanding of the importance of these systems. The village tank (Wewa) is a subsystem of the whole system called the Ellangava (Tennakoon 2015). The Ellangava system is a series of tanks hanging next to one another (Fernando 2019; Sylva and Sylva 2021; Tennakoon 2015). The subsystem decisions are taken in a decentralized manner without violating the needs of water of the whole system in the Ellangava system. The Ellangawa system is the cascade system of tanks filling smaller to medium to large tanks in the dry zone of Sri Lanka. The holistic view is dominant in planning these tanks, although the village decisions were made in a decentralized manner. Each subsystem has been planned to preserve water catchment by avoiding settlements in the natural forests and releasing water downstream after cleaning through natural biological cleaning ponds or wetlands. Diverting water from areas of excess water to areas with less rainfall has been done by constructing channels along the contour lines with slight gradients. One side of most of these channels has been bounded by forest strips, allowing wild animals to access safe drinking water to sustain and maintain the rich biodiversity in the country (Sylva and Sylva 2021).

The KFGs are another decentralized EM system that operates under holistic thinking. The home environment of a family is managed to ensure that it does not disrupt the functioning of the ecosystem while still meeting the family's food needs. This approach is recognized as a sustainable agricultural ecosystem in Sri Lanka (Gunawardena and Fernando 2022). While utilizing food, timber and medicines from KFGs, the members of the families also protect the soil and hydrological profile to ensure the fullfillment of the needs of the subsequent sub-systems. According to Landreth and Saito (2014), KFGs are a balance between human well-being and EM, providing many ecosystem services. They state, "Kandyan home gardens are socio-ecological systems characterized by deliberate human interactions balancing household livelihoods with the environmental function."

14.3.8 Conceptual Framework for Sustainable Ecosystem Management

Isolated small-, medium-, and large-scale development projects without landscape-level thinking have resulted in a mosaic of land use types in Sri Lanka. Presently, fragments of natural ecosystems at different successional stages are embedded in a mosaic of man-made ecosystems such as forest plantations (Pinus, Eucalyptus, Teak), cash crop plantations (Tea, Rubber, Coconut, Oil Palm), agricultural fields (Paddy, Vegetable plots), TCSs, KFGs, and urban areas. In the country, biodiversity is increasingly threatened by habitat fragmentation, pollution, overexploitation of natural resources, the spread of invasive species, and the impacts of climate change. Moreover, Sri Lanka has consistently been placed among the top ten countries at risk of extreme weather events by the Global Climate Risk Index. For instance, it was placed as the second most affected countries in 2017 (Eckstein et al. 2018). Thus, it is high time policymakers, scientists, educators, and communities effectively engage to take adaptation and mitigation actions to develop climate-resilient communities in the near future. Since massive knowledge of sustainable EM is embedded in local people, it must be harnessed through studying, understanding, and appreciating it. Integration of indigenous knowledge hidden in ancient EM systems can be updated using modern sustainable land management approaches. This will promote ecological resilience, socio-economic development, and cultural preservation amidst climatic challenges.

In sustainable EM, it is crucial to comprehend the environmental, social, cultural, and economic contexts of diverse stakeholders before implementing any new external interventions. Guided by local needs and contexts, this approach strives to attain the SDGs outlined by the United Nations in 2015, which include safeguarding the environment, eradicating poverty, and ensuring universal peace and prosperity by 2030. Despite the fragmentation of natural ecosystems in many parts of the world, both natural and man-made ecosystems share many common components, processes, and functions. Presently, these common elements are interconnected or isolated at the landscape level. Linking natural and man-made ecosystems using ecological links is vital to mixing genetic material between them for the sustenance of biodiversity on Earth. For instance, collaborative approaches involving all stakeholders in different land parcels are essential to promote coexistence between humans and wildlife, such as wild elephants, leopards, wild boars, peacocks and Toque monkeys in human-dominated landscapes. Thus, it is crucial to engage stakeholders during the project development phase to tap into the indigenous knowledge systems by creating platforms that encourage meaningful contribution of local communities in natural resource management, especially in biodiverse-rich regions. Consequently, the valuation of ecosystem services, assessment of ecosystem's carrying capacity, evaluation of environmental and social risks, identification of sustainable livelihood opportunities, enhancement of livelihood resilience, empowerment and engagement of stakeholders (especially next generation) in decision making, and restoration of ecosystems will collectively contribute to sustainable economic growth and well-being of all organisms on Earth in future.

Since many cross-sectional interactions between environment, livelihoods, development, and economy at the landscape scale challenge sustainable EM, it is essential to establish platforms that promote effective engagement among all stakeholders (Ghormare et al. 2024). These platforms must include local communities (women, marginalized groups, and indigenous peoples), government institutions, private sectors, academic institutions, and development practitioners at all levels representing different age groups, education literacies, and ethnic groups, to enable governments to capture and update indigenous knowledge to initiate sustainable EM at landscape level aligning with the SDGs. We propose a framework for sustainable EM using the Strengths, Opportunities, Aspirations, and Results (SOAR) approach (Fig. 14.3). The SOAR approach explained by Stavros et al. (2003) starts with an appreciative inquiry leading to discovering strengths (S) and opportunities (O) available for sustainable EM. Then, platforms must be created to enable all stakeholders to share their aspirations (A) and co-construct their most preferred sustainable future vision that aligns with SDGs. Finally, all stakeholders must recognize and design reward programs to inspire stakeholders to achieve measurable results (R).

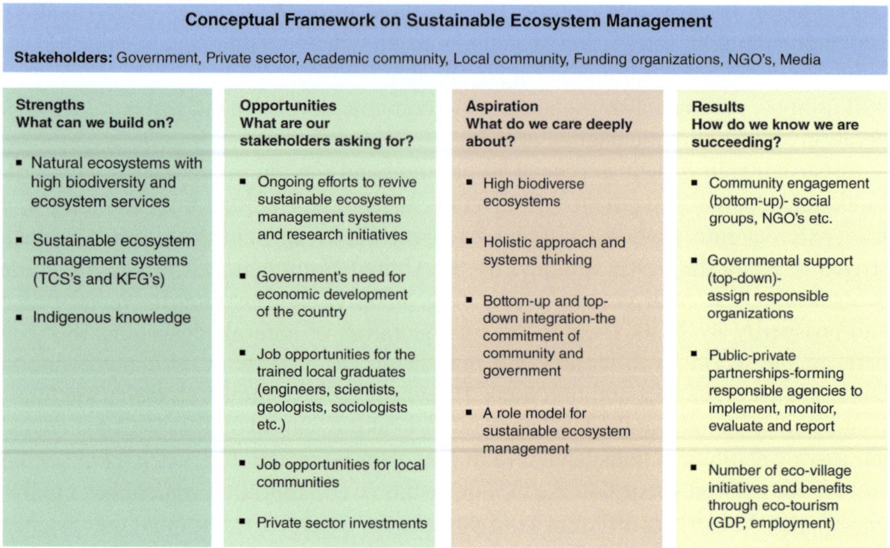

Fig. 14.3 Framework for sustainable EM

The proposed framework emphasizes the importance of multi-stakeholder platforms with effective and transparent governance mechanisms to build trust among all core stakeholders. This trust encourages active participation in designing, implementing, monitoring, evaluating, and reporting on projects, ensuring sustainability at every phase from inception onwards.

14.4 Conclusions

Indigenous knowledge and community engagement are deeply intertwined with cultural identities and comprehensive understandings of ecosystems and natural resource management in ancient Sri Lanka. Through case studies, theoretical frameworks, and governance structures, we advocate for the recognition and empowerment of indigenous communities as key stakeholders in environmental governance and advocate for a holistic approach towards achieving SDGs amidst climate change.

References

Agee J, Johnson D (1988) Ecosystem management for parks and wilderness. https://api.semanticscholar.org/CorpusID:129784702

Bandara CM (2010) Village tank cascade systems: a traditional approach to drought mitigation and rural wellbeing in the Purana villages* of Sri Lanka. In: Indigenous knowledge for disaster risk reduction. ISDR, European Union, p 68

Dayaratne R (2018) Toward sustainable development: lessons from vernacular settlements of Sri Lanka. Front Archit Res 7:334–346

De Lucia V (2015) The ecosystem approach between anthropocentrism and ecocentrism. Working paper, January 2014. https://doi.org/10.13140/RG.2.1.3247.6889

Dharmasena PB (1995) Kattakaduwa; a potential land for agro-forestry system development in Sri Lanka. In: Gunasena HPM (ed) Proceedings of the 6th regional workshop on multipurpose trees, 17th–19th August 1995. University of Peradeniya, Peradeniya, pp 96–104

Dharmasena PB (2016) The cascaded tank-village system (CTVS) in the dry zone of Sri Lanka. Report: Food and Agriculture Organization of the United Nations, Sri Lanka

Dharmasena PB (2020) Cascaded tank-village system: present status and prospects. In: Agricultural research for sustainable food Systems in Sri Lanka: Volume 1: A historical perspective. Springer, Singapore, pp 63–75

Dharmasena PB, Kadupitiya HK (2021) UNEP-GEF project: managing agricultural landscapes in socio-ecologically sensitive areas to promote food security, wellbeing and ecosystem health in Sri Lanka

Eckstein D, Hutfils ML, Winges M (2018) Global Climate Risk Index 2019. Who suffers most from extreme weather events, p 36

Fernando S (2019) Tank cascade system (ellanga system). Culture & Heritage, Ceylon Digest, 7 August 2019

Ghormare RR, Fatima S, Grover P, Phutela N, Kandpal V, Gonzalez EDS (2024) Exploring the paradigm shift towards sustainability: a systematic literature review on circular economy and eco-innovation. AIMS Environ Sci 11(6):940–959

Gunawardena AR, Fernando TT (2022) The Kandyan home garden – sustainable agricultural ecosystem in Sri Lanka. In: Vadrevu KP, Le Toan T, Ray SS, Justice C (eds) Remote sensing of agriculture and land cover/land use changes in South and Southeast Asian countries. Springer, Cham

Inglezakis V, Poulopoulos S, Arkhangelsky E, Zorpas A, Menegaki A (2016) Aquatic environment. In: Environment and development. Elsevier, Amsterdam, pp 137–212. https://doi.org/10.1016/b978-0-444-62733-9.00003-4

Kaufmann MR, Graham RT, Boyce DA Jr et al (1994) An ecological basis for ecosystem management. USDA Forest Service, Rocky Mountain Forest and Range Experiment Station, Fort Collins

Kekulandala B, Jacobs B, Cunningham R (2021) Management of small irrigation tank cascade systems (STCS) in Sri Lanka: past, present and future. Clim Dev 13(4):337–347

Köpke S, Withanachchi SS, Pathiranage R, Withanachchi CR, Ploeger A (2019) Social–ecological dynamics in irrigated agriculture in dry zone Sri Lanka: a political ecology. Sustain Water Resour Manag 5(2):629–637

Lackey R (1998) Ecosystem management: in search of the elusive paradigm. Hum Ecol Rev 4:107–113

Landreth N, Saito O (2014) An ecosystem services approach to sustainable livelihoods in the homegardens of Kandy, Sri Lanka. Aust Geogr 45(3):355–373. https://doi.org/10.1080/00049182.2014.930003

Mendis DLO (2002) Water heritage of Sri Lanka. Pugwash Group, Colombo. ISBN 955-647-003-3

Mendis D (2007) Ancient water and soil conservation ecosystems of Sri Lanka. In: Conference paper presentation in international history seminar on irrigation and drainage. Tehran, Iran

Mendonce S, Borelli T, Hunter D, Silva A, Bandara T, Ratnayake SS (2022) Healthy landscapes: managing agricultural landscapes in socio-ecologically sensitive areas to promote food security, wellbeing and ecosystem health. Global Environment Facility

Narasimhan T (2009) Encyclopedia of inland waters. Academic Press

Nicholas CW (1960) A short account of the history of irrigation works up to the 11th century. J Ceylon Branch R Asiat Soc Great Brit Irel 7(1):43–69

Panabokke CR (2002) Small tanks in Sri Lanka: evolution, present status, and issues. IWMI

Paranage K (2018) Understanding the relationship between water infrastructure and socio-political configurations: A case study from Sri Lanka. Water 10(10):1402

Punyawardena BVR (2010) Climate of the dry zone of Sri Lanka. In: Soils of the dry zone of Sri Lanka, Special Publication No 7. Soil Science Society of Sri Lanka, pp 9–26

Semali LM, Kincheloe JL (2002) What is indigenous knowledge? Voices from the academy. Routledge
Stavros J, Cooperrider D, Kelley DL (2003) Strategic inquiry appreciative intent: inspiration to SOAR, a new framework for strategic planning. AI Pract 11:1–21
Sylva K, Iyer-Raniga U (2023) Partnership in the built environment for realizing the 2030 agenda: a soft systems model incorporating systems theory and the circular economy. In: Walker T, Cucuzzella C, Goubran S, Geith R (eds) The role of design, construction, and real estate in advancing the sustainable development goals. Sustainable development goals series. Palgrave Macmillan, Cham. https://doi.org/10.1007/978-3-031-28739-8_9
Sylva KKK, Sylva KKLA (2021) Sustainable city planning and management strategies in vernacular settlement patterns in Sri Lanka. In: Iyer-Raniga U (ed) Sustainability in the built environment in the 21st century: lessons learned from India and the region. SDRAP 2019. Environmental science and engineering. Springer. https://doi.org/10.1007/978-3-030-61891-9_9
Tansley AG (1935) The use and abuse of vegetational concepts and terms. Ecology 16(3):284–307. https://doi.org/10.2307/1930070
Tennakoon MUA (2015) Traditional dry zone irrigation—ecosystematic concepts revisited and reconceptualized. South Asia Partnership Sri Lanka, Sri Jayawardenepura Kotte. https://sapsri.lk/. Accessed 2 Oct 2023
Thomas JW (1993) Forest ecosystem management: an ecological, economic, and social assessment. Report of the Forest Ecosystem Management Assessment Team. US Government Printing Office
Weber EP (2000) A new vanguard for the environment: grass-roots ecosystem management as a new environmental movement. Soc Nat Resour 13(3):237–259
Weber EP (2003) Bringing society back. In: Grassroots ecosystem management, accountability, and sustainable communities, American and comparative environmental policy. The MIT Press, Cambridge, MA/London. ISBN: 026223226X, 9780585446721, 9780262232265, 0262731517, 9780262731515
Weerahewa J, Hemachandra D, Pushpakumara G (2021) Land regulations, tenure, and land use of Sri Lanka: a historical analysis on agricultural lands. In: Examining international land use policies, changes, and conflicts. IGI Global, pp 139–161
Wickramagamage P (2017) Role of human agency in the transformation of the biogeography of Sri Lanka. Ceylon J Sci 46:19
Wickramasinghe A (1995) The evolution of Kandyan home-gardens. In: Conserving biodiversity outside protected areas: the role of traditional agro-ecosystems. IUCN/AMA, pp 164–182
Xu Y, Yang L, Hossain ME, Haseeb M, Ran Q (2024) Unveiling the trajectory of corporate green innovation: the roles of the public attention and government. J Clean Prod 444:141119

Restoring Ecosystems: Case of Private Sector and Best Practice Through Community Forestry Approaches

15

Nabin Bhattarai, Tulasi Sangraula, Shivang Sinha, Teiji Watanabe, Ram Avtar, Rajesh Bahadur Thapa, Ranjit Singh Rawat, Kripa Shrestha, Bikas Adhikari, Chanda Khadka, and Rabindra Adhikari

Abstract

The degradation of ecosystems worldwide is a significant environmental concern, with far-reaching consequences. Human-induced habitat destruction and climate change are some of the inherent causes for the loss of habitat and biodiversity. This causes cascading effects on ecosystem functioning, resilience, and services essential for human well-being, such as pollination, water purification, and climate regulation. The degradation of ecosystems not only impacts the local environment but also has broader implications for global biodiversity, climate

N. Bhattarai (✉)
Armale Innovation Pvt. Ltd, Balkot, Bhaktapur, Nepal

T. Sangraula
Federation of Community Forestry Users, Changunarayan, Nepal

S. Sinha · T. Watanabe · R. Avtar
Hokkaido University, Sapporo, Japan

R. B. Thapa
International Centre for Integrated Mountain Development, Lalitpur, Kathmandu, Nepal

R. S. Rawat
Indian Council for Forestry Research and Education, Dehradun, India

K. Shrestha
Montclair State University, Montclair, NJ, USA

B. Adhikari
Dhankuta Municipality, Dhankuta, Nepal

C. Khadka
Hetauda Sub-Metropolitan City, Hetauda, Nepal

R. Adhikari
University of Tubingen, Department of Geosciences, Chair of Soil Science, Tubingen, Germany

S. Singh, M. K. Gautam (eds.), *Biodiversity Conservation and Land Degradation Neutrality*, https://doi.org/10.1007/978-981-96-7210-3_15

stability, and human societies. The loss of forests, wetlands, grasslands, and other ecosystems contributes to the depletion of natural resources, loss of carbon sinks, and exacerbation of environmental problems, such as air and water pollution. The decade 2021–2030 is also considered the decade of ecosystem restoration, whose main agenda is to prevent, halt, and reverse the degradation of ecosystems worldwide.

Nepal is a pioneer in community forests (CFs) due to its success in the restoration of degraded forest areas. This chapter highlights the restoration efforts conducted by the private sector and CFs in Nepal. These restoration efforts have provided multiple benefits to local communities, such as easy and equitable access to forest produce, better economic conditions through livelihood diversification options, capacity building, and awareness.

Keywords

Ecosystem · Biodiversity · Restoration · Private sector · Communities

15.1 Introduction

The ecosystem forms the foundation of life for all living beings. A fully functioning ecosystem is essential for maintaining the life cycles of organisms. Understanding ecosystems is crucial for conservation efforts, resource management, and addressing global environmental challenges. Many countries are experiencing extreme temperatures and erratic precipitation leading to landslides, floods, and droughts. These changes not only affect human lives but also disrupt entire ecosystems. Plants are shifting their ranges, and fruit trees are flowering earlier, also resulting in biodiversity loss (Bartomeus et al. 2013). This loss affects both people and animals, depriving them of ecosystem services such as wild edible fruits, freshwater fisheries, and agricultural crops. This scarcity leads to food shortages and intensifies competition for resources, particularly impacting mountain communities that rely heavily on ecosystems. Moreover, these changes erode the cultural and spiritual significance of communities, contributing to the loss of traditional knowledge (Ebel et al. 2021). Additionally, pollution, lack of clean water, and insufficient nutritious food are causing frequent health issues.

Therefore, restoring ecosystems is crucial not only for current human lives but also for preserving the environment and ecosystems for future generations (Singh et al. 2024). Research indicates that approximately 3.2 billion people worldwide are adversely affected by land degradation, and up to $577 billion in annual crop production is at risk due to pollinator loss (Chalise et al. 2019; Potts et al. 2016). These issues are exacerbated by climate change and natural disasters, causing cascading impacts across global systems.

15.2 Ecosystem Restoration

The exploitation of forest resources in unplanned or unsustainable ways has significantly increased the degradation of forest landscapes. While some use forests for daily subsistence, others exploit them for economic gain. Studies estimate that approximately 10 million square kilometers of forest have been lost since the twentieth century. To address these degraded landscapes, approaches to forest restoration have evolved. The main objective of the United Nations (UN) Decade of Ecosystem Restoration (2021–2030) is to restore 350 million hectares of barren and degraded land (UNEP/FAO 2019). Achieving this target could generate USD 9 trillion in ecosystem services and sequester an additional 13–26 gigatons of greenhouse gases from the atmosphere (Lim et al. 2022).

15.3 Major Restoration Initiatives in the World

Several countries have their own restoration projects and activities aimed at benefiting ecosystems and livelihoods. Additionally, the UN has recognized ten transformative efforts worldwide (Table 15.1) in restoring degraded lands (UNEP and FAO 2024). These projects were designated as World Restoration Flagships and are eligible for UN support, which includes promotion, guidance, and funding. They were selected as part of the UN Decade on Ecosystem Restoration, a global initiative led by the UN Environment Programme (UNEP) and the Food and Agriculture Organization (FAO). This initiative seeks to halt and reverse the degradation of natural environments worldwide. These flagship projects aim to restore over 68 million hectares of land and create employment opportunities for 15 million people (UNEP 2022).

15.4 Role of Private Sectors

Restoring degraded landscapes has predominantly relied on government investments. However, the private sector's participation has been limited due to challenges in making a business case for investing in nature. While humans often exploit land until it becomes unusable, the recovery of degraded lands takes time, exacerbating global concerns about land degradation (Weeraratna 2022). Addressing this urgent issue requires actions beyond governments' capabilities alone, requiring the involvement of the private sector. It is crucial to note that the term "private sector" encompasses not only large corporations but also individual farmers, cooperatives, and social groups, all of whom can contribute to increasing productivity on degraded lands. By raising productivity, supporting local communities, and ultimately enhancing business growth, strategic alignment for land restoration becomes central.

Table 15.1 Restoration initiatives in the world

Restoration initiatives	Country	Aim	Area
Trinational Atlantic Forest Pact	Brazil, Paraguay, and Argentina	Create wildlife corridors for endangered species, like Jaguar and golden lion tamarin Secure water supplies for people and nature Creating employment opportunities Countering and building resilience to climate change	1 Mha by 2030 and 15 Mha by 2050
Abu Dhabi Marine Restoration	United Arab Emirates	Restore beds of seagrass Restore coral reefs and mangroves along the Gulf coast Conservation of turtles and dolphin Generating livelihood opportunities	12,000 ha by 2030
Great Green Wall for Restoration and Peace	Across Africa	Restore savannas, grasslands, and farmlands Transform the lives of millions in the Sahel region by creating a belt of green and productive landscapes across 11 countries Create 10 million jobs	100 Mha by 2030 Sequester 250 million tons of carbon
Ganges River Rejuvenation	India	Rejuvenating, protecting, and conserving the Ganges and its tributaries Reducing water pollution Rebuilding forest cover Reviving key wildlife species such as dolphins, softshell turtles, otters, and the hilsa shad fish Bringing benefits to 520 million people living around the basin	2525 km course
Multi-Country Mountain Initiative	Serbia, Kyrgyzstan, Uganda, and Rwanda	Protection of endangered mountain gorillas Conservation and management of grasslands for both livestock and Asiatic ibex	
Small Island Developing States Restoration Drive	Vanuatu, St Lucia, and Comoros	Ridge-to-reef restoration Harnessing blue economic growth to assist island communities in rebounding from the COVID-19 pandemic Reducing pressure on coral reefs	

(continued)

Table 15.1 (continued)

Restoration initiatives	Country	Aim	Area
Altyn Dala Conservation Initiative	Kazakhstan	Restore steppe, semi-desert, and desert ecosystems Conservation of wetlands Conservation of Saiga antelope and migratory birds	
Central American Dry Corridor	Costa Rica, El Salvador, Guatemala, Honduras, Nicaragua, and Panama	Tapping traditional farming methods to enhance landscape productivity, including biodiversity Boosting soil fertility and water availability Restoring biodiversity in tropical forests Creating 5000 permanent jobs	100,000 ha by 2030
Building with Nature in Indonesia	Indonesia	Constructing fence-like structures with natural materials along the shore to calm waves and trap sediment Creating conditions for mangroves to naturally rebound Increasing livelihood options	
Shan-Shui Initiative in China	China	Combining 75 large-scale projects within the country to restore ecosystems, from mountains to coastal estuaries across the nation	10 Mha by 2030

The economic benefits of such restoration efforts are substantial: nature underpins half of the world's gross domestic product, and every dollar invested in restoration can yield up to USD 30 in returns (Gupta et al. 2023; Johnson et al. 2021). Despite this economic significance, Sustainable Development Goal (SDG) 15, Life on Land, remains one of the least funded SDGs (Krauss 2022). The private sector holds a fundamental role in achieving the ambitious target of restoring 1.5 billion hectares of degraded land by 2030. It can drive investments, introduce sustainable technologies, and facilitate a shift toward responsible land use, thereby promoting sustainable consumption and healthy ecosystems.

Discussions at the Paris meeting underscored the importance of innovative financing models, emphasizing private sector investments in land restoration and productivity enhancement. The United Nations Convention of Combat Desertification (UNCCD) highlighted opportunities in carbon sequestration and natural-based solutions, demonstrating the economic feasibility of sustainable land projects and encouraging entrepreneurs to explore untapped potentials (Chausson et al. 2020; Creutzig et al. 2022).

15.5 Restoration of Degraded Land by the Private Sector in Nepal

In central Nepal, the private sector has demonstrated that the restoration of degraded land is beneficial for both businesses and the environment. Plantec Coffee Estate (P) Ltd. leased 68 hectares of land from the government for 40 years, utilizing 38.5 hectares for shaded coffee cultivation. Established in 1994, the estate focuses on 100% organic coffee production. Since its inception, the estate has planted trees such as *Dalbergia sissoo*, *Leucaena leucocephala*, *Michelia champaca*, *Melia azedarach*, and *Albizia lebbek* alongside traditional agricultural crops (Timalsina et al. 2017).

Initially degraded due to leftover stones and gravel from nearby hydropower construction, the land has undergone reclamation. Soil fertility has been enhanced by adding humus, organic fertilizer, and soil from external sources. This initiative has not only transformed the landscape but also boosted the organic coffee business and provided employment opportunities for local communities. Furthermore, the forest now hosts various birds, wildfowl, and mammal species such as barking deer, common leopards, and jackals.

In this area, Timalsina et al. (2017) conducted research to observe changes in land use after the intervention done by the private sector. Their study showed significant land cover transformation over the years. In 1991, the land cover analysis indicated that the area was predominantly cultivated land (73.2%), followed by barren areas (18.1%) and shrubs (6.3%). By 2004, the cultivated land area had decreased to 4.1%, with almost 79% of the land covered by forests, including bushes. In 2015, there was an increase in agricultural land due to the conversion of bushy areas where agricultural practices could be carried out. Table 15.2 and Fig. 15.1 show the land use and land cover changes from 1991 to 2015.

Further analysis using Google Earth images reveals noticeable changes in the area. Cultivated land, which expanded in 2015, has since been converted into forested areas (Fig. 15.2). Imagery from Google Earth from 2010, 2015, and 2020 illustrates these landscape changes over time. These observations are supported by

Table 15.2 Land use of Plantec Coffee estate in 1991, 2004, and 2015

Land cover (Area in ha)	1991 (ha)	1991 (%)	2004 (ha)	2004 (%)	2015 (ha)	2015 (%)
Closed forest	0	0	8.6	22.3	20	52
Open forest	0	0	9.7	25.2	11.3	29.3
Shrub/bushes	2.4	6.3	12	31.2	1.4	3.7
Plantation	0	0	0	0	0	0
Barren land	7	18.1	4.3	11.3	0.1	0.2
Cultivated land	28.1	73.2	1.6	4.1	4.3	11
Sand	0.2	0.4	1.5	3.9	0.7	1.9
Waterbodies	0.8	2	0.8	2	0.7	1.9
Total	38.5	100	38.5	100	38.5	100

Source: Timalsina et al. (2017)

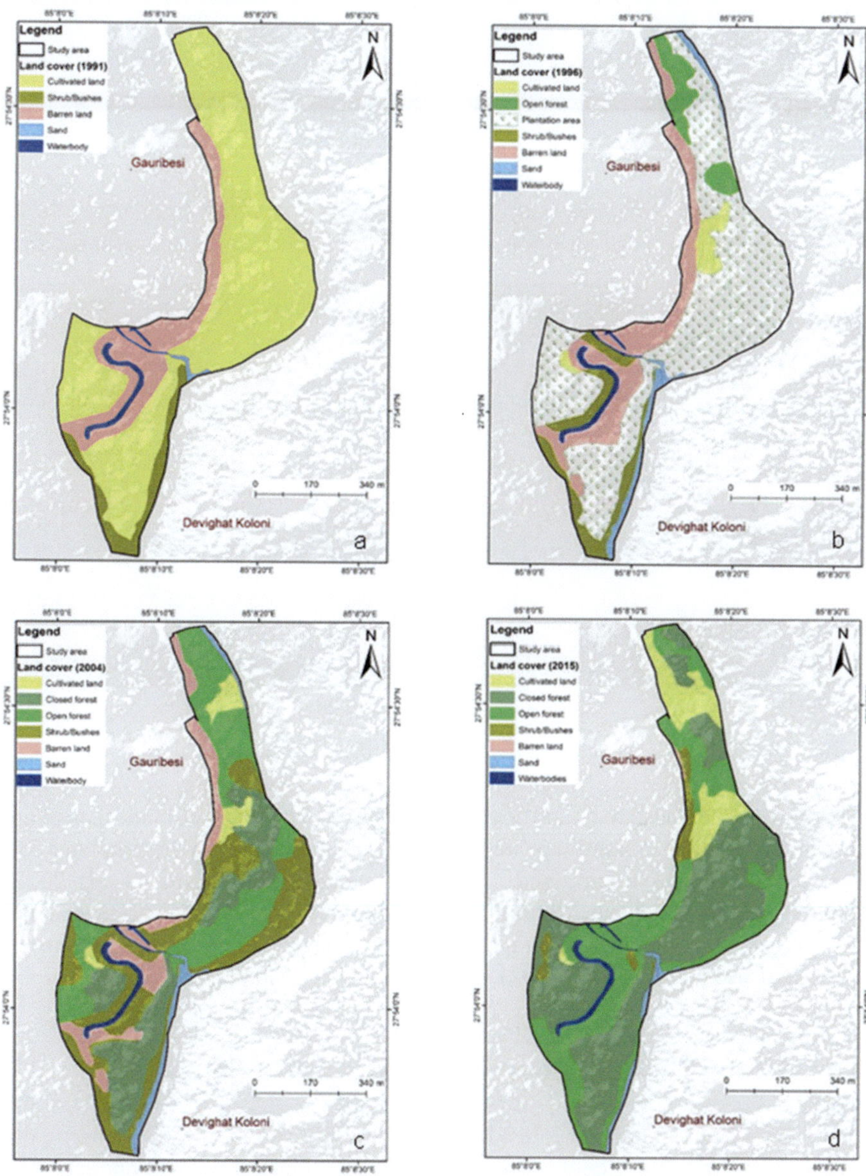

Fig. 15.1 Land cover and land use maps (**a**) 1991, (**b**) 1996, (**c**) 2004, and (**d**) 2015. (Source: Timalsina et al. 2017)

comparing repeated photographs taken from the same location in 1995 and 2016, which clearly demonstrate the area's transformation (Fig. 15.3).

Repeat photography is an effective method for documenting landscape changes by capturing multiple images from the same spot over time (Klett 2020). This

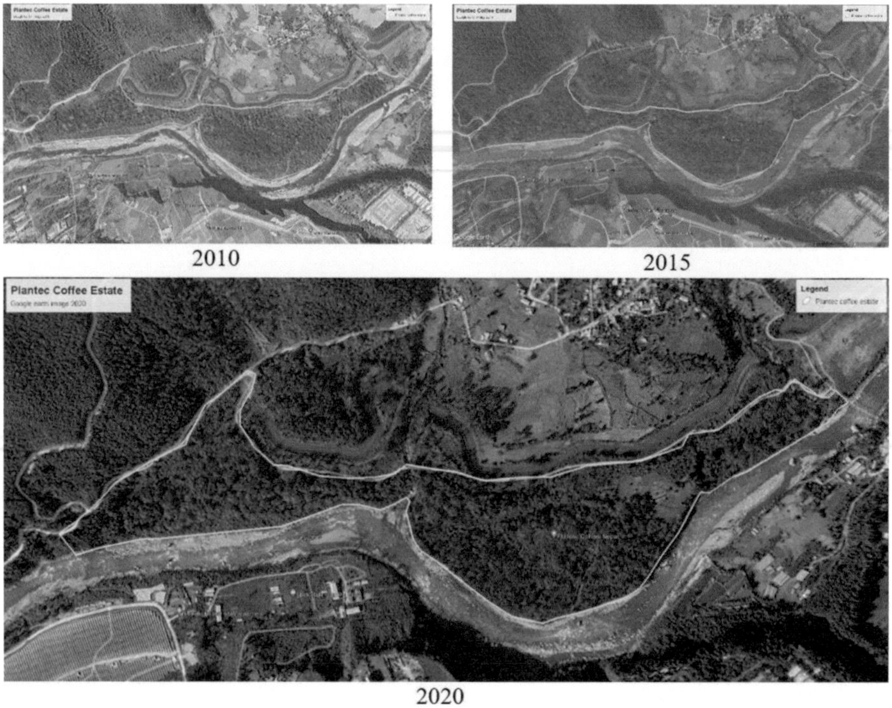

Fig. 15.2 Google images of Plantec coffee estate—2010, 2015, and 2020

technique has been used to quantify vegetation shifts induced by climate change and to reconstruct trends in landscape vegetation, such as tree encroachment, land conversion, field abandonment, and forest fragmentation (Depauw et al. 2022). Evaluating landscape change through repeat photography is valuable because it provides precise spatial and temporal information about ecological succession and human land-use patterns (Bayr 2021).

15.6 Efforts Through Community Forestry in Nepal

Nepal experienced significant forest loss from the 1850s to the 1950s primarily due to expanding agricultural lands, government revenue generation, and uncontrolled, unsustainable harvesting and utilization of forest resources (Bhattarai et al. 2021). Nepal's rich forest cover was once considered its greatest wealth, as reflected in the saying "Hariyo Ban Nepalko Dhan" (Green Forest is Nepal's wealth). However, unsustainable management and extraction practices exacerbated deforestation. Deforestation accelerated in the 1950s, particularly in southern Nepal, coinciding with malaria eradication and government resettlement programs. This period witnessed a massive population entry into the Terai region (Chapagain and Aase 2020). Recognizing the critical importance of forests, a movement emerged in the 1970s

Photo location: Lat. 27.8961° and Long 85.1348°) Above: 1995 and Below: 2016

Photo location: Lat. 27.8961° and Long 85.1348°) Above: 1995 and Below: 2016

Fig. 15.3 Repeated photographs illustrating changes in land use over time

Table 15.3 Forest cover of Nepal

Types	FRS (1963/64)	NRSC (1984)	NFI, DFRS (1999)	DFRS (2015)
Forest	45.5	35.9	29	40.36
Shrub			10.6	4.38
Total	45.5	35.9	39.6	44.74

Adopted from: Chapagain and Aase (2020)
FRS Forest Resources Survey, *NRSC* National Remote Sensing Center of Nepal, *NFI* National Forest Inventory, *DFRS* Department of Forest Resource and Survey

advocating for community involvement in forest protection and management, laying the groundwork for community forestry.

Following the establishment of community forestry regulations in 1990, the government began transferring forest areas to local communities for conservation, management, and sustainable use (Kusters et al. 2022). Since then, the number of community forests (CFs) has increased significantly, leading to the restoration of substantial areas of degraded land. By the fiscal year 2079/2080, Nepal had 23,682 registered CFs covering 2,490,194 hectares (MoFE 2024). Approximately 40% of the total forest area is now under community management, either in the form of CFs or collaborative efforts. This remarkable achievement has earned Nepal global recognition for its success in restoring degraded lands through the community forestry approach. The country has now nearly reached a forest cover percentage reminiscent of the 1960s, as shown in Table 15.3.

This section presents the success of forest restoration in three CFs—one from the mountains, one from the hills, and one from the Terai—located in Nepal's Koshi Province. As stated in the previous section, the province's and country's success in restoring degraded areas is largely attributed to community efforts. With over 90% of its 32,000 hectares of forest under community management, Koshi Province personifies the potential of community forestry. Communities are responsible for sustainably managing forest resources based on operational plans approved by the divisional forest office. These plans, developed for 5- or 10-year periods depending on available funds, guide forest management. A primary benefit for communities is free or low-cost access to timber, fuelwood, fodder, and non-timber forest products (NTFPs). Income primarily comes from entrance fees, membership fees, and annual renewals. Each CF uniquely adapts its practices to local conditions, resulting in best practices for sustainable forest management. Conservation efforts include planting seedlings, thinning, pruning, and volunteer forest guard roles. Fire prevention measures such as fire lines and regular litter removal are implemented. To increase revenue, barren areas are often planted with broom grass or bamboo.

Operational plans, approved by community general assemblies and the Forest Department, ensure the sustainable availability of forest products. These plans prioritize the needs of the poorest community members and include benefit-sharing mechanisms, incentives, and penalties for rule violations. Community Forest Consumer Groups (CFUGs) prioritize inclusivity and equity, with women in leadership roles and representation of indigenous groups and marginalized castes. Financial transparency is maintained through regular audits and public disclosure of income and expenses. Beyond ecological restoration, CFs contribute to poverty reduction, livelihood improvement, and tourism development. The following sections delve into the specific restoration stories of the three selected CFs.

15.6.1 Shree Baluwabash Community Forest

This CF is situated in Athrai Tribeni Rural Municipality, Taplejung district, and was established in 1993, covering 4.8 hectares. The forest is home to 34 households. Before the 1980s, the forest was thriving, and local communities relied heavily on it for their livelihoods. However, political instability, migration, increasing poverty, social conflicts, and a lack of leadership led to severe degradation between 1987 and 1993. As a result, communities faced challenges accessing essential forest products and water resources. Recognizing the critical link between forest loss and declining water availability, they initiated conservation efforts to reduce encroachment and forest fire. Given the steep terrain and poor soil quality, initial forest regeneration was slow. Nevertheless, the forest has gradually recovered, with dense vegetation now covering the area.

Before the formal establishment of the CFUG in 1993, a local conservation committee was formed to oversee forest protection. The CFUG's operational plan outlines its core objectives: conserving and sustainably managing forest biodiversity while improving community livelihoods. By effectively managing forest resources, the CFUG aims to enhance the production of timber, fuelwood, and other forest products, contributing to social, economic, and environmental well-being. Collaboration with local governments and stakeholders ensures a continuous supply of forest products, strengthens community resilience, and preserves the forest's ecological value. The CFUG prioritizes the needs of marginalized groups, empowering them through forest-based opportunities while safeguarding the forest ecosystem.

The forest is primarily composed of Sal (*Shorea robusta*) trees, with other significant species including Chilaune (*Schima wallichii*), Katus (*Castanopsis* sp.), Simal (*Bombax ceiba*), and Mauwa (*Engelhardia spicata*). NTFPs such as Khar (*Andropogon pumilus*), Bhorla (*Bauhinia variegata*), and Babiyo (*Eulalipsos binata*) are also found. The forest supports a diverse range of wildlife, including mammals such as monkeys, deer, wolves, leopard cats, pangolins, and porcupines. Bird species observed include crows, peacocks, doves, mynas, koels, owls, kalij pheasants, bulbuls, and others. Reptiles such as snakes and golden lizards inhabit the area. To enhance biodiversity and community benefits, the CFUG has planted fruit trees like Chiuri (*Diploknema butyraceae*), mango, banyan, and pipal on barren land. Open grazing is strictly prohibited. However, invasive species such as Kirne Kande (*Lantana aculeate*) and Gandhe jhar (*Ageratum conyzoides*) pose significant challenges to forest health.

The CF's operational plan includes an assessment of the demand for and supply of timber, firewood, grass, and fodder. Despite successful forest restoration efforts, the community still faces a shortage of these resources, as detailed in Table 15.4. To meet their needs, community members rely on resources from their private lands or purchase them from the market.

To achieve success, CFUGs have implemented several effective practices. They have initiated financial support programs to encourage the adoption of improved cooking stoves, biogas, and bio-briquettes. To reduce dependence on the CF, the CFUG incentivizes tree planting on private land through cash prizes and recognition for exemplary agroforestry practices. The operational plan outlines the distribution of forest products such as fuelwood, timber, grass, leaf litter, bedding materials, green fodder, and medicinal plants among user groups. Product prices are determined by the general assembly every 2 years, reflecting market rates. While Sal timber prices range from Rs. 900 to Rs. 1300 per cubic foot based on size, Katus is priced at Rs 500 per cubic foot, and other species are valued at Rs. 200 per cubic foot. However, Sal felling is currently prohibited. Grass, fodder, and leaf litter are provided free to all users, and timber is offered free to disaster victims. Less

Table 15.4 Demand, supply, and deficit of forest resources for the user groups

Items	Unit	Demand	Supply	Deficient
Timber wood	ft^3	980	32.75	947.25
Firewood	ton	48	13.5	34.5
Grass	*(note: 1 bhari = 30 kg)*	47.79	17.19	30.6
Fodder		31.5	7.5	24

Source: Shree Baluwabash Community Forest (2021)

valuable timber is sold within the community after committee approval, prioritizing the poor and marginalized. Wood for social and cultural functions is priced at Rs. 200 per quintal.

The CFUG's current working committee comprises nine members: five women and four men, with seven from indigenous groups, one from the Brahman/Chhetri caste, and one from the Dalit community. The general assembly meets annually, while user groups hold regular monthly meetings to make decisions by consensus. The CFUG employs a robust financial management system. Funds are managed through a bank account requiring joint signatures from the Chairperson and Treasurer. Internal audits are conducted by a financial and monitoring committee, and an external audit is performed by a registered auditor or firm annually. The CFUG maintains a revolving fund of NPR 200,000, offering loans to the poor at a 12% interest rate. Financial assistance for healthcare is also provided to needy members. Membership fees contribute to income, with new members paying NPR 551, separated households paying NPR 100, and an annual renewal fee of NPR 100.

The CFUG imposes penalties for violations of its constitution and forest management plan, considering the severity of the offense. Penalties for outsiders are double those for members. Specific penalties include fines ranging from NPR 200,000 to 1,000,000 per hectare for mortgaging, selling, or transferring CF land; fines of NPR 10,000 and potential seizure or destruction of structures for unauthorized construction or land degradation; fines of NPR 50–500 for entering the CF with fire hazards; and fines of NPR 100,000–1,000,000 per hectare for causing forest fires. Other violations are addressed through internal discussions and agreements.

15.6.2 Keureni Community Forest

This CF is located in Ward 10 of Myanglung Municipality, Terathum District. Established in 1994, it encompasses 883.94 hectares and serves 116 households. Prior to its establishment, the Kureni Forest was owned by local Tumuhamphe/ Subba families under the Kipat land tenure system. Following the cadastral survey in the 1970s, the Subba continued to manage the forest for a short period before government restrictions led to degradation. Illegal logging of Khayer (*Acacia catechu*) intensified, and forest authorities struggled to prevent locals from accessing grass, firewood, and other forest products. Annual forest fires exacerbated the situation, threatening both the forest and nearby settlements. To address these challenges and promote sustainable forest use, the community established the CF in 1994.

The CF aims to conserve forests, promote sustainable use of forest products, improve livelihoods, and protect biodiversity. Specific objectives include enhancing environmental quality, supporting marginalized groups, and building community capacity. The management plan outlines strategies for fire control, rainwater harvesting, and institutional development.

The forest is predominantly composed of Sal (*Shorea robusta*), Khote Sallo (*Pinus roxburghii*), Khayer (*Acacia catechu*), and other valuable species. A rich diversity of NTFPs is also present, including Kurilo (*Asparagus officinalis*), Amala

(*Phyllanthus emblica*), and Thakal (*Cirsium wallichii*). The forest supports a variety of wildlife, including mammals, reptiles, and birds. A recent forest inventory revealed an average stock of 56 cubic meters per hectare. The management plan recommends an annual allowable extraction of 7656 cubic feet for timber and 172.08 tons for firewood. However, current demand exceeds supply for both timber and firewood, as well as for grass and litter.

The user group has undertaken significant efforts to regenerate, conserve, and manage the CF. These activities include seedling planting, thinning, pruning, and clearing invasive species like Kirnekada (*Lantana aculeate*), which hampered grass growth. To mitigate the risk of forest fires, fire lines have been established and maintained, and restrictions on carrying lighters have been imposed on visitors. While small streams (Kholchi) provide water during the rainy season, the region's arid climate precludes the formation of wetlands. Given the prevalence of animal husbandry, open grazing remains a challenge. However, the construction of a rainwater collection pond has benefited both wildlife and domestic animals, leading to increased biodiversity.

The CF provides user groups with access to forest products such as fuelwood, timber, grass, fodder, medicinal plants, and bedding materials, as outlined in the management plan. Grass, fodder, leaf litter, and dead firewood are distributed free of charge, with priority given to the most vulnerable community members. The cost price of the forest produces is provided in Table 15.5.

Despite financial constraints, the CFUG prioritizes forest conservation activities in its latest operational plan, similar to other mountain CFs. The CFUG has also contributed to infrastructure development, including power, roads, and water, as well as livelihood improvements in ginger, walnut, poultry, and weaving. Additionally, the CFUG supports education and skills training and provides financial aid to community members in need.

Table 15.5 Costings of the forest produce

Species	Particular	Unit	Rate (Rs)		Remarks
			Poor	Others	
Sallo	Timber	ft^3	15	20	
Sal (*Shorea robusta*)	Timber	ft^3	50	75	
Karam (*Adina cordifolia*) and Phadir	Timber	ft^3	35	50	
Fire wood from standing tree		30 kg	5	10	
Khayar (*Acacia catechu*)		kg	20		Procedure should be followed
Agricultural tool (Hallo)		2 Piece	20	40	
Agricultural tool (Haris)		Piece	20	40	
Shed building materials (Khamo)		Piece	200–500		
Resign collection		kg	12		Given to private company
Coal		As per decision of CFUG			

Source: Keureni Community Forest (2023)

Despite the CFUG's dedicated efforts in forest conservation and management, several natural and human-made challenges continue to drive deforestation and forest degradation. The major obstacles include open grazing, logging, and the illegal trade of forest products. Inadequate awareness about forest conservation, out-migration of users, and the drying up of drinking water resources also contribute to the problem. Additionally, forest fires and the destruction of biodiversity exacerbate the situation. Other challenges include the need for Environmental Impact Assessments every 5 years, the complexities of the three-tier tax system, and the deteriorating accountability of newly elected leadership in CF management.

15.6.3 Hariyali Community Forest

This CF, located in Dharan Municipality, was established in 2006 and covers an area of 197.38 hectares with 496 households. It is part of the Charkoshe Jadi region in Nepal. Locals recall that, before the 1970s, the forest was dense and teeming with wild animals, making it largely inaccessible to all. Following the eradication of malaria, Dharan became a top destination, leading to significant migration from the mountainous regions of eastern Nepal. The rapid expansion of settlements around Dharan resulted in severe forest degradation from the 1970s through the 1990s.

Efforts to conserve the forest began in 1987 with the establishment of a range post nearby, but these efforts were thwarted by limited staffing and local resistance. As forest resources decreased, local residents initiated a conservation campaign, recognizing the need to support government efforts in forest protection. Consequently, residents from Dharan came together to manage the nearby forest to ensure a continuous supply of firewood, grass, timber, and other forest products. This led to the forest being formally handed over to the community in 2006. The forest was named "Hariyali" due to its loamy black soil, which supports rapid greenery growth.

The CF aims to conserve and manage the forest for essential products such as grass, firewood, fodder, litter, and timber while promoting biodiversity through sustainable practices. Their objectives include advancing ecotourism, supporting the livelihoods of the poor, single women, minorities, and disadvantaged groups through community forestry, and enhancing environmental conservation and natural beauty. They plan to increase income by establishing a commercial nursery and creating a safe house for injured and orphaned wild animals. Short-term goals focus on supplying essential forest products, removing dead and infected trees, planting trees along riversides and barren land, and maintaining a clean environment. Additionally, they aim to enhance the participation and leadership capacities of women and disadvantaged groups, manage grazing and fire risks, develop inclusive leadership skills, and attract tourists by developing the CF as a resource center.

The forest has broad-leaf vegetation. Local communities believe that the forest has improved greatly after being transferred into the hands of the CF, due to the collective efforts of the UG.

The major tree species found in the CF are Sal (*Shorea robusta*), Satisal (*Dalbergia latifolia*), Karam (*Adina cordifolia*), Bot dhaiyanro (*Lagerstroemia parvifolia*), Jamun (*Eugenia jambolana*), Sirish, Harro (*Terminalia chebula*), Barro (*Terminalia bellirica*), Bajhi, Hade, Aaulo, Seto Sirish (*Albizia procera*), Rajbriksha (*Cassia fistula*), Simal (*Bombax ceiba*), Kusum (*Schleichera oleosa*), Teak (*Tectona grandis*), Tatari (*Dillenia indica*), Rudraksha (*Elaeocarpus ganitrus*), Amba (*Psidium guajava*), Bhalayo (*Rhus wallichii*), Tanki (*Bauhinia purpurea*), Kalikath (*Myrsine semiserata*), Aashare, Kumbhi (*Careya arborea*), Amala (*Phyllanthus emblica*), and others. Similarly, climbers include Bhorla (*Bauhinia variegata*), Debre lahara (*Spathlobus parviflorus*), Charchare, *Michenia micrantha*, Ban tarul (*Dioscorea hamiltonii*), Cucurdyno (*Smilax* sp.), Githha (*Dioscorea bulbifera*), and others. In the shrub/bushes and grass, there are Banmara (*Eupatorium adenophorum*), Arare Kanda, Gelani, Titepati (*Artemisia vulgaris*), Kurilo (*Asparagus officinalis*), Sarpagandha (*Rauvolfia serpentina*), Niguro (*Tectaria coadunate*), Unyu (*Dryopteris* sp.), and others which are found in forest.

Leopard (*Panthera pardus*), Chittal (*Axis axis*), Barking deer (*Muntiacus vaginalis*), Wild boar (*Sus scrofa*), Monkey (*Macaca mulatta*), Dude, Porcupine (*Hystrix indica*), Fox, Khirkhire, Rabbit (*Oryctolagus cuniculus*), and others are some of the animals residing in the forest. Likewise, the bird species include Parrot, Myna (*Acridotheres tristis*), Sparrow (*Passer* sp.), Dove (*Spilopelia chinensis*), Bulbul (Pycnonotus sp.), Rani Chari, Peacock (Pavo cristatus), Owl (Tyto longimembris), Koel (*Eudynamys scolopaceus*), Kotero, Hornbill, and others. Various types of snakes, golden lizards, lizards, and others are also found in the CF.

The CFUG provides essential forest products, including fuelwood, timber, grass, leaf litter, bedding materials, green fodder, and medicinal herbs, according to their management plan. Firewood collection is permitted one day a week, with collection prohibited on other days. Sal (*Shorea robusta*) and Bhorla (*Bauhinia variegata*) leaves are available free on specified days. About 50% of users rely on firewood for cooking. The CFUG allocates 25% of surplus wood and 40% of surplus firewood to the District Forest Products Supply Committee. The CFUG prioritizes poorer households, categorizing them as 15 middle-class, 351 lower-middle-class, 116 poor, and 14 extremely poor households.

In terms of social development, the CFUG offers skill development training, assists with business ventures, provides scholarships, supports critically ill patients, funds youth clubs, promotes sports, and offers immediate relief during calamities. They have discontinued the practice of constructing permanent graves, replacing it with Smriti Chautaris—small, zinc-roofed structures for picnics that generate additional income. Other initiatives include a children's garden and a small zoo to educate youth on wildlife conservation.

The CFUG's efforts in forest management and social development have led to the construction of a new office, training room, and residential facilities, providing employment for ten people. The venue, known for its attractive environment and organic food, is popular for events, studies, and research. Increased funding has improved livelihoods, reduced poverty, and enhanced access to health and education while continuing to protect forests and biodiversity. The CFUG also operates a nursery, safe houses for wild animals, and various community amenities. Poor households benefit economically from making Duna and Tapari from free Sal and Bhorla leaves.

Despite these efforts, potential threats to the forest remain, including illegal logging, human–wildlife conflicts, invasive species, and challenges related to the three-tier government tax system. The costly and complex process of rescuing and rearing injured wildlife, along with fluctuating policies and waning volunteer enthusiasm, complicates ongoing conservation efforts.

Clearly, the combination of every aforementioned community-led effort is integral in each of the success stories of the CFUGs. This publication includes another noteworthy achievement of a community-led effort that is the development of a river forest corridor for disaster reduction. The Mawa khola flows in the boundary between Jhapa and Morang districts. In 1977, Mawa khola converged with Nansari khola that caused severe flooding that even displaced people. The locals then decided to take the matter into their own hands to mitigate the channel migration of Mawa khola. Under local leadership, the villagers and the then-Panchayat commenced the physical channelization of Mawa khola in 1984. They protected the channel bend migration zone with quality bioengineering through Gabain walls and bamboo on both banks of the Mawa simultaneously. Since then, soil fertility and green belt successfully resumed, and land has been reclaimed from the riverbanks.

Despite the best practices in place, the CFUGs do face some natural and anthropogenic challenges. Forest fire is still one of the major concerns yet to be fully addressed. Illegal logging and illicit trade of forest products like timber and herbs are on the rise. Spread of new invasive species that worsens grass availability and ecosystem resilience, and the three-tier taxation system are some of the concerns of the local communities managing forests. Drying of springs in forested lands, open grazing, and increasing human–wildlife conflict with increasing forest cover are some of the major challenges. And due to out-migration from the hills, there are fewer people to manage the increasing forest area.

Overall, when community forestry is practiced in a sustainable manner with adequate safeguards in place, there is evidently an opportunity for nature-based solutions that can improve the resilience of landscapes and the local communities to climate change.

15.7 Conclusion

Ecosystem conservation is essential for sustaining life on Earth due to the planet's limited capacity to grasp degradation and pollution. The accelerating exploitation of ecosystems poses serious threats to the food security, health, and cultural practices of vulnerable mountain communities. International frameworks, such as the UN Decade of Ecosystem Restoration, highlight the urgent need to restore ecosystems for economic and ecological benefits.

The private sector plays an important role in achieving restoration goals, as demonstrated by initiatives like the Plantec Coffee Estate in Nepal. Such initiatives show that rehabilitating degraded land enhances biodiversity, sustains economies, and promotes sustainable development. Additionally, Nepal's community forestry programs have proven that local communities can effectively manage resources to achieve significant ecological and social improvements. Collaborative efforts involving local governments, private sectors, and communities are essential to combat land degradation and climate change impacts. Innovative funding mechanisms and renewable energy solutions can support restorative projects, promoting more efficient land use and maximizing ecosystem benefits.

In conclusion, global initiatives, private sector investments, and community-led projects offer promising pathways to a sustainable future. Effective coordination and collaboration among local communities, who are the agents of change, are crucial for the success and sustainability of restoration activities. Furthermore, governments and development partners need to leverage funds to support local communities in economic generation activities. By prioritizing ecosystem restoration, it will benefit to safeguard the natural environment, preserve biodiversity, and ultimately enhance human well-being.

References

Bartomeus I, Park MG, Gibbs J, Danforth BN, Lakso AN, Winfree R (2013) Biodiversity ensures plant--pollinator phenological synchrony against climate change. Ecol Lett 16(11):1331–1338

Bayr U (2021) Quantifying historical landscape change with repeat photography: an accuracy assessment of geospatial data obtained through monoplotting. Int J Geogr Inf Sci 35(10):2026–2046

Bhattarai K, Conway D, Bhattarai K, Conway D (2021) Forestry and environment. In: Contemporary environmental problems in Nepal: geographic perspectives, pp 663–754

Chalise D, Kumar L, Kristiansen P (2019) Land degradation by soil erosion in Nepal: A review. Soil Syst 3(1):12

Chapagain PS, Aase TH (2020) Changing forest coverage and understanding of deforestation in Nepal Himalayas. Geogr J Nepal 13(10.3126):1

Chausson A, Turner B, Seddon D, Chabaneix N, Girardin CAJ, Kapos V, Key I, Roe D, Smith A, Woroniecki S et al (2020) Mapping the effectiveness of nature-based solutions for climate change adaptation. Glob Chang Biol 26(11):6134–6155

Creutzig F, Niamir L, Bai X, Callaghan M, Cullen J, Diaz-José J, Figueroa M, Grubler A, Lamb WF, Leip A et al (2022) Demand-side solutions to climate change mitigation consistent with high levels of well-being. Nat Clim Chang 12(1):36–46

Depauw L, Blondeel H, De Lombaerde E, De Pauw K, Landuyt D, Lorer E, Vangansbeke P, Vanneste T, Verheyen K, De Frenne P (2022) The use of photos to investigate ecological change. J Ecol 110(6):1220–1236

Ebel R, Menalled F, Ahmed S, Gingrich S, Baldinelli GM, Félix GF (2021) How biodiversity loss affects society. In: Handbook on the human impact of agriculture. Edward Elgar Publishing, pp 352–376

Gupta AK, Goyal MK, Singh SP (2023) Ecosystem restoration: towards sustainability and resilient development. Springer Nature

Johnson JA, Ruta G, Baldos U, Cervigni R, Chonabayashi S, Corong E, Gavryliuk O, Gerber J, Hertel T, Nootenboom C, et al (2021) The economic case for nature: a global earth-economy model to assess development policy pathways

Keureni Community Forest (2023) Keureni Community Forest Operational Plan 2079/80

Klett M (2020) Rephotography in landscape research. In: The Sage handbook of visual research methods, pp 114–128

Krauss JE (2022) Unpacking SDG 15, its targets and indicators: tracing ideas of conservation. Globalizations 19(8):1179–1194

Kusters K, De Graaf M, Ascarrunz N, Benneker C, Boot R, Van Kanten R, Livingstone J, Maindo A, Mendoza H, Purwanto E et al (2022) Formalizing community forest tenure rights: a theory of change and conditions for success. Forest Policy Econ 141:102766

Lim CH, Lim BS, Kim AR, Kim DU, Seol JW, Pi JH, Lee H, Lee CS (2022) Climate change adaptation through ecological restoration. In: Natural resources conservation and advances for sustainability. Elsevier, pp 151–172

MoFE (2024) Annual progress report 2079/80

Potts SG, Imperatriz-Fonseca V, Ngo HT, Biesmeijer JC, Breeze TD, Dicks LV, Garibaldi LA, Hill R, Settele J, Vanbergen AJ (2016) The assessment report on pollinators, pollination and food production: summary for policymakers. Secretariat of the Intergovernmental Science-Policy Platform on Biodiversity~…

Shree Baluwabash Community Forest (2021) Shree Baluwabash Community Forest Operational Plan

Singh S, Kumar A, Kumar A (2024) Harmony in extraction: navigating mining, land degradation and ecological restoration for sustainable futures. In: Forests and climate change. Springer Nature Singapore, pp 513–534. https://doi.org/10.1007/978-981-97-3905-9_25

Timalsina N, Bhattarai N, Karky B, Gilani H, Windhorst K, Pant B (2017) Contributions by the private secctor to climate change mitigation: lessons from the Plantec Coffee Estate in Nepal. https://lib.icimod.org/record/32714

UNEP (2022, Dec13) UN recognizes 10 pioneering initiatives that are restoring the natural world. UNEP. https://www.unep.org/news-and-stories/press-release/un-recognizes-10-pioneering-initiatives-are-restoring-natural-world

UNEP, & FAO (2024) World Restoration Flagships. https://www.decadeonrestoration.org/world-restoration-flagships

UNEP/FAO (2019) The UN decade on ecosystem restoration 2021–2030. UNEP/FAO Factsheet, February 2020. www.unep.org

Weeraratna S (2022) Understanding land degradation: an overview. Springer

Ecosystem Restoration: Success Stories and Best Practices

16

Meribeni Tsopoe, Arghya Chakrabarty,
Kambam Boxen Meetei, Krishna Giri, Gaurav Mishra,
and Sanjeev Kumar

Abstract

Ecosystem degradation caused by both natural and anthropogenic factors has resulted in severe biodiversity loss, reduced ecosystem services, and changes in the global climate system. Ecosystem restoration is now recognized as an important global method for mitigating environmental degradation and biodiversity loss. This chapter presents success stories and best practices of proven restoration initiatives, such as Chilika Lagoon and Vembanad-Kol Wetland, highlighting the critical importance of multi-stakeholder collaboration and customized solutions. These initiatives are guided by the principles of international efforts such as the Bonn Challenge and the United Nations Decade on Ecosystem Restoration, which integrate ecological, social, cultural, and economic components to improve resilience and functionality. Frameworks such as Biodiversity Ecosystem Functioning offer a scientific basis for understanding the relationship between biodiversity and ecosystem functions, further guiding restoration efforts. The chapter also delves into the implications of documenting efficacious techniques and promoting public involvement. A comprehensive account of the restoration outcomes aids in evaluating the efficacy, costs, and socio-economic implications, offering useful insights for future initiatives. Giving due importance to ecosystem restoration can revive degraded environments, promote sus-

M. Tsopoe · K. B. Meetei
ICFRE-Rain Forest Research Institute, Jorhat, Assam, India

A. Chakrabarty
ICFRE-Rain Forest Research Institute, Jorhat, Assam, India

Wetlands International South Asia, New Delhi, India

K. Giri (✉) · G. Mishra · S. Kumar
Centre of Excellence on Sustainable Land Management, Indian Council of Forestry Research and Education, Dehradun, India

S. Singh, M. K. Gautam (eds.), *Biodiversity Conservation and Land Degradation Neutrality*, https://doi.org/10.1007/978-981-96-7210-3_16

tainable development, and lessen the effects of climate change by setting clear goals, utilizing science-based methodologies, tracking progress, and encouraging long-term commitment and collaboration among stakeholders. Future efforts should continue emphasizing community involvement and evidence-based strategies to enhance ecosystem resilience and sustainability.

Keywords

Ecosystem degradation · Ecosystem restoration · Ecosystem services · Community involvement

16.1 Introduction

The United Nations Environment Programme (UNEP) defined ecosystem degradation as the decline in an ecosystem's ability to provide essential services to society due to anthropogenic or natural shifts. The ecosystem degradation is declining health, biodiversity loss, compromised water purification, and climate regulation functions. Human-induced activities such as habitat destruction, pollution, resource overexploitation, and climate change are principal catalysts for this degradation, disrupting ecological processes and jeopardizing ecosystems' capacity to continue effective functioning. Viewing ecosystem degradation through the lens of natural capital reveals a critical oversight within conventional national accounting systems (Dasgupta 2010). Ecosystem degradation, often arising from malpractices like deforestation or excessive fishing, introduces a paradox where economic advancements measured by gross domestic product (GDP) fail to acknowledge the concurrent depletion of natural wealth (Costanza et al. 1997). According to the World Bank (2006), a few nations experience wealth devaluation despite showing growth in net savings due to the degradation of natural resources.

Understanding ecosystem degradation requires thoroughly examining the driving forces that bring about negative change in ecosystems. According to the Millennium Ecosystem Assessment (MA), a driver is any natural or human-caused factor that directly or indirectly causes changes in an ecosystem (MA 2005). The direct drivers impact ecosystem processes, while indirect drivers operate more broadly by modifying one or more direct drivers. According to Ramsar, direct drivers are natural or human-caused factors that lead to changes in local and regional biophysical conditions (Van Asselen et al. 2013). On the other hand, indirect drivers have a broader impact, primarily by influencing direct drivers, and are often associated with institutional, socio-economic, demographic, and cultural processes. The MA identifies several direct drivers, such as climate change, nutrient use, land conversion, habitat alteration, invasive species, and diseases. Significant indirect drivers include demographic, economic, socio-political, scientific, technological, cultural, and religious factors (MA 2005). Distinguishing between natural and human-induced factors helps identify the drivers behind ecosystem changes. Natural drivers include solar radiation, weather fluctuations, seismic events (such as earthquakes and volcanic eruptions), pests, diseases, and ecological processes such as flood cycles and ecosystem succession. However, human-induced drivers encompass changes in land use, climate change, rising sea levels, water extraction, species introductions or removals, and resource consumption (MA 2005).

While climate variability is a natural driver, human-induced climate change arises from increased emissions of greenhouse gases, which constitute a significant global trend (IPCC 2014). These drivers can have positive and negative impacts, with negative impacts often seen as a decline in biodiversity, habitat quality, ecosystem services, or cultural significance. Conversely, positive drivers, such as conservation management or control of invasive species, represent human responses aimed at mitigating these changes. Understanding these drivers is complicated by the complex pathways that lead to ecosystem loss and degradation. Interactions between multiple drivers occur at various scales and result in regional variations (Ward et al. 2016; Craig et al. 2017). For instance, climate change can directly affect temperature, water levels, and hydroperiods (Renton et al. 2015) while acting indirectly through mitigation efforts such as biofuel production and hydropower, which can further stress ecosystems.

Substantial transformations in the primary drivers of ecosystem dynamics have occurred in India over recent decades. Deforestation and forest degradation have profoundly impacted the nation's forests, affecting nearly half of India's extant forest cover, primarily in Western Ghats and Northeast India (Prasad 2024). In marine ecosystems, the foremost among the direct drivers of alteration is the intensive exploitation of marine resources through fishing activities, precipitating substantial depletion of fish biomass and consequential modifications to aquatic biodiversity and habitats along India's extensive coastline (Morley et al. 2020). Within freshwater ecosystems, pivotal drivers of change encompass alterations in water regimes, introduction of invasive species, and pollution from industrial and agricultural sources (Cantonati et al. 2020). The burgeoning concern of excessive nutrient loading due to excess synthetic fertilizer application in agriculture has emerged as a significant force of ecosystem alteration in India, contributing to algal blooms and eutrophication in aquatic environments (Shukla et al. 2021). The impacts of climate change on India's ecosystems are evident through observable shifts in species distributions and significant changes in biological events such as flowering and migration patterns. According to the Intergovernmental Panel on Climate Change (IPCC) Sixth Assessment Report, climate change has led to notable alterations in the phenology of various plant and animal species, affecting ecosystem dynamics and biodiversity. These changes are particularly pronounced in regions like the Himalayas and Western Ghats where temperature rise and altered precipitation patterns have resulted in shifts in the timing of flowering and fruiting in plants, as well as changes in the migration patterns of birds and other wildlife (IPCC 2023). Concurrently, India contends with the challenges posed by biological invasions, wherein invasive species threaten indigenous biodiversity, ecological integrity, and concomitant economic implications (Rai and Singh 2021).

Changes in significant indirect drivers encompass diverse aspects of human society and development. Over the past four decades, the global population has doubled, with 2 billion people added in the last 25 years. Developing countries have predominantly driven this growth, while demographic patterns exhibit significant variability across regions and nations (United Nations 2020). Urbanization has emerged as a prominent trend, with urban areas currently accommodating approximately half of the world's population, albeit covering less than 3% of the terrestrial surface (World Bank 2019). Economic transformations have been profound, as evidenced by a nearly sevenfold increase in global economic activity between 1950 and 2000 (World Bank 2019). Despite the demographic expansion, average income per capita almost doubled during this period, although marked regional disparities persisted (Floerkemeier et al. 2021). With rising per capita income, shifts in consumption patterns occur, influencing ecosystem conditions and services. Energy and material intensity tends to diminish as GDP per capita increases, yet aggregate energy and material utilization has escalated. Sociopolitical dynamics have transitioned toward better democratic governance, empowering local communities and civil society (UNDP 2020). Science and technology advancements have significantly and positively impacted ecosystem services, particularly food production. However, unintended consequences such as eutrophication of freshwater systems and depletion of marine fish stocks may arise (Gonçalves and Summers 2024; Oberdorff 2022).

Various interacting factors operating over temporal and spatial scales typically drive changes in ecosystem services. For instance, the nexus between population growth, economic development, and technological advancements influences climate change (Ruane 2024). These drivers operate intermittently, with events such as droughts, conflicts, and economic fluctuations exerting significant impacts. Moreover, feedback loops between changes in ecosystem services and drivers occur, wherein alterations in ecosystems can engender shifts in land use patterns, necessitate institutional responses to mitigate resource degradation, and influence socio-economic dynamics such as income distribution. In India, deforestation arises from a complex interplay of drivers, including commercial exploitation of timber resources, expansion of agricultural frontiers, livestock rearing, and infrastructure development, with regional nuances shaped by several economic, institutional, technological, cultural, and demographic factors (Chakraborty 2023). Agrarian intensification, characterized by the heightened use of fertilizers and the adoption of high-yielding crop varieties has been pivotal in bolstering food production in India, driven by advancements in agricultural technologies, market dynamics, and policy interventions (Bhattarai et al. 2021). Urbanization, characterized by the growth of urban population, expansion of built-up areas, and urban sprawl, profoundly impacts ecosystems within urban centers and their peri-urban hinterlands (UN-Habitat 2019).

Ecosystem degradation in India has reached alarming proportions, marked by substantial deforestation, soil erosion, and deterioration of aquatic ecosystems. Between 2015 and 2020, the country experienced a net loss of approximately 668,400 hectares of forest land, placing it second globally regarding deforestation rates (Singh 2024). This period underscores the significant pressures on forest ecosystems, primarily due to logging, agricultural expansion, and infrastructural development. Soil degradation has emerged as a critical issue driven by the overuse of irrigation, pesticides, chemical fertilizers, and heavy machinery. These practices have depleted essential micronutrients such as zinc, sulfur, manganese, iron, and copper, which were not previously a concern half a century ago (Menon 2023). The degradation of soil health not only diminishes agricultural productivity but also affects the livelihoods of millions of small farmers who depend on the land for sustenance. Aquatic ecosystems are equally threatened, with excessive nutrient loading from agricultural runoff leading to eutrophication and harmful algal blooms. These phenomena disrupt aquatic life and degrade water quality (Menon 2023). According to various estimates, India has experienced a significant decline in its wetland areas, with nearly 30% of these ecosystems being lost over the past 30 years. This reduction can largely be attributed to factors such as rapid urbanization, land-use change, unregulated construction activities, and pollution (Kumar et al. 2021). Climate change has exacerbated these issues, manifesting in observable shifts in species distributions and alterations in phenological events such as flowering and migration patterns. These changes are particularly pronounced in ecologically sensitive regions such as the Himalayas and Western Ghats (IPCC 2021). Comprehensive policies integrating sustainable agricultural practices, robust reforestation efforts, and climate adaptation strategies are imperative to address these multifaceted challenges. Such measures are essential to restoring and protecting India's ecosystems, ensuring their resilience and sustainability for future generations.

16.2 Principles of Ecosystem Restoration

United Nations (UN) Decade on Ecosystem Restoration 2021–2030, declared by the UN General Assembly in 2019, highlights the global importance of ecological restoration. This initiative aims to combat ecosystem degradation, bolster biodiversity, and enhance ecosystem services. It emphasizes the necessity for overarching principles encompassing a wide array of restoration efforts outlined by the UN Decade, including ten principles aligned with its goals, guiding restoration endeavors across diverse biomes, sectors, and regions (Fischer et al. 2021). Effective ecosystem restoration is fundamentally underpinned by principles integral to global policy frameworks and practices of inclusive engagement. These principles are enshrined in international policies and standards, and their implementation necessitates the active participation of a diverse range of stakeholders, including local communities, governmental bodies, and organizations. This approach ensures that restoration efforts are comprehensive, considering multiple viewpoints and needs aligning with established global guidelines (Díaz et al. 2019). These principles advocate for a continuum of restoration activities that maximize ecological and socio-economic benefits (Keenleyside et al. 2012; Cowie et al. 2018). They stress proactive intervention to address root causes of degradation, harnessing diverse knowledge systems to inform restoration strategies (Hallett et al. 2013; Shackelford et al. 2013). Restoration efforts encompass ecological health, cultural values, and socio-economic considerations tailored to local contexts (SER Code of Ethics 2012), with rigorous monitoring and adaptive management crucial for achieving sustainable impacts (Suding et al. 2015).

Ecological restoration aims to autonomously restore ecosystem structure, processes, and functions, promoting resilience and fostering interactions with adjacent ecosystems (Nelson et al. 2024). Assessing restoration progress involves diverse measurable attributes, including maintaining species assemblages and representing essential functional groups (Hobbs et al. 2006). Effective ecosystem restoration necessitates that the ecosystem supports self-sustaining populations, ensuring that its species can persist and maintain themselves independently of continuous external intervention. The restored ecosystem must also integrate seamlessly into the broader landscape, contributing to ecological connectivity and overall environmental health. Mitigating threats such as pollution, invasive species, and habitat destruction is imperative for long-term resilience and sustainability (Nelson et al. 2024).

From a social perspective, effective restoration entails comprehensive stakeholder engagement and empowerment through participatory decision-making processes. It integrates various sectors and knowledge systems, delivering short-term benefits aligned with long-term ecological objectives. Restoration initiatives strive to balance human activities with environmental processes, engaging stakeholders in setting priorities and implementing strategies (Palmer et al. 2022).

Despite the challenges of degraded landscapes, ecological restoration is increasingly acknowledged as a fundamental tool for advancing conservation and sustainable development (Tomlinson et al. 2022). Recent years have witnessed a notable uptick in global organizations adopting restoration initiatives influenced by international policies and agreements (Aronson and Alexander 2013; Alexander et al. 2011; Suding et al. 2015; Cowie et al. 2018).

The Society for Ecological Restoration (SER) has played a pivotal role in shaping restoration policies from initial discussions to establishing international standards (Keenleyside et al. 2012). These frameworks enable practitioners to tackle ecological intricacies while addressing human livelihood concerns (Hallett et al. 2013; Shackelford et al. 2013). Clearly defined policy will be essential for navigating global challenges in restoration and capitalizing on opportunities in the future.

16.3 The Importance of Ecosystem Restoration

According to scientific data, extreme climate events such as heat waves, floods, droughts, landslides, and so on are increasing rapidly attributing to ongoing ecosystem degradation by natural hazards and aggravated by higher synthetic processes and exacerbating the ecologically fragile places. Parallelly, anthropogenic influences such as human encroachment in protected areas, reckless resource consumption, and weak laws and policies also fuel ecosystem degradation (Singh et al. 2022). Integrating ecosystem restoration into climate change adaptation strategies can reduce the impact and restore ecologically fragile regions by enhancing the ecosystem-based approaches, building resilience, and reducing vulnerability in communities and landscapes. The concept of restoring ecological systems has emerged as a critical adaptive management for addressing environmental degradation, biodiversity loss, and climate change impacts (Simonson et al. 2021), and it

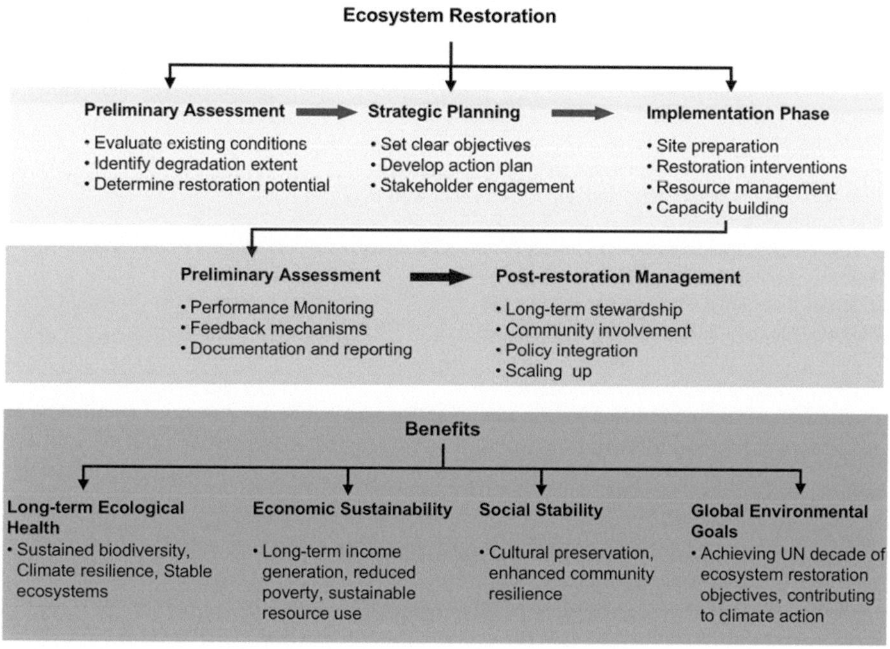

Fig. 16.1 Ecosystem restoration process and its multifaceted benefits

holds significance through the multidimensional benefits it offers in the most important sectors such as the ecological, economic, social, and policy dimensions (Wu et al. 2023) (Fig. 16.1).

A restored ecosystem enhances biodiversity by creating habitats for diverse plant and animal species, promoting ecological resilience, ecosystem functions, and services. It contributes to ecosystem stability, reduces the risk of species extinction, enhances ecosystem productivity, and becomes resilient to environmental stressors. Restored ecosystems enhance fundamental ecosystem functions including pollination, soil fertility, clean air and water, and carbon sequestration, thereby supporting human well-being and livelihoods (Haq et al. 2023).

Many international laws have been formulated to emphasize the significance of a healthy and restored ecosystem. Before the declaration of the UN Decade on Ecosystem Restoration in 2021, the Bonn Challenge was introduced in 2011, which set a global objective to restore about 150–350 Mha of degraded and deforested landscapes globally by 2020 and 2030, respectively. The International Union for Conservation of Nature (IUCN) complemented the Bonn Challenge 2011 as they shared common goals and collaboration in the areas comprising conservation and restoration of forests and other ecosystems. Initiatives for sustainable development must prioritize ecosystem restoration because it carries major significance toward realizing ecological preservation, adaptation and mitigation of climate change, sustainable economic growth, improved human well-being, coherent policy, global environmental goals, and building resilience against environmental threats (Table 16.1).

Table 16.1 Ecological restoration and its multi-dimensional benefits

Beneficial traits	Source
1. Ecological significance Improves biodiversity (both plants and animals) Promotes ecological resilience Restores ecosystem functions services and productivity Provides ecosystem services and supports human well-being and livelihoods	Bandyopadhyay and Maiti (2019)
2. Biodiversity conservation Restores degraded habitats and fragmented landscapes Restores threatened and endangered species Promotes genetic diversity within populations Enhances ecosystem functioning, resilience, and adaptation	FAO and UNEP (2020)
3. Climate change mitigation and adaptation Sequester carbon dioxide through afforestation, reforestation, wetlands, and coastal habitats Natural buffers against extreme weather events Reduces vulnerability in communities and landscapes	Goel and Rath (2022), Rawat et al. (2017)
4. Economic benefits Generates economic benefits (eco-tourism opportunities and sustainable resource management practices) Improves ecosystem services (water purification, flood control, and erosion control) Cost-effectiveness over traditional infrastructure-based approaches	Chase and Singh (2012), Pathak (2009)
5. Social and cultural importance Promotes social equity and environmental justice Provides cultural and spiritual values Enhances social cohesion Empowers marginalized groups	FAO and UNEP(2020), Singh et al. (2022)

The restoring ecosystem involves a dynamic healing process with shifting species distribution and structure; therefore, continuous intervention, management, and monitoring guide the restoring ecosystems into the intended balanced community. Identifying the cause and extent of the damaged or degraded ecosystems could help to understand the most suitable interventions required for restoring degraded ecosystems (Simonson et al. 2021). For example, in the traditional approach of forest restoration, ecological restoration is best suited in stable communities, because the interactions are predictable and direct changes in community structure called "ecological succession" which requires lesser external inputs. However, severely degraded systems require higher inputs, and it is difficult to restore already disturbed communities where a solo intervention could be inadequate to promote the regeneration of the forest. Performing these analyses leads to more effective utilization of restoration resources and enhances the overall success of restoration endeavors (Holl and Aide 2011). Protected areas, which primarily aim to minimize losses from ecological degradation, also help with restoration and are crucial to global efforts to safeguard biodiversity and ecosystem services for both aquatic and terrestrial structures (Wehi and Lord 2017). Regional restoration objectives vary throughout the world, for instance, in North America, most projects were conservation-focused, but, in Asia and Africa, restoration was through traditional knowledge and cultural harvesting (Wehi and Lord 2017). The Traditional Ecological Knowledge (TEK) practiced among indigenous communities stands as a backbone in many native or local restoration processes and their religious belief systems where nature is a sacred entity (Haq et al. 2023). However, TEK may not be self-sufficient to restore degrading ecosystems at a large scale. Local knowledge systems must be integrated with scientific and structured restoration approaches to enhance the promotion and protection of the natural ecosystem while also supporting allied sectors such as their livelihoods generation (through extracting food, fodder, medicines, firewood, timber, etc.), as well as culture beliefs and tradition setups (Haq et al. 2023).

Effective ecological restoration depends on maintaining species biodiversity in different ecosystem components including aquatic and terrestrial ecosystems. However, practices such as selective logging and deforestation alter the structure of plant communities in forests (Mansaurian and Vallauri 2014). Excessive and unethical fishing practices and exploitation of aquatic bodies endanger the innate ability of aquatic species to regenerate. (https://www.worldwildlife.org/threats/overfishing, accessed on 19-07-2024). In addition to these factors, the limitations in scientific understanding pose multiple challenges to the effective ecosystem restoration process. Therefore, research studies must largely be focused on addressing the knowledge gaps in Indigenous ecosystem knowledge that might aid ecosystem restoration successfully (Haq et al. 2023). By maintaining a balance between ecological and socio-economic welfare, a green and sustaining future could be achieved. Ecology Conservation experts suggest more efforts toward ecological restoration following a natural progression and assistive regeneration by working on the prevailing threats, viz., urbanization, policy threats, deforestation, selective logging, and so on to prevent further degradation of the forest after restoration goals (Goel and Rath 2022).

16.4 The Significance of Documenting Case Studies and Best Practices on Ecosystem Restoration

The degradation of forests, grasslands, and agricultural lands (terrestrial ecosystems), as well as both freshwater and marine water bodies, has been the primary cause of the alarming loss of biological diversity and biological wealth and in the long run, the absence of positive interventions could result in an irreversible and unrepairable loss in ecosystem (Singh et al. 2022). While various programs and policies have targets related to ecosystem restoration, policy actions and practices are guided toward achieving these goals by scientific knowledge, readiness, and critical analysis (Singh et al. 2022). The ecosystem restoration goals can be categorized under two broad methods. The first restoration method is passive restoration, also known as unassisted natural succession. It is inexpensive and non-interventionist and can be deemed appropriate in cases where a deteriorated system seems capable of recovering on its own. The need to alter biota or physical environmental conditions to hasten restoration and move it toward the intended outcome is typically the basis for the second approach, known as technical or assisted restoration (Da Silva et al. 2017).

Documenting successful case studies and practices on ecosystem restoration could benefit the conservation community at large through systematic collection and dissemination enhancing the broader conservation, development communities, and policymakers (Mansaurian and Vallauri 2014). Such activities provision to keep a record of the existing trend and pave the way for replicating the model in similar geographic, terrestrial, aquatic, and vegetation settings providing solutions for the restoration of degraded ecosystems. An ecosystem restoration study by Rawat et al. (2017) pointed out that forest restoration efforts ought to be a key element of national strategies and action plans for degraded land rehabilitation programs, provided that preventing deforestation and planting trees alone are insufficient to stop the loss of biodiversity and ecosystem services. Along with research and scientific interventions, effective policy frameworks and governance mechanisms are essential to promote ecosystem restoration at local, national, and international levels. Integration of ecosystem restoration into national biodiversity strategies, climate action plans, and sustainable development agendas enhances policy coherence and synergies across sectors (Rawat et al. 2017). Investing in capacity building, research, and monitoring efforts strengthens the evidence base for ecosystem restoration, informs policy decisions, and facilitates adaptive management approaches. Clewell and Aronson (2006) stressed the importance of pragmatic rationales for successful ecosystem restoration and claimed that a sound cohesive strategy is possible when the idealistic, pragmatic, and technical justifications are combined. For ecological restoration or maintenance, which goes beyond rearranging the prior biodiversity, it is necessary to have access to effective case studies and best practices. This can be done by carrying out regular surveys, keeping a close track of the presence and structure of forest cover, fragmenting land, maintaining habitat, and ascertaining that all government programs and regulations are implemented correctly. Effective restoration efforts require public participation in ecosystem maintenance and public awareness of the significance of maintaining a balanced environment (Goel and Rath 2022).

16.5 Successful Case Studies on Ecosystem Restoration

Ecological restoration is essential for safeguarding the critical ecosystems in coastal lagoons, vital where rivers meet seas, and contributes significantly to ecological, economic, and social dimensions.

Chilika Lagoon in Odisha, Asia's largest brackish water lagoon and a Ramsar site, exemplifies the importance of providing essential ecosystem services to local communities. Chilika Lagoon faced severe challenges, including siltation, invasive species, illegal aquaculture, and cyclonic disturbances, prompting the Chilika Development Authority to implement comprehensive restoration strategies. A pivotal intervention in 2000 involved constructing a new channel to enhance water exchange with the Bay of Bengal, restoring critical salinity levels for biodiversity and fisheries while mitigating environmental degradation. Authorities enforced regulatory measures to curb illegal aquaculture, ensuring water quality and habitat integrity. Conservation efforts focused on protecting habitats for migratory birds, endangered species, and commercially important fish, complemented by community engagement programs promoting sustainable practices and alternative livelihoods. The successful Restoration of Chilika Lagoon resulted in its removal from the Montreux Record in 2001 and earned the prestigious Ramsar Award in 2002, underscoring its role in preserving coastal ecosystems and fostering sustainable development (Finlayson et al. 2020; Giri et al. 2023).

The Vembanad-Kol Wetland in Kerala, spanning approximately 200 km^2, is another critical ecosystem facing environmental challenges such as pollution, encroachments, and altered hydrological patterns (Thomas 2019; Kumar et al. 2021). Restoration efforts since 2010 have focused on improving water quality through enhanced sewage treatment and organic farming practices to reduce chemical inputs (Kumar et al. 2021). These measures are essential for sustaining over 150 fish species and supporting the livelihoods for thousands of fisherfolk (Mathew 2018). Rejuvenating wetlands and mangrove forests along the wetland's edges has enhanced biodiversity and provided vital breeding grounds for fish and other aquatic organisms (Thomas 2019). Community engagement has been pivotal since the early 2000s, promoting sustainable practices among local stakeholders, including eco-friendly tourism and responsible fishing techniques (Mathew 2018). Besides, integrated wetland management plans and scientific monitoring have guided these restoration initiatives, ensuring sustainable development and biodiversity conservation in the Vembanad-Kol Wetland (Kumar et al. 2021).

Deepor Beel, the only Ramsar site in Assam, serves as a crucial stormwater reservoir for Guwahati City faced degradation from urbanization, pollution, invasive species, and unsustainable practices. Restoration efforts involved collaborative endeavors among government bodies, nongovernmental organizations (NGOs), academic institutions, and local communities. These initiatives included management of invasive species, habitat restoration through reforestation, and water quality management. Policy support and community participation have been integral to the success of these efforts, guided by comprehensive management plans and adaptive management strategies (Deka et al. 2020; Sharma et al. 2024).

Successful wetland restoration is an international example at the Djegbadji Lagoon Ramsar Site in Benin. The lagoon faced significant mangrove deforestation due to heavy dependence on mangrove wood harvesting for salt production, threatening its ecological integrity and essential ecosystem services. In response, the "Hydrological Restoration of Mangroves in Djegbadji Lagoon, Benin" project, led by the Coordination for Research and Development in the Environment (CRDE), implemented a pilot restoration effort. The restoration activities involved establishing earth channels based on indigenous community knowledge to re-establish tidal flows, setting up a community nursery, and successfully reforesting 30 hectares with a high success rate using more than 250,000 mangrove seedlings of *Avicennia germinans* and *Rhizophora racemose* (CRDE 2020). The restoration led to the recovery of crucial fish species, and supported endemic, resident, and migratory birds, demonstrating the effectiveness of collaborative restoration efforts in preserving biodiversity and supporting local communities.

The Restoration of Chilika Lagoon, Vembanad-Kol Wetland, Deepor Beel, and Djegbadji Lagoon illustrates the importance of multi-stakeholder engagement and tailored restoration strategies in addressing complex environmental challenges. These case studies provide valuable lessons and frameworks for replication across diverse international, national, and regional contexts, ensuring the sustainability of critical wetland ecosystems worldwide.

Forest restoration approaches formulated by the World Wildlife Fund (WWF): The international conservation group WWF implemented forest landscape restoration programs for 10 years in collaboration with several developed and developing nations. The study's key findings emphasized that the conservation community as a whole might benefit if lessons from these projects are systematically gathered and rationally conveyed to policymakers as well as other conservation and development groups (Mansaurian and Vallauri 2014). The sustainable and durable restoration process can only be achieved by garnering local communities' support and proactively involving important partners and stakeholders. For example, the involvement of international organizations such as WWF projects for ecosystem restoration cemented and supported engagements for sustainable and successful restoration programs at a large scale (WWF: A future for people and nature).

Creating and executing a successful monitoring system: WWF dedicated a significant amount of time and effort to work with local communities and the government in Vietnam, holding over 60 meetings at the national, province, district, and commune levels to define the success story in the restoration of Central Annamites region. The result was an extensive monitoring framework with critical indicators to monitor progress in the areas of (a) livelihoods; (b) forest ecosystem services (e.g., protecting water catchments); (c) biodiversity and forest condition; (d) the capacity to regulate ecosystem services sustainably; and (e) threats monitoring (Mansaurian and Vallauri 2014).

Safeguarding Indigenous species knowledge: Local communities in Madagascar's Fandriana-Marolambo environment exclusively cultivated exotic species before 2007. However, more than 80% of indigenous seedlings were planted by WWF by 2010 (268,416 out of 328,416). This substantial shift in local species is attributed to the work done by local groups and mediators to engage with the native communities, deepen their understanding of native species, and better define their needs in terms of agroforestry (Roelens et al. 2010).

Meeting socio-economic requirements: Large-scale forest restoration cannot sustain without local support, which can only be attained through understanding and consideration of local needs (Adams et al. 2016). The Fandriana-Marolambo forest restoration project in Madagascar included several local facilitators to increase community involvement (Roelens et al. 2010). It is imperative for both developing and developed countries to address local socio-economic issues. For instance, majority of restoration initiatives by the United States Collaborative Forest

Landscape Restoration Program in the United States included employment generation as a key social goal (USDA 2009).

Connecting demands for repair and restoration needs to a broad vision for conservation: The WWF assisted communities in Sichuan province, China, to increase their usage and understanding of native agroforestry species, conversely promoting the rehabilitation of the Giant panda's (*Ailuropoda melanoleuca*) habitat. By expanding the giant panda's habitat, the ecosystem restoration in Sichuan served both ecological and socio-economic goals by enhancing food security and income creation (Mansaurian and Vallauri 2014).

Implementing successful Payment of Environmental Services (PES) Scheme in Costa Rica: The Costa Rican PES program, which prioritized four ecological services—capturing and storing atmospheric carbon, protecting water resources, maintaining biodiversity, and maintaining scenic beauty—was among the first payment systems ever created globally (Porras and Chacon 2018). One of the goals of the PES plan which was adopted from several countries, and each had its specific applications to reward landowners for keeping their forests and other services that would otherwise be lost to other land uses. Costa Rica was among the nations that implemented this plan with the greatest degree of success. By giving landowners financial incentives to encourage either forest regeneration or actively planting trees, Costa Rica was able to defy the national trend of declining forest cover and begin expanding its forests in the 1980s. This was achieved in tandem with the country's efforts to conserve biodiversity and promote ecotourism on a large scale. The provisions of PES in Costa Rica successfully contributed to improving the forest cover in the country. The percentage of the country's territory covered by forests increased from 21% in the 1980s to almost 53% by 2020 (Costa Rica: A Remarkable Tale of Environmental Redemption area 2024).

Asola Bhatti Wildlife Sanctuary (ABWS): This sanctuary is situated on the southern Delhi Ridge of the Aravalli highlands covering 32.71 km^2. Asola Wildlife Sanctuary zone serves as Delhi's green lungs and potential carbon sink, mitigating local-level warming (Kushwaha et al. 2014). It was officially declared as a protected area in 1986 and is well-known for its abundant avian life, which is greater than that of mammals, as well as its medicinal plants, which comprise hundreds of different types of grasses, shrubs, and plants with 252 bird species and 246 animal species, including a variety of insects and mammals. The human–animal conflict has increased because of urbanization and an increase in agricultural activities, encroachment of forested land, deforestation, and the lack of safe wildlife corridors. Since then, several initiatives have been undertaken to restore and protect the region adjoining the Wildlife Sanctuary through organic farming, planting saplings, rainwater harvesting, restoring degraded land and habitat, agroforestry, herbal plantations, and harnessing solar energy. The repair has been effective as a result of the ongoing support and interventions from the governing bodies in collaboration with the local dwellers and the forest department stationed in the sanctuary (Goel and Rath 2022). Besides, the Wildlife Institute of India has prepared a wildlife management plan to ensure their protection and well-being. Some of the initiatives include: (i) regeneration projects, which began in 2020 and focused on the assistive regeneration of plants rather than the mass plantation of trees, thereby re-imposing the significance of grassland and shrubs; (ii) plantation drive, which involves around 34 lakh saplings and 14 varieties of grass; the grass species selected were those that helped increase the microbial population, which, in turn, will increase the availability of nutrients to the plants. During the last 10 years improvement in the green cover of 18 km^2 from 324 km^2 (21.9%) in 2019 to 342.2 km^2 (23.06%) has been observed in 2022, (iii) q 45 km patrol passage outside the sanctuary has been built to prevent trespassing and encroachment on forest land; (iv) abandoned mine pits have been used to harvest and store rainwater during monsoon season, which ensures water availability during dry season; and (v) government organized protection awareness events to raise public understanding of the value of wildlife conservation (ABWS).

Khonoma Nature Conservation and Tragopan Sanctuary (KNCTS): The KNCTS in Khonoma village is an in-situ protected area created in 1998, which is about 18 km far from Nagaland's capital city Kohima. The Angami ethnic community inhabits the village and the name Khonoma is derived from the vernacular name of a native plant called "Khwunoria" (Giri et al. 2018). Contracting its past practices of extensive hunting, the village councils through the support from the local community took the initiative to restore and conserve its natural wildlife which was becoming extremely vulnerable from the hunting culture that existed among the local community, transcending from once being a popular hunting area to being recognized as the first Green Village of Asia in 2005 (ENVIS 2007). Since then, ecosystem conservation and promoting a green lifestyle became the prime elements of social empowerment in the village and received formal recognition for overcoming ecological challenges and conserving its present status through community conservation efforts (Pathak 2009). Covering approximately 25 km^2, KNCTS is located close to the settlements of Khonoma, Mezoma, and Dzüleke. It is home to over 300 bird species, many of which are native to North East India including the Mountain bamboo partridge, Blyth's tragopan, Crested finch bill, Assam laughing thrush, Striped laughing thrush, Spot-breasted scimitar babbler, Flavescent bulbul, and Naga wren-babbler, as well as many endangered mammals, including Clouded leopard, Asiatic black bear, and Hoolock gibbon, and many more such as Pallas's squirrel, Himalayan striped squirrel, Barking deer, Binturong, and Serow. The KNCTS case study popularizes the local community's participation and involvement in overcoming issues such as deforestation and poaching by playing a pivotal role in preserving the natural ecosystem of the area.

Joint Forest Management (JFM) in Nagaland State: The Indian Himalayas spans approximately 59 million hectares, of which 13.5 million are government forests that have been degraded, 7.3 million are degraded community lands, and 1.2 million are private agricultural areas that have been abandoned (Rawat et al. 2017). Since the early 1980s, restoring degraded land has been a top priority in India. The Social Forestry Program in 1980 was initiated in India to recover degraded wasteland and to fulfill the demand for biomass. Subsequently, the JFM Program was introduced which involved greater community participation and involvement (Rawat et al. 2017). The 1988 forest policy acknowledged that community participation in forest management was critical to enhance livelihood systems and safeguard forest resources, and in 1990, the Government of India concurrently issued directives to state governments to legally establish "JFM" systems in nearly all states making it one of the most extensive JFM projects globally (Behera 2009). Indian rural communities responded well to the JFM program, which was introduced to promote community-led involvement in the management and restoration of forests nationwide (Meetei et al. 2023). The program had positive effects on the "Zeliang Naga" community. The settlement in the Eastern Himalayan region of India near "Old Jalukie," Nagaland. In Jalukei, the transition from vast jhum farming to ecological restoration started in 2002. The local communities and forest officials with the active involvement of the village councils established a 370-hectare community biodiversity reserve in the region. Both the local economy and the ecology of the region benefited from it and created opportunities for employment and livelihood while also restoring biodiversity through the cultivation of multipurpose trees, orchid species, native tree species, enhanced bird populations, and an increase in various herb and scrub species (Bhattacharjee 2020). The successful land restoration and biodiversity conservation efforts of JFM in Jalukei, Nagaland, are an excellent example of concerted efforts of local communities, village councils, and government (Bhattacharjee 2020; Meetei et al. 2023).

Biodiversity Ecosystem Functioning (BEF): Focusing on alternate aspects of forest restoration that are sometimes overlooked by more conventional approaches to restoration, the BEF approach offers a helpful framework for assessing forest restoration in the context of ecosystem functioning (Aerts and Honnay 2011). The BEF concept was developed on the premise that the diversity of all the biota, including species, genotypes, and functional groups, directly influences the levels of ecosystem functions (such as productivity, nutrient cycling, and decomposition) and the stability of those functions (Doherty and Zedler 2018). Reinstating the association between ecosystem functioning and biodiversity is an essential objective of restoration from the BEF perspective. For example, the initiatives to restore degraded forests and increase biodiversity have been implemented in the Western Ghats, India, which is a UNESCO World Heritage site and a hotspot for biodiversity. Here, native tree planting, agroforestry practices, and community involvement in conservation initiatives were highlighted through projects jointly coordinated by various government agencies, private companies, community partaking, and publicly funded organizations like the Nature Conservation Foundation (NCF), paving the way to restore ecosystem functions such as carbon sequestration, water regulation, and habitat provision for endangered species like the Malabar civet and lion-tailed macaque and protecting around 1075 ha in 35 rainforest fragments under restoration and recovery programs (Nature Conservation Foundation n.d.).

16.6 The Prevailing Best Practices on Ecosystem Restoration

Ecosystem restoration is an essential and ongoing process that helps maintain the long-term health and functionality of natural environments across diverse ecosystems (Ramsar Secretariat 2010). A holistic approach, emphasizing the interconnectedness of ecological components is pivotal in successful restoration efforts (Tedesco et al. 2023). This approach aims to restore multiple environmental benefits simultaneously reflecting the self-sustaining dynamics observed in natural ecosystems where vegetation, wildlife, and ecological processes interact synergistically.

Community engagement is fundamental to ecosystem restoration, and it involves local stakeholders in planning and implementation to ensure inclusive decision-making and sustainable outcomes (Jones et al. 2018). It is crucial to tackle the underlying degradation factors such as excessive water resource use and pollution from agricultural, industrial, and urban activities (Johnson and Brown 2016). Restoration initiatives should prioritize reintroducing native flora and fauna while managing and eradicating invasive species (Spatz et al. 2022). Cleaning up debris and waste from degraded areas not only aids in ecosystem recovery but also mitigates environmental hazards (Thompson et al. 2019). Proper zoning and access management are essential in delineating areas for human activities while protecting sensitive habitats (Duffy and John 2020). The principles of wetland restoration, as outlined by the Ramsar Secretariat, are highly applicable beyond wetlands due to their holistic and comprehensive nature (Ramsar Secretariat 2010). These practices promote ecological health, biodiversity conservation, and sustainable resource management across various landscapes and habitats globally. By adopting these best practices, restoration efforts can effectively contribute to ecosystems' resilience and long-term sustainability worldwide.

Adaptive management, community involvement, and scientific concepts serve as the foundation for best practices in ecosystem restoration. By putting these best practices into action, restoration projects can be made far more successful and result in resilient ecosystems that support biodiversity and offer vital services. Enlisted below are some current best practices for effective ecosystem restoration derived from various studies (Table 16.2).

16.6.1 Strategies for Upscaling Best Practices

Ecologically and economically sound strategies are essential to scale up optimal practices for ecosystem restoration. Leveraging evidence-based restoration, integrated methods, information exchange, and stakeholder participation are important factors in upscaling best practices. The effectiveness of initiatives aimed at restoring ecosystems depends heavily on knowledge exchange. Restoration activities can be more successfully scaled up internationally by sharing best practices and lessons discovered. Effective ecosystem restoration relies on integrated methods (Mostafa et al. 2022). In particular, wetland restoration initiatives can significantly benefit from applying integrated concepts, such as combining constructed/artificial

Table 16.2 Ecological restoration (ER) systems and best practices in India

S. No	ER systems	Best practices examples	Source
1.	*Promotion of traditional knowledge:* Promoting traditional knowledge helps communities become more engaged, improves ecological and cultural sustainability, and supports adaptive management techniques in the face of environmental problems like climate change. Such information is a priceless source of wisdom that, when combined with scientific methods, can result in restoration projects that are more robust and successful.	The "Zabo system" which was developed more than 80 years ago by the ingenious and skillful farmers of the Chakhesang tribe of the Kikruma village in Nagaland as an initiative to solve the water scarcity issues in the village vicinity is recognized for judicious land use system. The traditional technique involves multi-tier integrated farm enterprises such as forest conservation at the top-most, followed by the water catchment area, livestock rearing, and crop cultivation at the base. Besides conserving the ecological system of the villages, it also generates food and income for the farmers, efficiently.	Singh et al. (2018)
2.	*Community-led forest conservation:* Community participation in the restoration and management processes is deemed necessary for the accomplishment of sustainable forestry, agroforestry, land rehabilitation, and restoration initiatives in the Himalayas.	(i) Khonoma village in Nagaland is recognized for promoting ER practices by involving local communities led by the government, and NGOs. The village imposes a strict ban on activities such as hunting and logging in its forest areas. The village has been attributed as the first "Green Village" in Asia and recognized for its restoration and conservation practices such as the alder-based agroforestry which is a sustainable modification of the traditional jhum farming, community-managed forests, and KNCTS. (ii) The successful JFM implementation in Old Jalukie, located in the Eastern Himalayan region of India, exemplifies the transition from vast degrading lands to restoring a 370-hectare Community Biodiversity Reserve.	Pathak (2009), Chase and Singh (2012)
3.	*Shifting cultivation alternatives for ER in hilly regions:* Introducing sustainable agricultural practices as alternatives to traditional slash-and-burn cultivation (particularly with less fallow periods) to reduce deforestation, soil degradation, and water scarcity problems.	(i) Bamboo-based agroforestry systems in the Northeastern States of India such as Mizoram and Manipur, where bamboo is widely used for timber, handicrafts, and erosion control. In this system, bamboo cultivation is integrated with other tree species and crops to enhance soil fertility, productivity, biodiversity, resource conservation, sustainability, and economic livelihoods. (ii) Commonly practiced Agri + Horti + silvicultural systems in the Western Himalayas (Himachal and Uttarakhand). In addition to intercropping vegetable crops and fruit trees, it involves multipurpose trees, for tree fodder and fuelwood collection which constitute an integral part of the traditional farming systems across the Himalayas.	Tewari et al. (2015), Kumar et al. (2018)

(continued)

Table 16.2 (continued)

S. No	ER systems	Best practices examples	Source
4.	*Conservation of Wetlands:* The restoration and conservation of wetlands is a prerequisite for a healthy ecosystem by maintaining water quality, supporting aquatic biodiversity, and sustaining local livelihoods.	(i) Loktak Lake, a Ramsar site in Manipur harbors rich biodiversity. The lake's restoration was primarily focused on reducing pollution, restoring floating biomass (phumdis), and conserving biodiversity (aquatic plants and animals). (ii) The conservation efforts in Chilka Lagoon, Odisha, hold immense ecological significance due to its unique biodiversity hotspot, bird sanctuary, fisheries mangrove ecosystems ecosystem services, and so on. The lagoon emphasized protecting habitats for migratory birds, endangered species, and commercially important fish, complemented by community engagement programs promoting sustainable practices and alternative livelihoods.	Meetei et al. (2023), Giri et al. (2023)

wetlands with comprehensive management and evaluation programs, inclusive governance, and active public participation. Prioritizing the effectiveness and cost-efficiency of these restoration efforts is crucial, as emphasized by Vardi et al. (2021). Restoration initiatives may be broadened to include larger regions and create more substantial results within limited assets by prioritizing sustainability and affordability.

Assuring the stability and resilience of restored ecosystems, Heneghan et al. (2008) emphasized that the comprehension of soil biology is essential to restoration management. Understanding soil dynamics and biological processes can help overcome impediments to ecosystem recovery and make optimal use of resources in restoration activities. Addressing the multitude of associations that occur throughout ecosystems requires a multifaceted approach to restoration (Danovaro et al. 2021). Because species and ecological processes are interdependent, restoration efforts can be more comprehensive and effective in promoting ecosystem resilience and health. Furthermore, adopting evidence-based restoration approaches is essential for steering restoration projects with scientific rigor (Cooke et al. 2018). By ensuring that restoration plans are grounded on facts and research findings, evidence-based approaches guarantee that ecosystem recovery succeeds with greater efficiency. Additionally, community participation, incorporating local knowledge, and sustainable restoration techniques are all facilitated by integrating stakeholders and local communities in restoration initiatives (Eisinger 2022). By involving stakeholders, restoration initiatives are made sure to correspond with local demands and their values which promote sustained support and care for restored ecosystems.

16.7 Implications of Documenting the Studies on Ecosystem Restoration

Documenting research on ecosystem restoration is critical for promoting the field and directing restoration techniques. The reliability of empirical outcomes is improved by meticulously recording study findings, restoration aims, techniques, and costs (Krishnan and Osuri 2022). This documentation is critical for maintaining the pertinent nature of research in restoration contexts and supporting comparative and innovative research initiatives. Jones et al. (2018) underlined the necessity of documenting the restoration of ecosystems after anthropogenic perturbations to understand the efficacy of restoration procedures. Meta-analyses of research on ecosystem recovery provide insightful information on the efficacy of restoration efforts following immense disturbances such as intensive agriculture, logging, and mining operations.

Assessing the financial advantages and ecological efficacy of various restoration schemes requires extensive studies that examine the costs and benefits of restoration efforts over time (Groot et al. 2013). By utilizing these elements, the most economical methods for restoring different various ecosystems can be determined. The inclusion of socio-economic ramifications in restoration studies, especially in mangrove restoration programs, has shown that it is necessary to take into account wider social repercussions (Bayraktarov et al. 2020). Monitoring socio-economic variables alongside ecological successes facilitates decision-making easier and increases our understanding of what constitutes successful restoration. Moreover, recording the longevity of restored ecosystems is essential for evaluating their ability to sustain biodiversity and offer ecosystem services in the long run (Reid et al. 2017).

16.8 Conclusion

Reversing environmental deterioration and maintaining the long-term stability of natural ecosystems depend on ecosystem restoration. The efficaciousness of customized solutions and multi-stakeholder participation is exemplified by the successful stories of Chilika Lagoon, Vembanad-Kol Wetland, DeeporBeel, Djegbadji Lagoon, AsolaBhatti Wildlife Sanctuary, Khonoma Nature Conservation, and several forest restoration initiatives. These case studies demonstrate the value of combining sound policy frameworks, community engagement, and technological advances. Future projects may learn much from these best practices, including developing robust monitoring systems, preserving indigenous knowledge, attending to socio-economic requirements, and applying holistic restoration methodologies. By strongly emphasizing biodiversity, adaptive management, and community involvement, these strategies work together to create resilient ecosystems that promote human well-being and biodiversity.

To further advance the field of ecological restoration, it is necessary to carry out collaborative work based on scientific research, effective public engagement, and strict non-allowance provisions. Future efforts in restoration should be focused on restoring the afflicted ecosystems through the implementation of effective evidence-based strategies and methods presented by successful examples. The priority for restoration initiatives must be community engagement if the solutions are to be long-lasting and financially beneficial. Increasing the opportunity for interested parties to share information and opinions fosters teamwork, which promotes restoration exercises. For motivating effective and inspiring cooperation, the state goals of ecological recovery should be synchronized with the goals that are related to global and national programs, for instance, the tasks of reducing biodiversity losses and combating the consequences of climate change. Pursuing technological advances, flexibility at the management level, and constant supervision and evaluation are essential in monitoring success and tackling new challenges. Also, expanding the role of financial factors, for instance, PES can facilitate the involvement of landowners and communities in restoration programs. Therefore, the suggested option is the promotion of a comprehensive and integrated approach that should involve the usage of indigenous knowledge as well as the available scientific and modern technological points and international cooperation to improve biome sustainability and health.

References

Adams C, Rodrigues ST, Calmon M, Kumar C (2016) Impacts of large-scale forest restoration on socio-economic status and local livelihoods: what we know and do not know. Biotropica 48(6):731–744. https://doi.org/10.1111/btp.12385

Aerts R, Honnay O (2011) Forest restoration, biodiversity, and ecosystem functioning. BMC Ecol 11:1–10

Alexander S, Aronson J, Whaley O, Lamb D (2011) The relationship between ecological restoration and the ecosystem services concept. Ecol Soc 16(1):5

Aronson J, Alexander S (2013) Ecosystem restoration is now a global priority: time to roll up our sleeves. Restor Ecol 21(3):293–296

Bandyopadhyay S, Maiti SK (2019) Evaluation of ecological restoration success in mining-degraded lands. Environ Qual Manag 29(1):89–100. https://doi.org/10.1002/tqem.2164

Bayraktarov E, Brisbane S, Hagger V, Smith C, Wilson K, Lovelock C, Saunders M (2020) Priorities and motivations of marine coastal restoration research. Front Mar Sci 7. https://doi.org/10.3389/fmars.2020.00484

Behera B (2009) Explaining the performance of state–community joint forest management in India. Ecol Econ 69(1):177–185. https://doi.org/10.1016/j.ecolecon.2009.08.015

Bhattacharjee A (2020) Forest landscape restoration as an NbS strategy for achieving bonn challenge pledge: lessons from India's restoration efforts. In: Nature-based solutions for resilient ecosystems and societies. Springer, Singapore, pp 133–147

Bhattarai K, Conway D, Bhattarai K, Conway D (2021) Agriculture and environment. In: Contemporary environmental problems in Nepal: Geograph Persp, pp 335–445

Borawake N, Datta A, Naniwadekar R (2021) Tropical forest restoration in the eastern Himalaya: evaluating early survival and growth of native tree species. Ecol Restor 39(3):52–63. https://doi.org/10.3368/er.39.3.194

Cantonati M, Poikane S, Pringle CM, Stevens LE, Turak E, Heino J, Znachor P (2020) Characteristics, main impacts, and stewardship of natural and artificial freshwater environments: consequences for biodiversity conservation. Water 12(1):260

Chakraborty M (2023) An analysis of the causes of deforestation in India. In: The causes of tropical deforestation. Routledge, pp 226–239

Chase P, Singh OP (2012) People's initiative for conservation of forests and natural resources: a success story of Khonoma village forest, Nagaland. Nebio 3(3):61–67

Clewell AF, Aronson J (2006) Motivations for the restoration of ecosystems. Conserv Biol 20(2):420–428. https://doi.org/10.1111/j.1523-1739.2006.00340.x

Cooke S, Rous A, Donaldson L, Taylor J, Rytwinski T, Prior K, Bennett J (2018) Evidence-based restoration in the Anthropocene—from acting with purpose to acting for impact. Restor Ecol 26(2):201–205. https://doi.org/10.1111/rec.12675

Costa Rica: a remarkable tale of environmental Redemptionarea. MarttaPebet, 31st January, 2024. https://greencircleexperience.com. Accessed on 29-07-2024

Costanza R, d'Arge R, de Groot R, Farber S, Grasso M, Hannon B, Limburg K, Naeem S, O'Neill RV, Paruelo J, Raskin RG, Sutton P, van den Belt M (1997) The value of the world's ecosystem services and natural capital. Nature 387(6630):253–260

Cowie AL, Orr BJ, Sanchez VMC, Chasek P, Crossman ND, Erlewein A, Welton S (2018) land in balance: the scientific conceptual framework for land degradation neutrality. Environ Sci Policy 79:25–35

Craig LS, Olden JD, Arthington AH, Entrekin S, Hawkins CP, Kelly JJ, Wooten MS (2017) Meeting the challenge of interacting threats in freshwater ecosystems: a call to scientists and managers. Elem Sci Anth 5:72

CRDE (Coordination for Research and Development in the Environment) (2020) Hydrological Restoration of Mangroves in Djegbadji Lagoon, Benin. Report by CRDE, Benin

Da Silva AM, Bortoleto LA, Castelli KR, Silva RA, Mendes PB (2017) Prospecting the potential of ecosystem restoration: a proposed framework and a case study. Ecol Eng 108:505–513

Danovaro R, Aronson J, Cimino R, Gambi C, Snelgrove P, Dover C (2021) Marine ecosystem restoration in a changing ocean. Restor Ecol 29(S2). https://doi.org/10.1111/rec.13432

Dasgupta P (2010) Nature's role in sustaining economic development. In: Mäler KG, Vincent JR (eds) Handbook of environmental economics. Elsevier, pp 1457–1497

Deka M et al (2020) Policy framework and community participation in DeeporBeel restoration. Aust J Environ Manag 18(1):150–165

Díaz S, Settele J, Brondízio ES, Ngo HT, Guèze M (2019) The pervasive human-driven decline of life on Earth points to the need for transformative change. Science 366(6471):1377–1386

Doherty J, Zedler JB (2018) Biodiversity-ecosystem function (BEF) theory and wetland restoration. The Wetland Book: I: structure and function, management, and methods 1991–1996

Duffy JE, St. John MA (2020) Zoning and access management in ecosystem restoration: balancing human activities and conservation priorities. Mar Ecol Prog Ser 18(2):220–235

Eisinger A (2022) Restoration, land tenure, conflict, and opportunities in peacebuilding: 15 best practices for restoration. https://doi.org/10.7926/66030219

Environmental Information System, ENVIS (2007) Centre on Himalayan Ecology: Nagaland Institute of Health Environment and Social Welfare (NIHESW) Kohima, Nagaland, Fauna-Status of Environment and Related Issues. http://www.envis.nic.in

FAO and UNEP (2020) State of the world's forests: forests, biodiversity and people. https://openknowledge.fao.org. Accessed on 29-07-2024

Finlayson CM, Rastogi G, Mishra DR, & Pattnaik AK (Eds) (2020) Ecology, conservation, and restoration of chilika lagoon, India. Berlin/Heidelberg, Germany: Springer, Volume 6, pp 7–23

Fischer J, Riechers M, Loos J, Martin-Lopez B, Temperton VM (2021) Making the UN decade on ecosystem restoration a social-ecological endeavor. Trends Ecol Evol 36(1):20–28

Floerkemeier MH, Spatafora MN, Venables A (2021) Regional disparities, growth and inclusiveness. International Monetary Fund

Giri K, Mishra G, Jayaraj RSC, Kumar R (2018) Agrobio-cultural diversity of alderbased shifting cultivation practiced by Angami tribes in Khonoma village, Kohima, Nagaland. Curr Sci 115(4):598–599

Giri S, Panda S et al (2023) Ecological restoration strategies for Chilika Lagoon: a case study. J Coast Res 7(2):123–135

Goel V, Rath R (2022) Asola Bhatti wildlife sanctuary case study of ecological restoration: its success and present status. Vantage J Them Anal 3(2):104–116. https://doi.org/10.52253/vjta.2022.v03i02.09

Gonçalves AM, Summers JK (2024) Marine ecosystems: biodiversity, ecosystem services and human impacts. BoD–Books on Demand

Groot R, Blignaut J, Ploeg S, Aronson J, Elmqvist T, Farley J (2013) Benefits of investing in ecosystem restoration. Conserv Biol 27(6):1286–1293. https://doi.org/10.1111/cobi.12158

Hallett LM, Diver S, Eitzel MV, Olson JJ, Ramage BS, Sardinas H, Suding KN (2013) Do we practice what we preach? Goal setting for ecological restoration. Restor Ecol 21(3): 312–319

Haq SM, Pieroni A, Bussmann RW, Abd-ElGawad AM, El-Ansary HO (2023) Integrating traditional ecological knowledge into habitat restoration: implications for meeting forest restoration challenges. J Ethnobiol Ethnomed 19(1):33. https://doi.org/10.1186/s13002-023-00606-3

Heneghan L, Miller S, Baer S, Callaham M, Montgomery J, Pavao-Zuckerman M, Richardson S (2008) Integrating soil ecological knowledge into restoration management. Restor Ecol 16(4):608–617. https://doi.org/10.1111/j.1526-100x.2008.00477.x

Hobbs RJ, Arico S, Aronson J, Baron JS, Bridgewater P, Cramer VA, Epstein PR, Ewel JJ, Klink CA, Lugo AE, Norton D, Ojima D, Richardson DM, Sanderson EW, Valladares F, Vila M, Zamora R, Zobel M (2006) Novel ecosystems: theoretical and management aspects of the new ecological world order. Glob Ecol Biogeogr 15(1):1–7

Holl KD, Aide TM (2011) When and where to actively restore ecosystems? For Ecol Manag 261(10):1558–1563. https://doi.org/10.1016/j.foreco.2010.07.004

IPCC (2014) Climate change 2014. Synthesis report. . In: Pachauri RK, Allen MR, Barros VR, Broome J, Cramer W, Christ R, Church JA, Clarke L, Dahe Q, Dasgupta P, Dubash, NK

(eds) Contribution of Working Group I, II, and III to the Fifth Assessment Report of the Intergovernmental Panel on Climate Change (p 151).
IPCC (2021) Climate change 2021 the physical science basis. In: Masson-Delmotte VP, Zhai A, Pirani SL, Connors C, Péan S, Berger N, Caud Y, Chen L, Goldfarb MI, Gomis M, Huang K, Leitzell E, Lonnoy JBR, Matthews TK, Maycock T, Waterfield O, Yelekçi R, Yu A, Zhou B (eds) Contribution of Working Group I to the sixth assessment report of the intergovernmental panel on climate change. Cambridge University Press, Cambridge, UK/New York. https://doi.org/10.1017/9781009157896
IPCC (2023) Summary for policymakers. In: climate change 2023: synthesis report. In: Core Writing Team, Lee H, Romero J (eds) Contribution of Working Groups I, II and III to the sixth assessment report of the intergovernmental panel on climate change. IPCC, Geneva, pp 1–34. https://doi.org/10.59327/IPCC/AR6-9789291691647.001
Johnson EA, Brown S (2016) Addressing root causes of degradation in natural ecosystems: strategies and challenges. Environ Sci Pol 28(1):150–165
Jones H, Jones P, Barbier E, Blackburn R, Benayas J, Holl K, Moreno-Mateos D (2018) Restoration and repair of earth's damaged ecosystems. Proc R Soc B Biol Sci 285(1873):20172577. https://doi.org/10.1098/rspb.2017.2577
Keenleyside K, Dudley N, Cairns S, Hall C, Stolton S (2012) Ecological restoration for protected areas: principles, guidelines and best practices. IUCN
Krishnan A, Osuri A (2022) Beyond the passive–active dichotomy: aligning research with the intervention continuum framework of ecological restoration. Restor Ecol 31(4). https://doi.org/10.1111/rec.13828
Kumar BM, Handa AK, Dhyani SK, Arunachalam A (2018) Agroforestry in the Indian Himalayan region: an overview. In: Temperate agroforestry systems. CAB International, pp 153–172
Kumar R, Ganapathi H, Palmate S (2021) Wetlands and water management: finding a common ground. In: Water governance and management in India: issues and perspectives, vol 2. Springer, Singapore, pp 105–129
Kushwaha SPS, Nandy S, Gupta M (2014) Growing stock and woody biomass assessment in Asola-Bhatti Wildlife Sanctuary, Delhi, India. Environ Monit Assess 186:5911–5920. https://doi.org/10.1007/s10661-014-3828-0
Mansourian S, Vallauri D (2014) Restoring forest landscapes: important lessons learned. Environ Manag 53(2):241–251. https://doi.org/10.1007/s00267-013-0213-7
Mathew AT (2018) Community engagement and sustainable development in Vembanad-Kol Wetland. J Sust Dev 12(2):89–102
Meetei KB, Tsopoe M, Giri K, Mishra G, Verma PK, Rohatgi D (2023) Climate-resilient pathways and nature-based solutions to reduce vulnerabilities to climate change in the Indian Himalayan region. In: Climate change in the Himalayas. Academic Press, pp 89–119
Menon R (2023) Soil degradation in India spells doom for millions
Mongabay Millennium Ecosystem Assessment (2005). https://www.millenniumecosystemassessment.org. Assessed on 03-07-2024
Morley SA, Abele D, Barnes DK, Cárdenas CA, Cotté C, Gutt J, Constable AJ (2020) Global drivers on southern ocean ecosystems: changing physical environments and anthropogenic pressures in an earth system. Front Mar Sci 7:547188
Mostafa H, Attia S, Feisal Z (2022) A strategy for lakes ecological restoration by integrated constructed wetlands, case study: Lake Qaroun, Egypt. IOP Conf Ser Earth Environ Sci 1113(1):012008
Nature Conservation Foundation (NCF) (n.d.). https://www.ncf-india.org/western-ghats/reviving-the-rainforest. Accessed on 29-07-2024
Nelson CR, Hallett JG, Romero Montoya AE, Andrade A, Besacier C, Boerger V, Weidlich EWA (2024) Standards of practice to guide ecosystem restoration: a contribution to the United Nations decade on ecosystem restoration 2021–2030. FAO
Oberdorff T (2022) Time for decisive actions to protect freshwater ecosystems from global changes. Knowl Manag Aquat Ecosyst 423:19

Palmer CG, Fry A, Libala N, Ralekhetla M, Mtati N, Weaver M, Scherman PA (2022) Engaging society and building participatory governance in a rural landscape restoration context. Anthropocene 37:100320

Pathak N (ed) (2009) Community conserved areas in India - a directory. Kalpavriksh, Khonoma village: a case study. Kalpavriksh.EnvironmentalActionGroup, pp 453–456. https://kalpavriksh.org›khonoma-village

Porras I, Chacón-Cascante A (2018) Costa Rica's payments for ecosystem services program. International Institute for Environment and Development

Prasad R (2024) The future of Indian forests: the outlook 2050. In: Forest policies, laws, and governance in India: conservation challenges in the face of climate change. Springer Nature Singapore, Singapore, pp 383–422

Rai PK, Singh JS (2021) Plant invasion in protected areas, the Indian Himalayan region, and the North East India: progress and prospects. Proc Ind Nat Sci Acad 87(1):19–35

Ramsar Secretariat (2010) Guidelines for ecosystem restoration under the Ramsar convention. Ramsar Convention Secretariat

Rawat LS, Maikhuri RK, Dhyani D, Bahuguna YM, Pharswan DS (2017) Ecological restoration of village common degraded land through participatory approach for biodiversity conservation and socio-economic development in Indian Himalayan region. Acta Ecol Sin 37(4):240–252

Reid J, Wilson S, Bloomfield G, Cattau M, Fagan M, Holl K, Zahawi R (2017) How long does restoring ecosystems persist? Ann Missouri Bot Gard 102(2):258–265. https://doi.org/10.3417/2017002

Renton DA, Mushet DM, DeKeyser ES (2015) Climate change and prairie pothole wetlands: mitigating water-level and hydroperiod effects through upland management (No. 2015-5004). US Geological Survey

Roelens JB, Vallauri D, Razafimahatratra A, Rambeloarisoa G, Razafy FL (2010) Restauration des paysagesforestiers: Cinqans de re'alisation a` Fandriana-Marolambo

Ruane AC (2024) Synthesis report of the IPCC sixth assessment report (AR6)

Shackelford N, Hobbs RJ, Burgar JM, Erickson TE, Fontaine JB, Laliberté E, Ramalho CE, Perring MP (2013) Primed for change: developing ecological restoration for the 21st century. Restor Ecol 21(3):297–304

Sharma P, Sarkar R, Deka JP, Koley S, Saha B (2024) Assessing water quality of Deepor Beel, Assam, NE India, using water quality index: a case of Ramsar wetland. Arab J Geosci 17(1):20

Shukla AK, Behera SK, Chaudhari S, Singh G (2022) Fertilizer use in Indian agriculture and its impact on human health and environment. Indian J Fertil 18(3):218–237

Shukla AK, Singh A, Banerjee T, Chaudhary SK (2021) Impact of agricultural chemical inputs on human health and environment in India. Agric Chem Inputs 66

Simonson WD, Miller E, Jones A, García-Rangel S, Thornton H, McOwen C (2021) Enhancing climate change resilience of ecological restoration—a framework for action. Perspect Ecol Conserv 19(3):300–310. https://doi.org/10.1016/j.pecon.2021.05.002

Singh US (2024) Indian forests in comatose. Biophilia Insights 2(1) e202421001. https://doi.org/10.52679/bi.e202421001

Singh RK, Hanna K, Asangla A, Bharali R, Borkotoky D (2018) Zabo: a time-tested integrated farming system practiced by the Chakhesang Tribe of Nagaland. Ind J Hill Far 31(1):188

Singh K, Byun C, Bux F (2022) Ecological restoration of degraded ecosystems in India: science and practices. Ecol Engi 182:106708

Spatz DR, Holmes ND, Will DJ, Hein S, Carter ZT, Fewster RM, Russell JC (2022) The global contribution of invasive vertebrate eradication as a key Island restoration tool. Sci Rep 12(1):13391

Suding KN, Higgs E, Palmer M, Callicott JB, Hobbs RJ (2015) Editorial: Committing to ecological restoration. Restor Ecol 23(3):267–271

Tedesco AM, López-Cubillos S, Chazdon R, Rhodes JR, Archibald CL, Pérez-Hämmerle KV, Dean AJ (2023) Beyond ecology: ecosystem restoration as a process for social-ecological transformation. Trends Ecol Evol 38(7):643–653

Tewari S, Banik RL, Kaushal R, Bhardwaj D, Chaturved O, Gupta A (2015) Bamboo-based agroforestry systems. ENVIS center on forestry, National Forest Library and Information Centre, Forest Research Institute, Dehradun, p 24
Thomas R (2019) Marine biology: an ecological approach. Scientific e-Resources. ISBN 978-1-83947-453-8, 80-190
Thompson LA et al (2019) Debris Cleanup in ecosystem restoration: implications for habitat recovery and environmental health. Restor Ecol 12(3):180–195
Tomlinson S, Tudor EP, Turner SR, Cross S, Riviera F, Stevens J, Lewandrowski W (2022) Leveraging the value of conservation physiology for ecological restoration. Restor Ecol 30:e13616
UN-Habitat (2019) World cities report (2018) learning from city dynamics. UN-Habitat
United Nations (2020) World population prospects: the 2019 revision. Department of Economic and Social Affairs, Population Division
United Nations Development Programme (UNDP) (2020) Human development report 2020: the next frontier - human development and the anthropocene
USDA (2009) Collaborative forest restoration program: lessons learned. United States Department of Agriculture, Forest Service, Washington, DC
Van Asselen S, Verburg PH, Janse JH, Perez-Soba M (2013) Drivers of wetland conversion: a global meta-analysis. PLoS One 8(11):e81292
Vardi T, Hoot W, Levy J, Shaver E, Winters R, Banaszak A, Montoya-Maya P (2021) Six priorities to advance the science and practice of coral reef restoration worldwide. Restor Ecol, 29(8). https://doi.org/10.1111/rec.13498
Vitousek PM, D'Antonio CM, Loope LL, Westbrooks R (1997) Biological invasions as global environmental change. Am Sci 84(5):468–478
Ward TD, Algera DA, Gallagher AJ, Hawkins E, Horodysky A, Jørgensen C, Cooke SJ (2016) Understanding the individual to implement the ecosystem approach to fisheries management. Conserv Physiol 4(1):cow005
Wehi PM, Lord JM (2017) Importance of including cultural practices in ecological restoration. Conserv Biol 31(5):1109–1118
World Bank (2006) Where is the wealth of nations? Measuring capital for the 21st century. World Bank
World Bank (2019) World development indicators 2019. World Bank
Wu X, Lü Y, Zhang J, Lu N, Jiang W, Fu B (2023) Adapting ecosystem restoration for sustainable development in a changing world. Innovation 4(1) https://doi.org/10.1016%2Fj.xinn.2023.100375

Supporting Landscape Restoration for Resilient Ecosystems

17

Seema Bhatt, Konda Chavva, and Jasmeet Kaur

Abstract

This chapter explores the pressing issue of land degradation, a global crisis characterized by soil erosion, deforestation, and desertification. The far-reaching consequences of this environmental challenge include declining agricultural productivity, biodiversity loss, and exacerbated climate change. However, the chapter posits that land restoration offers a viable solution to mitigate these impacts. By restoring degraded lands, we can enhance food security, regulate climate through carbon sequestration, protect biodiversity, and improve human well-being. The chapter emphasizes the economic benefits of ecosystem restoration and its role in achieving sustainable development goals. It advocates for an integrated approach to land management that prioritizes restoration and sustainable practices.

Keywords

Land degradation · Ecosystem restoration · Sustainable development goals (SDGs) · Carbon sequestration · Biodiversity conservation

17.1 Land Degradation: A Global Challenge

Diversity is the basis of life on Earth. Earth's ecosystems, which range from wetlands and coastal regions to mountains, forests, and deserts, provide the complex web services necessary to keep life on the planet alive. However, these services are increasingly being disrupted by escalating threat of deterioration of land resources.

S. Bhatt (✉) · K. Chavva · J. Kaur
Food and Agriculture Organization of the United Nations (FAO), New Delhi, India
e-mail: seema.bhatt@fao.org; konda.chavva@fao.org; jasmeet.kaur@fao.org

S. Singh, M. K. Gautam (eds.), *Biodiversity Conservation and Land Degradation Neutrality*, https://doi.org/10.1007/978-981-96-7210-3_17

Land degradation potentially leads to the long-term reduction in a land's biological or economic productivity (United Nations Convention to Combat Desertification (UNCCD) 1993). It includes harmful processes such as salinization, deforestation, soil erosion, and water depletion.

17.1.1 Scope of the Problem

Data on land degradation presents an alarming picture. Over 16 million square kilometers are said to be continuously degraded, according to the UNCCD (2022) report. As a result, vegetative productivity has been reduced over a large area of the planet; estimates indicate that natural areas, pastures, grazing lands, and agricultural land have declined by 12–14%. Sub-Saharan Africa is most affected by this (UNCCD 2022).

17.1.2 Crisis Beyond Productivity Loss

While productivity loss is the immediate effect of land degradation, there are far more serious repercussions. According to the UNCCD report, from 2015 to 2050, land use change and soil degradation are expected to add an extra 69 gigatons of carbon emissions to the atmosphere. This leads to a significant rise in greenhouse gas emissions, which are equal to 17% of the current yearly emissions. According to estimates by the Food and Agriculture Organization (FAO), land degradation plays a significant role in the loss of biodiversity worldwide. If the present trajectory persists, a 10% decline in ecosystem services by 2050 is anticipated (IPBES 2018).

The disturbance of water cycles and erosion-based sedimentation caused by unsustainable land use practices and deforestation impacts terrestrial and aquatic ecosystems and the biodiversity therein. Sedimentation brought on by erosion impacts essential habitats and disrupts food webs in waterways, causing further disruption of the aquatic ecosystem. This loss of biodiversity jeopardizes the complex web of life that sustains all lifeforms by undermining the basis of ecosystem resilience and health.

17.2 Toward Restoration

The scientific consensus on land degradation presents a disturbing scenario. The basis for both our environmental and food security is at risk. A complex web of life that supports all living things is in danger of being destroyed as a result of desertification, soil erosion, salinization, and deforestation, which could turn once-fertile landscapes into barren wastelands. However, land restoration provides a beacon of hope amidst this bleak situation. This strategic approach presents a vital chance to restore the planet's ecosystems while safeguarding humanity's future. Large-scale restoration projects should be given top priority so that we can start a revolutionary path toward a future that is more robust and sustainable.

17.2.1 Safeguarding Food Security: A Lifeline for a Growing Population

The world faces two unprecedented challenges. On the one hand, the basic framework of our architecture for food security is being threatened by the progressive decline in agricultural output brought about by land degradation. On the other hand, the global population is projected to reach 10 billion (United Nations, Department of Economic and Social Affairs, World Population Prospects 2022). This combination of circumstances makes things perilous, particularly for the most vulnerable regions that are already combating hunger and malnutrition.

A stable food supply is based on robust and healthy ecosystems. Degradation of the land impedes essential ecological functions, resulting in a reduction of water and soil fertility. Lower agricultural yields as a result put food security in jeopardy, especially for communities that primarily rely on rain-fed agriculture. In disadvantaged areas, where food insecurity is already a persistent problem, the effects are felt most strongly. Efforts to achieve food security promoted the adoption of chemicals in agriculture and put the focus on a handful of crops and livestock. Additionally, the adoption of industrial processes to achieve food security undermined agro-biodiversity.

Our ability to enhance food production and build a more resilient food system for coming generations will be greatly enhanced by rehabilitating degraded lands and encouraging sustainable land management (SLM) techniques. The World Resources Institute's research (2014) indicates that restoring just 12% of the degraded agricultural land worldwide may result in a noteworthy 10% rise in food production. This equates to feeding 200 million more people, which is an essential step in guaranteeing fair access to the most basic human right—the right to food—and subsequently establishing global food security (Blazhevska 2019).

Healthy soils, nurtured through restoration practices, retain water more effectively, mitigating the impact of drought on crops. Additionally, restored ecosystems support natural pest control mechanisms, reducing reliance on harmful pesticides. This translates to increased and more sustainable food production, contributing to global efforts to achieve zero hunger by 2030. Healthy ecosystems support biodiversity, including pollinators essential for crop production. FAO estimates that pollinators contribute to over 35% of global crop production (FAO 2019). Additionally, restored ecosystems can help in the regulation of water cycles, minimize soil erosion, and improve soil fertility, leading to increased agricultural yields and greater food security. Beyond terrestrial ecosystems, coastal restoration also plays a crucial role. For instance, restoring mangroves can yield an astounding 60 trillion edible and commercially valuable fish and invertebrates annually (Worthington and Spalding 2019). This highlights the potential of ecosystem restoration for enhancing food security and livelihoods, particularly in coastal communities.

17.2.2 Toward Climate Restoration

Establishing land restoration as the highest priority will enable us to address many interrelated problems and take a holistic approach to environmental sustainability. Restored ecosystems can serve as strong carbon sinks, absorbing and storing carbon dioxide from the atmosphere to lessen the impacts of climate change. Moreover, by offering healthy habitats for a large number of species of flora and fauna, restoration initiatives can support the conservation of biodiversity. Healthy ecosystems also play a significant role in regulating water cycles, ensuring clean water for both human consumption and agricultural production.

According to the United Nations Environment Programme-World Conservation Monitoring Centre (UNEP-WCMC's 2020) report, the restoration and rehabilitation of 12 million hectares of degraded land annually could contribute to a 25% decline in global emissions by 2030. This leads to a significant reduction in atmospheric carbon dioxide levels, slowing the rate of climate change and reducing its potentially devastating impacts.

The Intergovernmental Panel on Climate Change (IPCC) (2019) reports that land degradation contributes to approximately 23% of global human-caused greenhouse gas emissions. Ecosystems are essential for mitigating and adapting to climate change. Restoration initiatives such as tree planting and reforestation create effective carbon sinks, by absorbing atmospheric carbon dioxide, a key greenhouse gas. Restoring mangroves and wetlands can provide coastal communities with protection against sea-level rise and storm surges. For instance, mangroves can sequester carbon at a rate up to four times faster than tropical rainforests (GEF 2021). The restoration of 100 million hectares of degraded land in the Sahel region through the Great Green Wall initiative could potentially contribute to one-third of the climate change mitigation needed by 2030 (FAO 2019).

Additionally, healthy ecosystems function as natural buffers against severe weather events such as floods and droughts. Forests, for example, help regulate water flow and reduce the risk of flash floods. These combined factors make ecosystem restoration a critical tool in mitigating climate change and its associated impacts.

Restored ecosystems serve as natural buffers against climate-related hazards. For instance, the restoration of oyster reefs in Mobile Bay, Alabama, decreased average wave heights and shoreline energy by 53–91% (Fodrie et al. 2017; Griscom et al. 2019; Abell et al. 2016), offering protection from coastal erosion and flooding. Coastal wetlands, when restored, can reduce flood damage by up to 66% (WRI 2020).

17.2.3 Safeguarding Biodiversity

Healthy habitats and ecosystems are the foundation of a biodiverse planet. A decrease in biodiversity and a chain reaction of unfavorable effects result from the disruption of essential habitats and ecological processes caused by land degradation. Numerous species face extinction due to unfavorable environmental conditions such as soil erosion, fragmented landscapes, and altered habitats. In addition to diminishing the intrinsic value of the natural world, declining biodiversity also causes disruptions in the complex web of life that supports all living things.

Land restoration addresses this threat. As part of sustainable development goal (SDG) 15: Life on Land, we have the power to safeguard biodiversity while offering secure habitats for wildlife by restoring degraded landscapes. A study that emphasizes the importance of restoration efforts was published in the prestigious journal *Science Advances* (2020). According to the study, there could be a 500% increase in bird populations worldwide if degraded forests are restored (Potts et al. 2010). Despite being an estimate, this measures the possible benefit for a specific group of organisms. An increase in the bird population could potentially revitalize entire ecosystems. Healthy forests provide vital habitats for a diverse range of species, from insects and amphibians to mammals and reptiles. Restoration efforts that prioritize specific ecosystems, such as wetlands and grasslands, can further amplify the positive impacts on biodiversity conservation, supporting the objectives of SDG 14: Life Below Water, and SDG 13: Climate Action.

The Convention on Biological Diversity (CBD) estimates that up to one million species could be driven to extinction by 2050 due to habitat loss and degradation (UNEP CBD 2020). Ecosystem restoration serves as a lifeline for biodiversity conservation. By restoring lost habitats and fostering natural regeneration, we provide a safe haven for a diversity of plant and animal species to thrive. This not only protects valuable biodiversity but also safeguards the ecosystem services these species provide, such as water purification and pollination. According to a study by the UNEP, effective restoration of just 15% of converted lands globally could prevent 60% of projected species extinctions (UNEP 2024). By restoring habitats for these species, we could create corridors for wildlife movement, allowing species to adapt to changing environmental conditions. Studies have proven that the restoration of 15% of converted lands could help avoid up to 60% of projected species extinctions (IPBES 2019).

17.2.4 Securing Human Well-Being

Ecosystem health is inextricably connected to human well-being. Land degradation has far-reaching consequences, impacting everything from water security to disaster risk and population displacement. Degraded landscapes often struggle to retain water, leading to droughts and unreliable water supplies for communities and agriculture, jeopardizing progress toward SDG 6: Clean Water and Sanitation and SDG 2: Zero Hunger. Furthermore, barren landscapes are more susceptible to soil erosion, which can exacerbate flooding and sedimentation in waterways, undermining progress toward SDG 11: Sustainable Cities and Communities. These environmental challenges force communities to migrate in search of fertile lands and reliable water sources, disrupting livelihoods and social structures and hindering progress toward SDG 10: Reduced Inequalities.

Land restoration offers a powerful solution to these interconnected problems. By prioritizing large-scale restoration efforts, we can promote environmental health, improve livelihoods, and foster sustainable development in alignment with the overarching goals of the 2030 Agenda. The World Bank estimates that the benefits of restoring degraded lands can outbalance the costs by a factor of 10–30 (World Bank 2020). These benefits include the following:

- *Enhanced Water Security*: Restored ecosystems have a remarkable ability to absorb and retain water, leading to more secure water supplies for communities and agricultural production, supporting progress toward SDG 6: Clean Water and Sanitation (UNEP).
- *Reduced Disaster Risk*: Healthy landscapes with intact vegetation are better equipped to withstand natural disasters such as floods and landslides, mitigating the risk of damage to infrastructure and loss of life, furthering progress toward SDG 11: Sustainable Cities and Communities (FAO).
- *Increased Livelihood Opportunities for Local Communities*: The restoration process itself can create jobs and income opportunities in rural areas, promoting economic growth and supporting SDG 8: Decent Work and Economic Growth. Furthermore, restored ecosystems can support sustainable livelihoods, such as ecotourism and the harvesting of non-timber forest products, contributing to SDG 1: No Poverty (International Labor Organization).

The benefits of land restoration extend beyond these immediate improvements. By restoring degraded landscapes, we are investing in a more sustainable future for all. Healthy ecosystems support essential services such as clean air and water and regulating the climate by absorbing carbon dioxide, supporting progress toward SDG 13: Climate Action, and offer invaluable cultural and recreational opportunities.

Land restoration is not just about environmental protection; it is a strategic intervention critical for achieving the SDGs and safeguarding the well-being of present and future generations. By prioritizing restoration efforts and promoting a holistic approach, we can create a healthier planet teeming with life, ensuring the security of vital resources for all.

Ecosystem restoration offers substantial health and welfare benefits. An analysis reveals that investing a mere US$4 per resident in urban tree planting can significantly improve the health of millions. Trees, it seems, are not only nature's beauty but also a cost-effective solution for enhancing public health.

Aligned with this, the World Health Organization (WHO) 2022 report underscores the profound link between healthy ecosystems and human well-being. Restored ecosystems, such as wetlands and forests, act as natural filters, purifying both air and water. Forests and wetlands act as natural filters, improving air and water quality. Studies have shown that increased green spaces in urban areas can reduce respiratory illnesses by up to 15% (WHO 2018). Moreover, access to nature has been linked to reduced stress, anxiety, and depression, contributing to overall mental health and well-being. This reduction in pollutants and waterborne diseases directly enhances human health. Moreover, the psychological benefits of green spaces are undeniable, as access to nature has been shown to reduce stress and promote overall mental well-being.

17.2.5 Water Security and Quality

Healthy ecosystems are critical for water security. Forests, for example, act like sponges, absorbing and slowly releasing rainwater. This helps regulate water flow and prevents floods. Additionally, restored wetlands can filter pollutants from water and store excess water during heavy precipitation, releasing it gradually during dry spells. The Ramsar Convention on Wetlands emphasizes the restoration of wetlands. Globally, 81% of cities could address sediment or nutrient pollution in the water used through initiatives of forest protection and restoration and improvements in agricultural practice (Abell et al. 2016).

17.2.6 Economic Benefits

The economic value of ecosystem services is significant. According to a study, approximately $44 trillion of economic value generation—over half of the world's gross domestic product—relies moderately or heavily on nature and its services (IPBES 2019). The degradation of the ecosystems and the resulting loss of their services could lead to substantial economic losses. Conversely, investing in ecosystem restoration can generate new economic opportunities and contribute to sustainable development.

Investing in ecosystem restoration is not just good for the environment; it is also a sound economic decision. Restored ecosystems contribute to economic well-being through increased productivity (higher yields and catches), reduced disaster risks (protecting infrastructure), ecotourism opportunities, and improved human health (lower healthcare costs). Africa's Great Green Wall initiative seeks to restore 100 million hectares of degraded land in the Sahel while also fostering peace and security in the region (Mirzabaev et al. 2022). Investing in ecosystem restoration offers a cost-effective and sustainable solution to a multitude of environmental challenges. The benefits extend far beyond ecological health, encompassing food security, climate change mitigation, conservation of biodiversity, and human well-being.

17.3 Guiding Principle for Landscape Restoration

Successful landscape restoration requires a holistic and adaptive approach that considers ecological, social, and economic factors. Figure 17.1 provides a framework for effective restoration efforts.

17.4 FAO's Work in India and Land Degradation

The FAO is a key partner for India in addressing agricultural and environmental challenges, particularly land degradation. FAO collaborates with the Government of India (GoI) to develop policies, build capacity, and promote SLM practices.

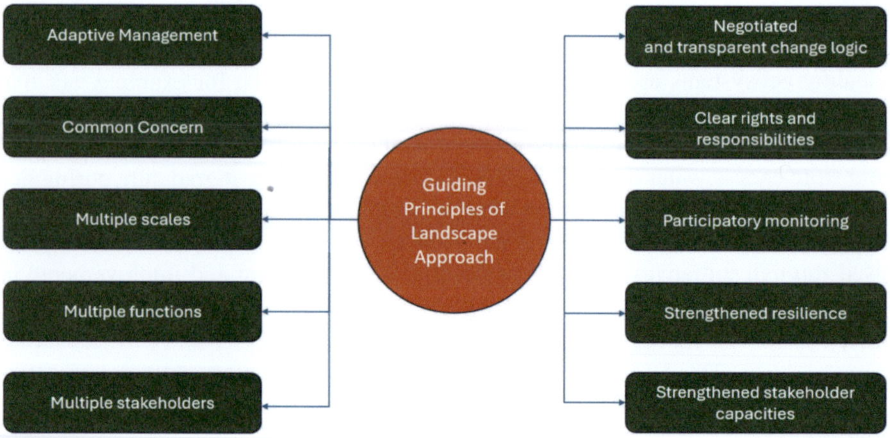

Fig. 17.1 The ten guiding principles of the landscape approach. (Adapted from Sayer et al. (2013))

Among FAO's current projects, many in collaboration with the Global Environment Facility (GEF), are attempting to address land degradation in multiple Indian states. Under the Sixth GEF cycle, FAO in collaboration with the Ministry of Agriculture and Farmers' Welfare, and the Ministry of Environment, Forest and Climate Change, is working on a 7-year project (2019–2026) titled, "Green Agriculture: Transforming Indian Agriculture for Global Environmental Benefits and the Conservation of Critical Biodiversity and Forest Landscapes." The project is working in five states: Madhya Pradesh, Mizoram, Odisha, Rajasthan, and Uttarakhand. This project aims to mainstream biodiversity, climate change, and SLM objectives and practices into the Indian agricultural sector. Additionally, the project seeks to align the priorities and investments between India's agricultural and environmental sectors, thereby fostering national and global environmental benefits while ensuring that the nation's development aspirations remain uncompromised.

FAO's project: "Promotion of Sustainable Food Systems in India through Transforming Rice-Wheat Systems in Punjab, Haryana, Odisha and Chhattisgarh" was launched under GEF's seventh funding cycle. This 5-year project supports the GoI, in association with key stakeholders (ranging from civil society and the private sector), in designing, implementing, validating, and scaling up a transformative model of sustainable and integrated food systems. This project aims to deliver significant global environmental benefits, including reductions in land degradation, biodiversity loss, and greenhouse gas emissions. It also seeks to enhance the environmental sustainability of food systems while improving farmers' incomes and resilience. Furthermore, the project offers additional advantages for other food system stakeholders. These include: greater access to safe and nutritious food and increased employment opportunities. The project follows a "whole food system" approach. The basis of this approach is that farmers are supported in managing their farming and livelihood systems in an integrated manner aligned with the principles of sustainability, resilience, and agroecology. These include: locally adapted best practices that could best deliver multiple and linked environmental, social, and health benefits.

The Green Climate Fund project, implemented across the states of Odisha, Chhattisgarh, Madhya Pradesh, and Punjab, focuses on several key objectives: enhancing capacity for coordinating climate finance; developing strategic frameworks to support low-emission investments; strengthening adaptation planning processes; creating a transformative pipeline for climate-resilient initiatives; and fostering knowledge sharing and learning to promote sustainable practices.

Beyond specific projects, FAO is actively involved in the UN Decade on Ecosystem Restoration and provides platforms for knowledge sharing and experience exchange on land restoration, desertification, and drought resilience. Through these efforts, FAO is contributing significantly to India's goal of building a sustainable future. FAO continues to stand committed to supporting sustainable food systems in India through promoting agro-ecology for healthy landscapes and ecosystems; supporting policy design for food systems and landscape management; and encouraging participatory land use planning to improve governance and address environmental degradation.

17.5 Conclusion

Land restoration is a crucial strategy for addressing the global problems of biodiversity loss, climate change, and land degradation. By giving ecosystem restoration top priority, we can protect biodiversity, improve climate resilience, and secure food and water supplies. These initiatives provide solutions that combine social well-being, economic growth, and environmental health, all of which are in line with the SDGs. We can create the foundation for a resilient and sustainable future that will benefit both people and the environment by managing land in a comprehensive and flexible manner.

References

Abell R, Lehner B, Thieme M, Linke S (2016) Looking beyond the fenceline: assessing protection gaps for the world's rivers. Conserv Lett 10(4):384–394). Wiley. https://doi.org/10.1111/conl.12312

Blazhevska V (2019) It is everybody's business to put degraded lands back to life, UN says. United Nations Sustainable Development, July 17, 2019. Retrieved December 11, 2024, from United Nations Sustainable Development website: https://www.un.org/sustainabledevelopment/blog/2019/06/it-is-everybodys-business-to-put-degraded-lands-back-to-life-un-says/

FAO (2019) The state of the world's biodiversity for food and agriculture. Bélanger J, Pilling D (eds). FAO Commission on Genetic Resources for Food and Agriculture Assessments, Rome, 572 p

Fodrie FJ, Rodriguez AB, Gittman RK, Grabowski JH, Lindquist NL, Peterson CH, Piehler MF, Ridge JT (2017) Oyster reefs as carbon sources and sinks. Proc R Soc B Biol Sci 284(1859):20170891). The Royal Society. https://doi.org/10.1098/rspb.2017.0891

Global Environment Facility (GEF) (2021) Mangrove restoration: a guide for practitioners

Griscom BW, Lomax G, Kroeger T, Fargione JE, Adams J, Almond L, Bossio D, Cook-Patton SC, Ellis PW, Kennedy CM, Kiesecker J (2019) We need both natural and energy solutions to stabilize our climate. Glob Chang Biol 25(6):1889–1890). Wiley. https://doi.org/10.1111/gcb.14612

IPBES (2018) The IPBES assessment report on land degradation and restoration. Montanarella L, Scholes R, Brainich A (eds). Secretariat of the Intergovernmental Science-Policy Platform on Biodiversity and Ecosystem Services, Bonn, 744 p

IPBES (2019) Global assessment report on biodiversity and ecosystem services of the Intergovernmental Science-Policy Platform on Biodiversity and Ecosystem Services. Brondizio ES, Settele J, Díaz S, Ngo HT (eds). IPBES Secretariat, Bonn, 1148 p. https://doi.org/10.5281/zenodo.3831673

IPCC (2019) Climate change and land: an IPCC special report on climate change, desertification, land degradation, sustainable land management, food security, and greenhouse gas fluxes in terrestrial ecosystems [P.R. Shukla, J. Skea, E. Calvo Buendia, V. Masson-Delmotte, H.-O. Pörtner, D.C. Roberts, P. Zhai, R. Slade, S. Connors, R. van Diemen, M. Ferrat, E. Haughey, S. Luz, S. Neogi, M. Pathak, J. Petzold, J. Portugal Pereira, P. Vyas, E. Huntley, K. Kissick, M. Belkacemi, J. Malley (eds.)]. https://doi.org/10.1017/9781009157988.012

Mirzabaev A, Sacande M, Motlagh F, Shyrokaya A, Martucci A (2022) Economic efficiency and targeting of the African Great Green Wall. Nat Sustain 5(1):17–25

Potts SG, Biesmeijer JC, Kremen C, Neumann P, Schweiger O, Kunin WE (2010) Global pollinator declines: trends, impacts and drivers. Trends Ecol Evol 25(6):345–353). Elsevier BV. https://doi.org/10.1016/j.tree.2010.01.007

Sayer J, Sunderland T, Ghazoul J, Pfund J-L, Sheil D, Meijaard E, Venter M, Boedhihartono AK, Day M, Garcia C, van Oosten C, Buck LE (2013) Ten principles for a landscape approach to reconciling agriculture, conservation, and other competing land uses. Proc Natl Acad Sci 110(21):8349–8356. https://doi.org/10.1073/pnas.1210595110

UNCCD (1993) Land degradation. United Nations Convention to Combat Desertification (UNCCD).. Accessed 10 Oct 2024

UNCCD (2022) Chronic land degradation: UN offers stark warnings and practical remedies in Global Land Outlook 2. Retrieved from UNCCD website: https://www.unccd.int/news-stories/press-releases/chronic-land-degradation-un-offers-stark-warnings-and-practical

UNEP (2024) World Environment Day 2024: accelerating land restoration, drought resilience & desertification progress. Retrieved from UNEP website: https://www.unep.org/news-and-stories/statements/world-environment-day-2024-accelerating-land-restoration-drought

UNEP CBD, S. B. S. T. T. A (2020) Post-2020 global biodiversity framework: scientific and technical information to support the review of the updated goals and targets, and related indicators and baselines. Provisional Agenda CBD. SBSTTA/24/3, United Nations Environment Program (UNEP)

UNEP-WCMC (2020) 10 years to boost ecosystem restoration for people and planet. Retrieved from UNEP-WCMC website: https://www.unep-wcmc.org/en/news/10-years-to-boost-ecosystem-restoration-for-people-and-planet

World Bank (2020) Resource mobilization and the status of funding of activities related to land degradation. World Bank Group

United Nations Department of Economic and Social Affairs, Population Division (2022) World Population Prospects 2022: Summary of Results. UN DESA/POP/2022/TR/NO. 3

World Health Organization (WHO) (2018) Air pollution and child health: prescribing clean air: summary (No. WHO/CED/PHE/18.01). World Health Organization

World Health Organization (WHO) (2022) Compendium of WHO and other UN guidance on health and environment (No. WHO/HEP/ECH/EHD/22.01). World Health Organization

World Resources Institute (WRI) (2020) Restoring nature: paths to recovery. WRI

Worthington T, Spalding M (2019) Mangrove restoration potential: a global map highlighting a critical opportunity. Apollo – University of Cambridge Repository. Retrieved from: https://doi.org/10.17863/CAM.39153

MIX
Papier aus verantwortungsvollen Quellen
Paper from responsible sources
FSC® C105338

If you have any concerns about our products,
you can contact us on
ProductSafety@springernature.com

In case Publisher is established outside the EU,
the EU authorized representative is:
Springer Nature Customer Service Center GmbH
Europaplatz 3, 69115 Heidelberg, Germany

Printed by Libri Plureos GmbH
in Hamburg, Germany